IV International Forum on Nanotechnology 2011

(RUSNANOTECH 2011)

Journal of Physics: Conference Series Volume 345

Moscow, Russia
26-28 October 2011

ISBN: 978-1-62276-691-8
ISSN: 1742-6588

Printed from e-media with permission by:

Curran Associates, Inc.
57 Morehouse Lane
Red Hook, NY 12571

CURRAN ASSOCIATES INC. proceedings .com

Some format issues inherent in the e-media version may also appear in this print version.

Printed by Curran Associates, Inc. (2013)

Institute of Physics
Dirac House, Temple Back
Bristol BS1 6BE UK

Phone: 44 1 17 929 7481
Fax: 44 1 17 920 0979

techtracking@iop.org

Additional copies of this publication are available from:

Curran Associates, Inc.
57 Morehouse Lane
Red Hook, NY 12571 USA
Phone: 845-758-0400
Fax: 845-758-2634
Email: curran@proceedings.com
Web: www.proceedings.com

TABLE OF CONTENTS

NANOMATERIALS

The RUSNANOTECH 2011 International Forum on Nanotechnology was held from 26–28 October 2011, in Moscow, Russia. It was the fourth forum organized by RUSNANO (Russian Corporation of Nanotechnologies) since 2008. In March 2011 RUSNANO was established as an open joint-stock company through the reorganization of the state corporation Russian Corporation of Nanotechnologies. RUSNANO's mission is to develop the Russian nanotechnology industry through co-investment in nanotechnology projects with substantial economic potential or social benefit. Within the framework of the Forum Science and Technology Program, presentations on key trends of nanotechnology development were given by foreign and Russian scientists, R&D officers of leading international companies, universities and scientific centers.

The science and technology program of the Forum was divided into four sections as follows (by following hyperlinks you may find each section's program including videos of all oral presentations):

- Nanoelectronics and Nanophotonics
- Nanomaterials
- Nanotechnology and Green Energy
- Nanotechnology in Healthcare and Pharma (United business and science & technology section on 'RUSNANOTECH 2011')

The scientific program of the forum included more than 50 oral presentations by leading scientists from 15 countries. Among them were world-known specialists such as Professor S Bader (Argonne National Laboratory, USA), Professor O Farokzhad (Harvard Medical School, USA), Professor K Chien (Massachusetts General Hospital, USA), Professor L Liz-Marzan (University of Vigo), A Luque (Polytechnic University of Madrid) and many others. The poster session consisted of over 120 presentations, 90 of which were presented in the framework of the young scientists' nanotechnology papers competition. This volume of *Journal of Physics: Conference Series* includes a selection of 47 submissions.

Section editors of the proceedings:

Nanoelectronics and nanophotonics Corresponding Member of Russian Academy of Sciences, Professor Anatoly Dvurechenskii (Institute of Semiconductor Physics, RAS).

Nanomaterials Member of Russian Academy of Sciences, Professor Mikhail Alfimov (Photochemistry Center, RAS), Professor Igor Suzdalev (Semenov Institute of Chemical Physics, RAS), Member of Russian Academy of Science, Professor Vyacheslav Osiko (Prokhorov General Physics Institute, RAS), Member of Russian Academy of Science, Professor Aleksey Khokhlov (Physical department of Moscow State University).

Nanotechnology and green energy Corresponding Member of Russian Academy of Sciences, Professor Eduard Son (Joint Institute for High Temperatures, RAS).

Nanotechnology in Healthcare and Pharma Member of Russian Academy of Sciences, Professor Konstantin Skryabin (Bioengineering Center, RAS), Member of Russian Academy of Sciences, Professor Rem Petrov (RAS), Corresponding Member of Russian Academy of Sciences, Professor Sergey Deev (Institute of Bioorganic Chemistry).

Journal of Physics: Conference Series **345** (2012) 012001 doi:10.1088/1742-6596/345/1/012001

Sensors and filters based on nano- and microchannel membranes for biomedical technologies

S I Romanov[1], D V Pyshnyi[2], P P Laktionov[2]

[1] A V Rzhanov Institute of Semiconductor Physics SB RAS

[2] Institute of Chemical Biology and Fundumental Medicine SB RAS

E-mail: romanov@isp.nsc.ru

Abstract. A new technology is presented in a concise form which enables the silicon membranes to be produced over a wide range of channel dimensions from a few nanometers to tens of micrometers. There is good reason to believe that this method based on rather simple technical processing is competitive with other technologies for fabricating nanofluidic analysis systems. Some of the completed developments involving microchannel membranes, namely, the optical DNA-sensor and the human cell separation system are demonstrated without going into details. The other applications of micro- and nanochannel membranes, namely, the electrical sensor and electrokinetic filters for detecting and separating liquids and biomolecules are shown with the first results and are in progress.

1. Introduction

Genetic roots of diseases are the mainstream of current biomedical research and development. A new generation of fast, cheap sequencing technologies and nano- and microfluidic analysis systems is needed to achieve personalized medicine in the near future. Rapidly emerging developments in microfabrication technologies have enabled a variety of microfluidic systems consisting of valves, pumps, mixers, and reactors to be utilized effectively in biology and medicine for drug delivery, DNA analysis, and biological/chemical agent detection sensors. These microfluidic systems require integration of sample separation, collection, detection units with fluid pumping, flow control elements, and the necessary electronics on a single microchip [1].

Fluidic nanochannel structures have been also used for chemical and biomolecular sensing, separation of charged analytes, and other unique modes of molecular manipulations. These applications have been facilitated by the significant increase in the range of advanced nanofabrication techniques. Nanofluidic systems are fabricated using bulk and surface nanomachining based on electron beam and interferometric lithography or self-assembly patterning [2].

No technology for producing versatile membranes with channels the dimensions of which may be varied over a wide range however exists. This circumstance has threatened to retard biomedical research and development. In this paper we briefly describe a new approach to fabricate silicon nano- and microchannel membranes which can form the basis for advanced fluidic systems and can be operated as "fluidic transistor", optical DNA-sensor, electrical sensor, and as filters for mammalian cell separation.

2. Technology for producing membranes

A new technology of silicon nanochannel membrane fabrication has been developed. Conceptually the structure of these membranes comprises a monolithic connection of a nanochannel array with rigid supporting microchannel one and has an ordered profile of open through channels. Coupled with a fluidic system, the device which is performed in a monocrystalline silicon wafer is intended for separation, concentration, and detection of nanoparticles, cells, polynucleotide and protein molecules, their conjugates and associates [3-6].

The technology has been realized by two different techniques. In technique referred to as the "top-down" method, a nanochannel part of membranes is formed through series of plasmochemical and pyrolytical depositions of silicon and silicon oxide films directly on a microchannel array with overlapping microchannels. The lateral size of these microchannels can be decreased gradually from several micrometers to tens of nanometers, figure 1.

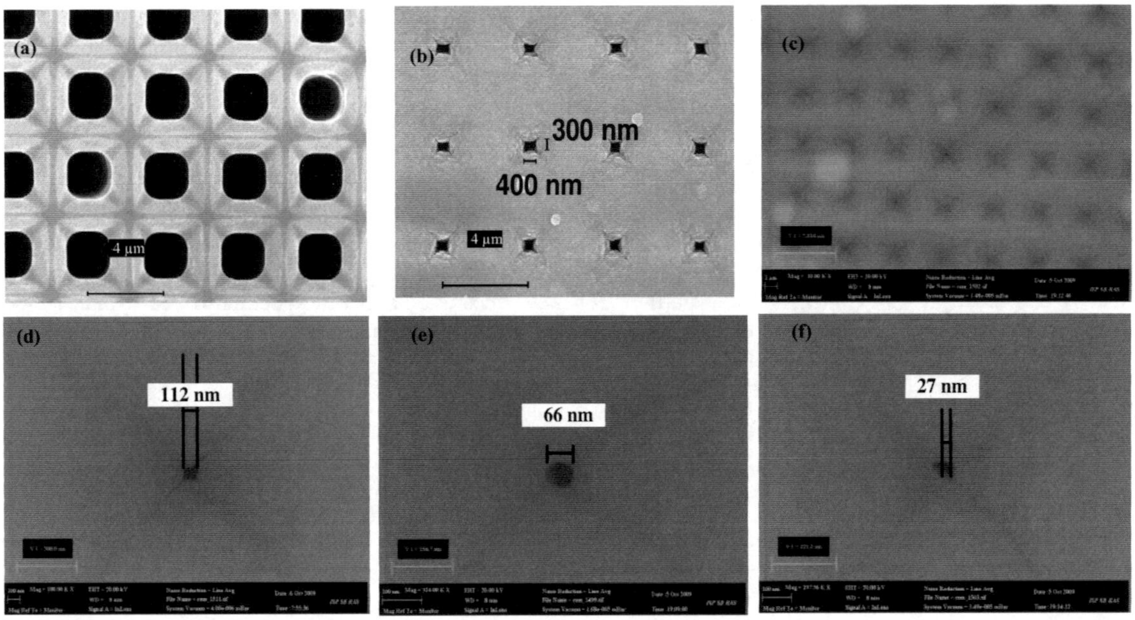

Figure 1. SEM images of the nanochannel membrane surface after different fabrication stages: (a) the starting microchannel membrane, (b) plasmochemical deposition of silicon film, (c) pyrolitic deposition of silicon oxide film – the final structure, (d-f) the separate channels.

In the "bottom-up" method, channels of microchannel array are sealed with SiO_2-nanospheres having a diameter of 50-100 nm. Then, a porous amorphous silicon film is plasmochemically deposited on the solid surface. After removal of silica plugs a molecular filter is obtained. This filter consists of the silicon film 40-400 nm in thick which is suspended over microchannels. Pore size in the amorphous silicon is less than a few nanometers. During subsequent processing these pores become through and expand in dimensions. Their diameter can be controllably increased to tens of nanometers, figure 2.

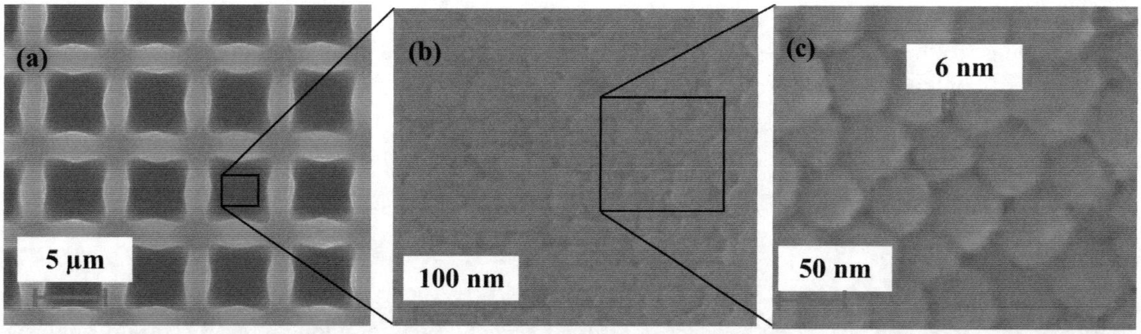

Figure 2. SEM images of the nanochannel membrane surface after different fabrication stages: (a) a porous silicon film deposited on silica plugs, (b) a fragment of the membrane over microchannel after removal of silica plugs, (c) a fragment of the membrane after electrochemical etching.

The thickness of nano- and microchannel parts of the membranes (length of the channels) is 0.1-6 μm and 100-250 μm, respectively. The lateral size of the channels changes in a wide range from a few nanometers to a few micrometers. Their surface densities are 10^5-10^8 cm^{-2} and 10^{10}-10^{11} cm^{-2} for the 1-st and 2-nd methods, respectively.

The nano- and microchannel membranes made of biocompatible silicon by the present technology hold much promise for their integration with advanced electronic, optical, and biochemical devices.

3. Microchannel sensors

3.1. Optical DNA-sensor

We propose a simple optical method for hybridization analysis of DNA. This method is based on overlapping silicon microchannels with microspheres consisted of light scattering and light absorbing materials. DNA targets immobilized on the spheres bind them to microchannel surface due to specific hybridization reactions with complementary oligonucleotide probes covalently bound to an internal surface of this microchannel sensor. As a result, the registered intensity of transmitted light decreases, figure 3.

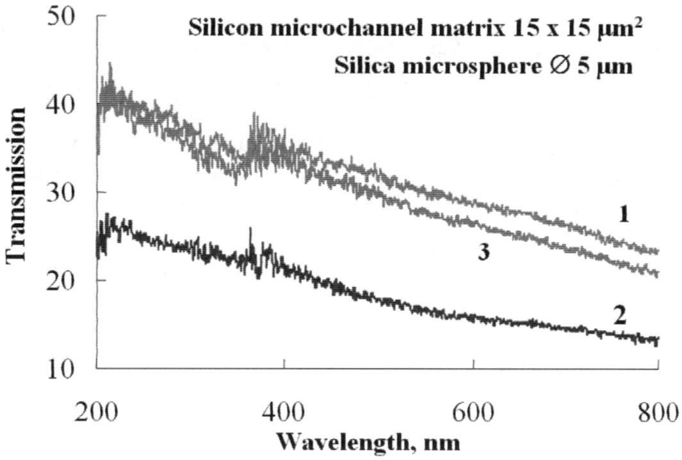

Figure 3. Transmission spectra of silicon microchannel matrix: 1) before and 2) after hybridization reactions, 3) the restored state of the matrix with oligonucleotide probes after denaturation of complementary complexes and removal of microspheres with DNA targets.

On the basis of the present method it becomes possible to develop a simple optical device for detection of DNA and other biological substances. This sensor can be designed of light source, microchannel matrix, photodetector, and several lenses. Multiplexing these elements on a silicon substrate enables a three-dimensional optical biochip to be constructed.

3.2. Electrical sensor

We have developed an electrical sensor based on microchannel semiconductor for liquid analysis. Silicon microchannel matrix acts as a sensitive element the conductivity of which changes specifically with the type of liquid coming to contact, figure 4.

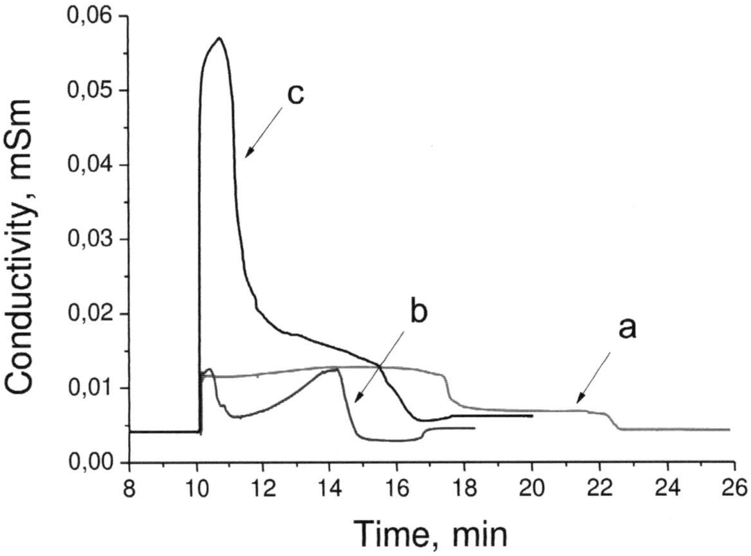

Figure 4. Responses of the electrical sensor: (a) deionized water, (b) water-ethanol solution, 1:1; (c) aqueous solution of ammonium hydroxide, 18 w.%; fluid volume of 3 μl.

Specific and reproducible responses of the sensor for every kind of liquid make it possible to distinguish biological solutions and to register specific biochemical reactions in the interior of microchannels. Multiplexing these sensors on a silicon substrate permits an electrical biochip to be designed.

3. Human cell separation filters

A cell separation system is proposed which comprises a pile of the microchannel membranes with different channel lateral sizes of 2-30 μm. This microfluidic device is intended for size-selective and receptor-specific separation of viable rare cells circulating in human blood. In the second case, receptor specific antibodies are immobilized onto an internal surface of microchannel filter. It was found that the silicon membranes were easily permeated with even undiluted blood and did not damage human cells. This microfluidic system has provided its high efficiency for separating fetal cells from mother's blood and tumour cells from blood of patients with breast cancer [7] or multiple myeloma. The possibility to isolate tumour cells from patient's blood is especially required for monitoring the tumour reoccurrence after therapy and our microfluidic device based on silicon microchannel membranes presents this opportunity, figure 5.

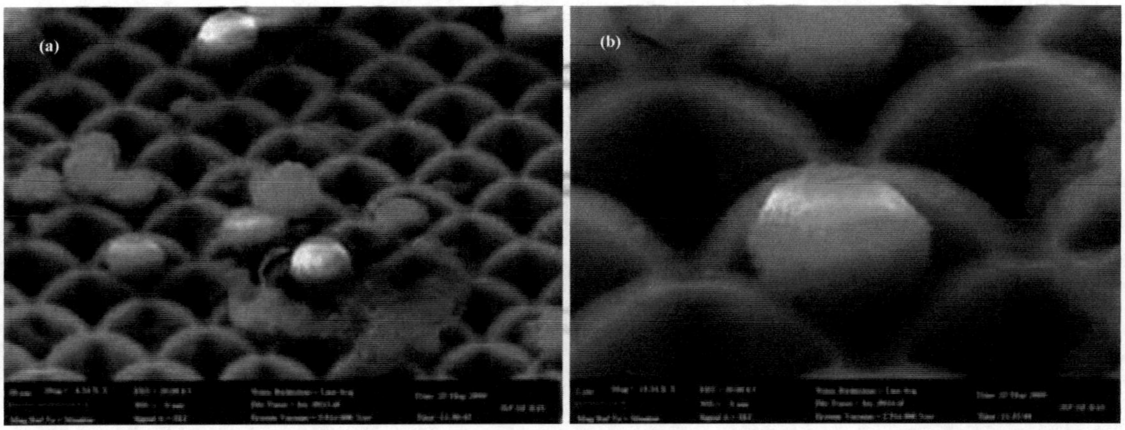

Figure 5. SEM images of oncotransformed HeLa cells separated from blood (in model experiment) on the surface of silicon microchannel filter with a pitch of 10 μm and channel openings of ~8.5 μm.

4. Electrokinetic silicon channel filters

Our silicon membranes provide a basis for separating and detecting different biological substances such as DNA, proteins, and cells around directed electrokinetic controlled transport of fluids and dissolved analytes through micro- and nanochannels. Using the silicon filters as a "fluidic" transistor, we can control over electroosmotic and electrophoretic flows by applying external biases to the semiconductor (as a "gate") and biological solutions (as a "source" and a "drain"). This electrokinetic flow control scheme which does not require any moving components has been tested with profit to transport of buffered solutions, proteins, and CdS nanoparticles. The electroosmotic phenomenon in the silicon nano- and microchannel membranes has received much consideration in our experiments.

5. Conclusions

The technology has been developed which enables the silicon membranes to be formed over a wide range of channel dimensions from a few nanometers to tens of micrometers. We believe that this new method based only on simple technical processing is quite competitive with present-day technologies for production of nanofluidic analysis systems controlled with low-driving electric fields.

This technology has been applied advantageously to the completed developments involving the microchannel membranes, namely, the optical DNA-sensor and the cell separation system. The other applications of micro- and nanochannel structures for the electrical sensor and electrokinetic filters are still being developed.

Acknowledgments

This research has been supported by Russian Government (P1073), Siberian Branch of Russian Academy of Sciences, and the program of Russian Academy of Sciences "Molecular and Cellular Biology".

References

[1] Albala J S *et al.* 2006 *BioMEMS and Biomedical Nanotechnology (Micro/Nano Technology for*

Genomics and Proteomics vol II) ed M Ozkan and M J Heller (New York: Springer Science +Business Media LLC) p 537

[2] Agrawal A *et al.* 2006 *Nanotechnology in biology and medicine: methods, devices, and applications* ed Tuan Vo-Dinh (New York: Taylor&Francis Group)

[3] Vandysheva N V, Romanov S I 2009 *Patent RF 2388109*

[4] Vandysheva N V, Kosolobov S S, Romanov S I 2009 *Patent RF 2410792*

[5] Romanov S I, Vandysheva N V, Semenova O I, Kosolobov S S 2010 *Patent RF 2428763*

[6] Romanov S I, Vandysheva N V, Daniluk A F, Semenova O I, Kosolobov S S 2010 *Patent RF 2433502*

[7] Laktionov P P, Vainer O B, Zaporozhchenko I A, Pyshnaya I A, Pyshnyi D V, Dmitrienko E V, Romanov S I, Vlassov V V 2009 *Patent RF 2423698*

HgCdTe nanostructures on GaAs and Si substrate for IR and THz radiation detecting

M V Yakushev[1], **V S Varavin, V V Vasil'ev, S A Dvoretsky, N N Mikhailov, I V Sabinina, Yu G Sidorov, V A Shvetz, A L Aseev**
A.V. Rzanov Institute of Semiconductor Physics Siberian Branch of RAS,
Av. c. Lavrent'eva, 13, Novosibirsk, 630090, Russia

E-mail: yakushev@isp.nsc.ru

Abstract. All-round studies of heteroepitaxial HgCdTe nanostructures (NS) growth on GaAs and Si substrates by molecular beam epitaxy have been carried out. In case of Si substrate HgCdTe NS's is very perspectives for IR detectors because of equal thermal expention coefficient with silicon read-out circuits. The problems of HgCdTe conjugations with Si at epitaxy connected with large differences in lattice mismatch and differences in chemical bonding that leads to antiphased domains. We found that the precise formation of transition layer (2 nm in thickness between Si substrates and first ZnTe buffer layer leads to growth HgCdTe NS's without antiphrasis domains. V-defects and etch pits densities are equal to 10^3 cm^{-2} and 10^7 cm^{-2} respectively. The HgCdTe/Si were used for fabrication photovoltaic 640×512 MWIR focal plane arrays. The operability for $\lambda_{1/2}$= 4.2 μm (77K) was over 97%. The response (Sv) and NETD were as 1,5×10^9 V/W and less 20 mK respectively. We developed the precise growth of symmetric and antisymmetric HgTe QW. We found the following effects: the presence 2D electron gas with high mobilities over 5×10^5 cm/V×s in doped HgTe QW, the presence @D holes and electrons in undoped HgTe QW and high sensitivity to linear and circular polarized IR and THz radiation in 6 – 400 μm region.

1. Introduction

A trend of modern infrared (IR) thermal-imaging systems is developing large format focal plane array (FPA). Mercury cadmium telluride (MCT) alloys have an advantage over other photosensitive materials because of unique physical properties [1]. To realize large format IR FPA based on MCT it is necessary to grow material with high lateral composition uniformity that leads to wavelength uniformity. Nanostructures, such as HgTe/CdTe quantum walls (QW) or superlattices (SL) are an alternative material for IR detection with better wavelength cutoff uniformity than in alloys [2]. Recently, considerable effort has been directed to the development of the growth of heteroepitaxial MCT structures by molecular beam epitaxy (MBE) on Si large in diameter substrates [3]. MCT/Si solved the problem of thermal cycling in temperature range 80-300 K of cooled large format IR FPAs as a hybrid assemblage because of equal thermal expansion coefficients of the photosensitive element and a Si read-out circuit (ROIC). The substrate orientation is essentially influence on quality MCT MBE. It was found [4] that MCT MBE growth on the (112)B substrate surface is carried out at low Hg vapor pressures without twin formation that leads to high material quality. However, the (112) orientation is very sensitive to insignificant variations in the growth conditions. That determines a narrow range of optimal growth condition for fabrication MCT MBE with a minimal defect [5]. But intensive development of

[1] To whom any correspondence should be addressed.

equipment is being carried out for MCT MBE growth on large in diameter (up to 6") (112) Si substrates. It was shown that MCT/Si can be used for fabrication of high quality photonic IR FPAs in a spectral range up to 10 μm [6]. The MCT MBE growth on (013) GaAs substrates [7] have been studied. It was shown that the empirically determined substrate orientation (310) provides growth of high quality MCT/GaAs without stacking faults and twin lamellae also at low Hg pressures in essentially broad interval of optimal growth conditions. The successes in MCT MBE alloys growth on GaAs allows to develop and grow high quality HgTe QW. Recently a detector system for room temperature electric detection of the polarization state of laser beams at terahertz frequencies by photogalvanic effects was realized using low symmetry GaAs and SiGe QWs [8]. In this study, we examined the growth of MCT heterostructures on Si (310) substrates with a diameter as large as 100 mm for IR FPAs of the spectral range of 3–5 μm and the formation and characteristics of p–n junctions and photoelectric parameters of IR photodetectors with different formats. We present the results of reproducible growth of single and multiple (up to 30) HgTe QWs and photogalvanic effect in IR and THz spectral range.

2. MCT MBE growth on Si(310)

2.1 Experimental procedure

MCT MBE on (0133)Si had were grown at multichamber ultrahigh vacuum MBE set "Ob" [9]. For monitoring *in situ* growth processes reflection high-energy electron diffraction (RHEED) and single-wave ellipsometry at wavelength $\lambda = 632.8$ nm was used. (013)Si substrates 76 and 100 mm in diameter were treated by the standard RCA procedure [10] and then hydrogenated into a 1% aqueous HF solution [11]. After loading there is preepitaxial vacuum annealing at 550–600°C in the arsenic flow and cooling. The ZnTe layer 0.01 μm in thickness on Si surface was grown on Si at 200–240°C. The CdTe layer 6–8 μm in thickness was grown on ZnTe/Si at 280–320°C. MCT layers were grown on CdTe/ZnTe/Si(310) substrate by the process which was described in detail in [9]. Special the coaxial arrangement of the molecular sources is realized for growth MCT on GaAs or Si without substrate rotation with very uniform composition distribution over the surface area [12,13]. The calculation shows that the composition uniformity should not exceed 0.0002 cm^{-1}. Experimentally the composition gradient x/l over the surface for MCT layers (x = 0.3–0.35) less than 0.002 cm^{-1} that leads to wavelength cut-off accuracy at 77 K less than 0.1 μm. Fig. 1 shows MCT MBE growth on Si (a), composition (b) and wavelength cut-off (c) uniformity over the surface area

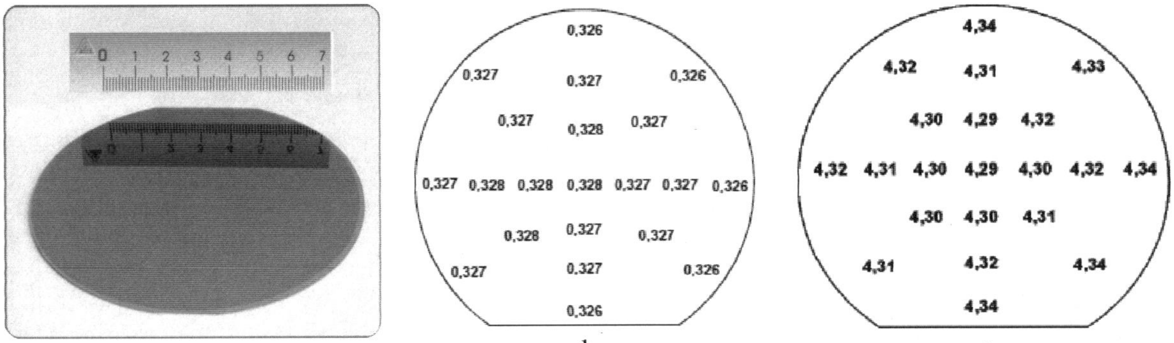

a b c

Figure 1. The MCT MBE on (013)Si (a) and composition (b, mole fr.) and wavelength cut-off (c, μm) uniformity over the surface area.

We developed passivation procedure of absorber layer by graded wide gap layers [14] at boundaries during the MCT MBE growth. Fig. 2 shows the composition distribution throughout the thickness measured by ellipsometry *in situ*.

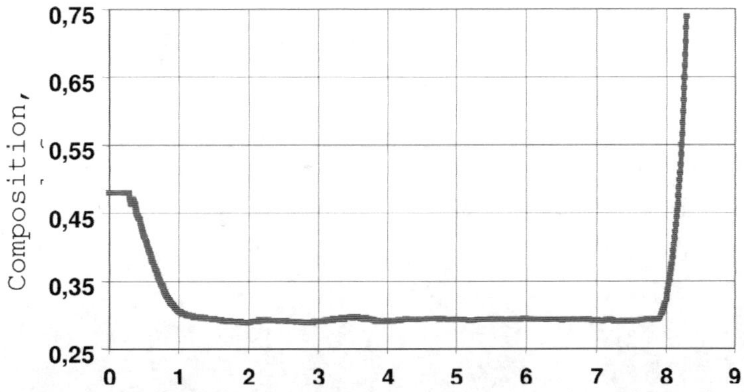

Figure 2. The MCT composition distribution throughout the thickness.

Such graded wide gap layers help to increase minority lifetime [15, 16] and to decrease or eliminate surface leakage [17] in PV FPA. There are the following defects in MCT MBE grown on Si: antiphase domains (APD), stacking faults (SF), threading dislocations (TD) and macroscopic morphological V-defects. Bulk defects were studied by transmission electron microscopy (TEM) and revealed by selective etching. TEM analysis form of thin foils parallel to the growth surface and transverse section was performed using the electron microscope JEM-4000EX (JEOL). We used etchants 10ml HNO3 + 20 ml H2O + 4g K2Cr2O7 + 1,5 mg AgNO3 (E-Ag1) [9] and 5g CrO3 + 3 ml HCl + 15 ml H2O (etchant Schaake) [10] for selective etching of CdTe and MCT layers respectively. The etching pits (ND) were as elongated triangles for CdTe and points for MCT. APD boundaries were identified as lines of arbitrary shape. SF was represented in the form of parallel straight lines. The surface defect (V-defect, etching pits and lines) densities and sizes were measured using optical microscope equipped by scanning system and built-in CCD camera.

2.2 Antiphase domains

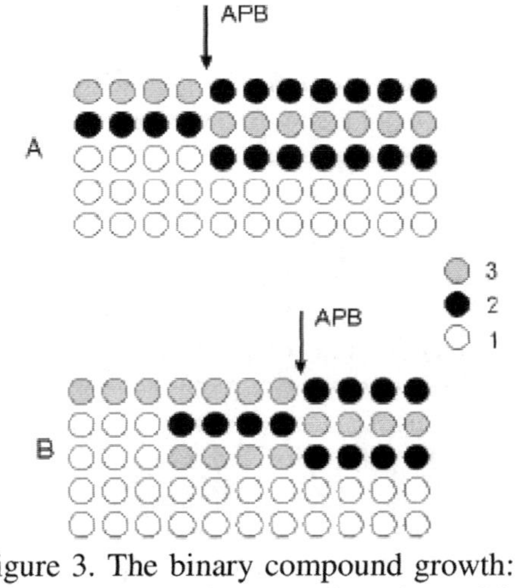

Figure 3. The binary compound growth: A – on monatomic steps Si surface with APD formation; B – on diatomic steps Si surface without APD formation. 1 – Si substrate, 2 – V or VI atoms, 3 – II or III atoms.

The detailed mechanism of the APD formation is developed in [11]. The APD formation at III-V or II-VI layers grown on Si substrates is due to the monatomic of silicon surface. Fig. 3a shows the scheme of the deposition on monatomic Si at first As and then Ga atoms which leads to APD. So to cancel the formation APD it is necessary diatomic Si surface. Fig. 3b represents the growth GaAs on Si without formation APD. But for growth ZnTe on Si the presence of diatomic steps on the silicon surface is not enough to grow films without APD. Really clean Si surface after preepitaxial annealing is interacted with species in residual vacuum created contamination centers which leads to defects formation. For elimination contamination at clean Si surface the passivation is carried out by As at preepitaxial annealing. It was experimentally shown that the absence of As on (211) Si leads to polycrystalline CdTe growth. It is known that Zn or Te is absorbed in the form of separate islands of As/Si(310) or clean Si(310) surfaces [18]. That can be the other reasons for APD formation at ZnTe growth.

We investigated the dependence of preepitaxial condition (different temperatures 500-1250 ^{0}C) on Si surface morphology by LEED and STM. We found the optimal condition of formation diatomic step Si surface, Nevertheless APD could occur in HgCdTe/CdTe/ZnTe/Si(310). Fig. 4 shows (100) TEM images of foil prepared from CdTe/ZnTe/Si(310) with APD formation. It can also be seen that APD gives under view structural damage. This allows us to identify APD by chemical etching in a selective E-Ag1 etchant. We investigated the growth CdTe (6 - 8 μm)/ZnTe (0.01 - 0.02 μm)/Si(310) for determination of APD appearance and growth conditions of ZnTe. The CdTe growth conditions and preepitaxial preparation processes were the same. It was found that the growth conditions of ZnTe at 200 – 220 ^{0}C and Zn/Te_2 ~ 5 – 20 leads to growth CdTe/ZnTe/Si without APD (see Fig. 5).

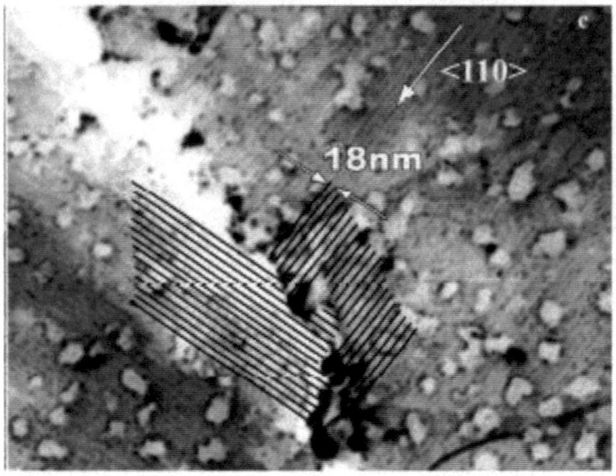

Figure 4. (100) TEM-images of APD in CdTe/ZnTe/Si(310).

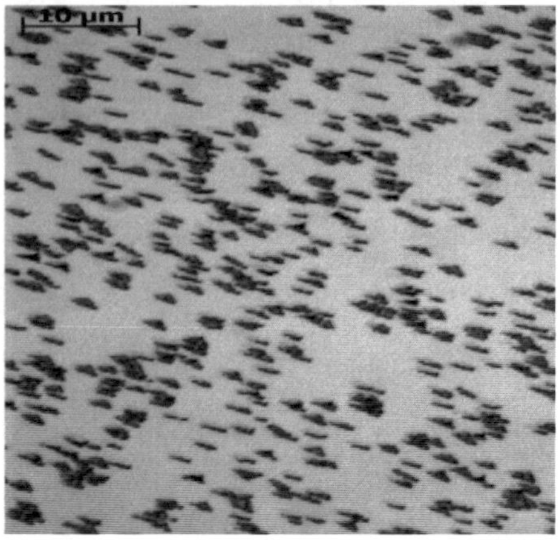

Figure 5. Etch pits on the surface of CdTe/ZnTe/Si(310) grown at 200 ^{0}C without APD

There are two APD types for ZnTe/Si. These are Si-As-Zn-Te and Si-As-Te-Zn configuration. To determine the probability of APD configuration it is necessary molecular flux of a one component higher in 1 – 2 orders over molecular flux of an other component. A series ZnTe/Si was grown under conditions where Te_2/Cd >> 1 to realized Si-As-Te-Zn configuration. However at these conditions ZnTe layers are grown as polycrystalline. So we think that Si-As-14 Zn-Te configuration is more preferably. The growth at Cd/Te_2>> 1 leads to crystalline ZnTe and in optimal condition without APD.

2.3 Stacking faults

SF in HgCdTe/CdTe/ZnTe/Si(310) are a subtracting type with the density of 10^5-10^7 cm^{-2} , lie parallel planes (111) intersecting the plane (310) at an angle of 68.58 degrees and nucleate at ZnTe/Si(310) interface and grow throughout the film up to surface (see Fig. 6). Selective etching revealed SF in the form of parallel straight lines. There are several of reasons for the formation of stacking faults in heteroepitaxy of semiconductors with a sphalerite lattice. It must be taken into account that SF occurs only in one of four possible {111} planes in CdHgTe/CdTe/ZnTe/Si(310). The proposed model must explain not only the formation of SF but also their anisotropy. Two {111} planes form an angle of 43.09^o with the (310)-plane and the other two - 68.58^o. SF lying to the boundary at a smaller angle will have a larger area than SF lying at a larger angle with the same thickness of the grown layer. Accordingly, they will have a large excess energy and their formation will not be profitable. The two remaining SF lie in (111) planes with different polarity. It remains to understand why one of the two different polar planes is preferable to another. The detailed investigation of formation of SF was studied in ZnSe/GaAs(100). The maximum SF observed at 3-D growth [19] and formed on the (111) facets that occur on the slopes of three-dimensional islands [20].

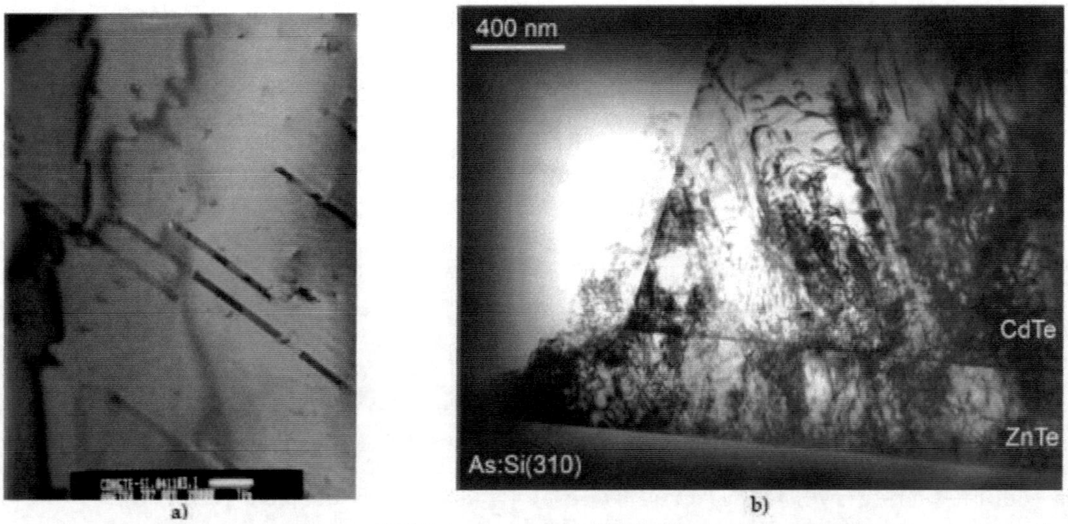

Figure 6. Stacking faults: a) planar TEM image of HgCdTe/CdTe/ZnTe/Si(310); b) cross-section TEM image CdTe/ZnTe/Si(310).

The lattice mismatch between film and substrate for ZnTe/Si (310) heterostructure is f = 12.3%. Therefore, the formation of SF can be caused by misfit strain in a heterojunction (formation of partial dislocations at the initial stage of stress relaxation). An indirect confirmation of this fact is, firstly, the type of SF (nearly all stacking faults have subtraction type and formed by sliding of Shockley partial dislocations) and, secondly, the formation of SF occurs in closely spaced parallel planes (111). In addition to the deformation mechanism there is also the growth mechanism of the formation of SF in ZnTe/Si. Such a mechanism is realized when a coalescence of three-dimensional islands occurs at the initial stage of growth. The confirmation of growth mechanism of SF formation in HgCdTe/CdTe/ZnTe/Si(310) is observed correlation between the SF density and ZnTe growth rate. We found the decreasing SF density in ZnTe/Si at decresing growth rate. It was shown that thermal annealing of HgCdTe/CdTe/ZnTe/Si(310) in inert atmosphere at a temperature of 200 - 250^{0}C for 5 - 10 hours leads to the disappearance of stacking faults in the whole volume of HgCdTe layer. Thermal annealing of CdTe/ZnTe/Si(310) in tellurium at 350^{0}C leads to SF disappearance.

2.4 Threading dislocations

The TD density (NTD) in the epitaxial film decreased from the interface to the surface and described by the equation:

$$\frac{dN_{DS}(x)}{dx} = -aN_{DS}(x) - bN^2_{DS}(x)$$

where x – coordinate (interface at x = 0 and surface at x = h), a and b-constants [21].

The interaction of dislocations (second term in (1) is the dominant process at small thicknesses (h <50 mm) typical for the MBE and the solution of equation expressed as:

Molecular-beam epitaxy, as a rule, provides the $N_{TD} \sim 10^8$-10^6 cm^{-2} values in surface regions of films at thickness more 5 microns. This expression is satisfactorily described experimental data with b = (7.0-9.0)×10^{-5}. That is the final density of threading dislocations in CdTe / Si (310) heterostructures is determined by reactions between pairs of dislocations with identical Burgers vectors. It was found that the dislocation density in the surface region $N_{TD} \sim 10^7$ cm^{-2} at the surface of CdTe layer h=5 μm grown on Si (310). Further reduction of N_{TD} without increasing the thickness of the film requires new decisions.

2.4 Morphological V-defects

V-defects (void) is specific defects may be called as a "visit card" of MBE MCT. The nature of this defect was connected with possibility of tellurium crystallization at initial stage of MCT growth and during the growth. The thermodynamic analysis shows that at MCT MBE growth condition (low growth temperature ~ 180-190 ^{0}C) there are two solid phases such as MCT and Te may be exist [19]. So at increasing temperatures and decreasing Hg pressure the formation of a polycrystalline film of tellurium is observed experimentally at these temperatures by RHEED *in situ*. The V-defect formation is due the Te nature at growth. The tellurium evaporates and reaches the MCT growing surface as a diatomic molecule Te_2. At low growth temperatures Te_2 which do not react with mercury and cadmium can not be evaporated and crystallized as solid phase. The formation of tellurium phase on the surface leads to avalanche multiplication of polycrystalline MCT growth and to V-defect formation. Fig. 7 shows AFM and TEM images of V-defects. It is well known that defects are as “killer” of diodes parameters fabricating on MCT MBE. The presence of V-defect in n-p junction gives linear volt-ampere characteristic that do not allow resisted radiation. Nevertheless, for practical device applications especially for miltielement pixels FPA MCT MBE with V-defects density lower 10^3 cm ^{2} is widely used. The process of growing MCT film with a low density of V-defects requires precise maintenance of the growth conditions and high surface quality of the buffer layer. When growth conditions deffer from optimal ones there is not possibility to grow MCT MBE od high crystal perfection with device production request. The results of selective etching allow to find correlation between V-defects density and AD density. To optimize the conditions of preepitaxial preparation processes of the substrate and growth of ZnTe and CdTe buffer layers it is obtained MCT on Si(310) without AD that reduced the V-defects density to values lower 2000 cm^{-2} (see Fig. 8).

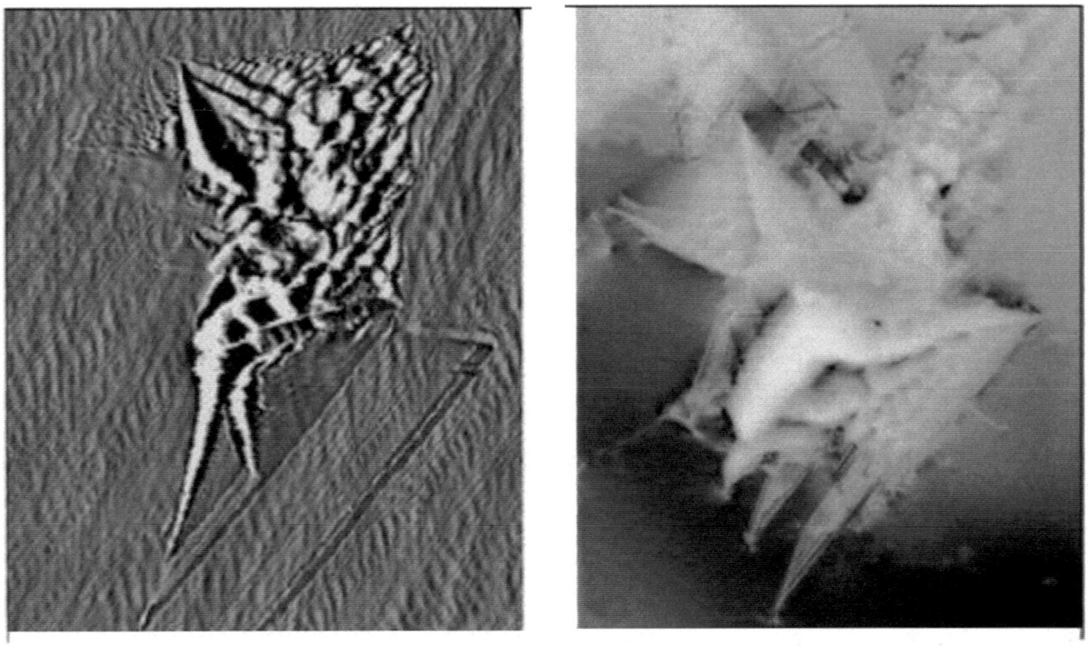

a b

Figure 7. Typical AFM (a) и TEM (b) 12×12 μm^2 images of a V-defect consisting of stacking faults, twin lamellaes и structure defects on the surface of MCT MBE on Si or GaAs (310).

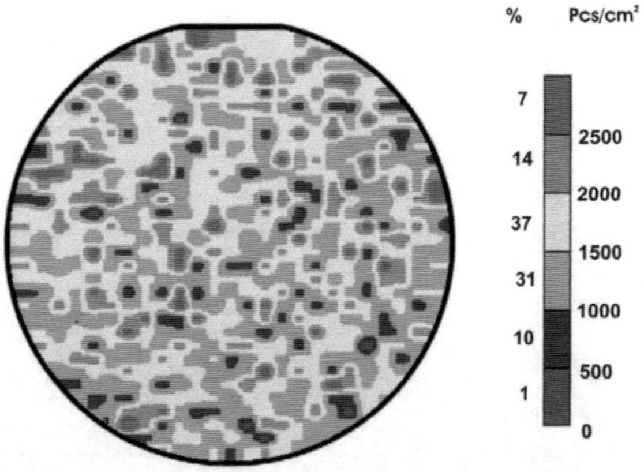

Figure 8. Typical distribution of V-defects over the surface area of MCT on Si(310)

2.5 Electrophysical characteristics

As-grown undoped MCT MBE on GaAs or Si substrates are n-type conductivity. P-type conductivity were obtained by thermal annealing in a helium atmosphere during 20 hours. The thermal annealing at Hg atmosphere gives again n-type conductivity. The carriers concentration in n-type films are $(1\text{-}10)\times10^{14}$ cm$^{-3.}$ We suggested that the residual donor at MCT MBE growth is anti-site tellurium atoms [22]. P-type MCT MVE after thermal anneling is determined by Hg vacancies. The majority mobility and minority lifetime of charge carriers in MCT/Si is lower than in the MCT/CdZnTe. Especially noticeable difference is observed for the n-type conductivity. We found that for MCT (x = 0.3) electrons mobility in n-type are weakly decreased with increasing of TD density. In case of SF electron mobility is close to the theoretical values 40000 cm^{2}V^{-1}s^{-1} at SF densities lower 2.5×10^6 cm^{-2}. At SF densities $(2.5\text{-}5.5)\times10^6$ cm^{-2} electron mobility varies from to high values. At SF densities more than 5.5×10^6 cm^{-2} electron mobility is low values. It is clear that it is possible to grow MCT MBE on Si with high carrier mobility close to the theoretical ones despite the presence of stacking faults.

2.6 Focal plane arrays

2.6.1 MWIR

MWIR photodiodes arrays 320×256 (pixel size 30 μm) and 640×512 (pixel size 25 μm) "N+-p" type were fabricated from the p-type MCT x=0.29–0.33 by B+ implantation. The current–voltage (I–V) characteristics, differential resistance, and ampere–watt sensitivity of photodiodes were measured at 77K. Fig. 9 shows the dependence of dark current (Id) under a bias voltage of –100 mV on the inverse temperature. It is seen that, in temperature 160–300 K, the dark current is proportional to n_i^2 and determined by the diffusion mechanism [23]. In temperature 140–160K Id is proportional to n_i and determined by generation–recombination processes in the depletion region. The photocurrent I_{ph} has a maximum value at 160–180 K. At temperatures 77-160 K I_{ph} increases with temperature increasing. It is explained by continuously increasing carriers diffusion length in temperatures 50-210 K due to an increase in the lifetime [24]. So we conclude that photodiodes characteristics are determined by background radiation up to 170K. The R_0A product is near 10^6 Ω×cm. FPA was fabricated on photodiode arrays and silicon ROIC by In bumps hybrid assembly. The measurements showed that FPA operability is more than more 98% (the beat more 99%). The defect pixels are uniformly distributed over the FPA area and do not form clusters in the central FPA (see Fig. 10 a). Fig. 10 b shows histogram of NETD. Fig. 11 shows the thermal image obtained FPA 640×512 visualized in real time

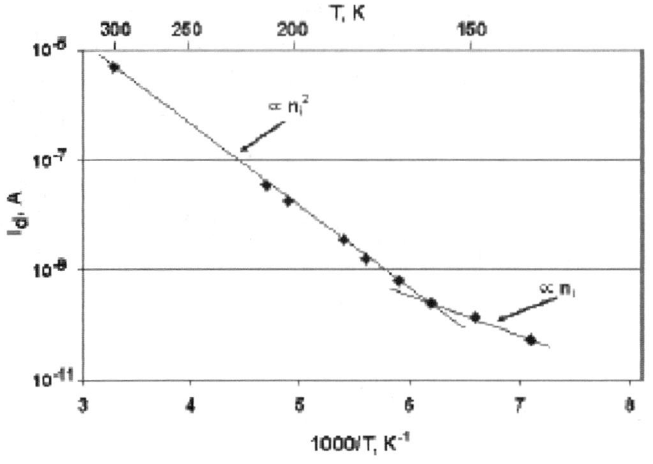

Figure 9. The dependence of dark current on temperature for MWIR photodiode. Bias voltage -100 mV, x=0.328: dots – experiment; lines –calculation.

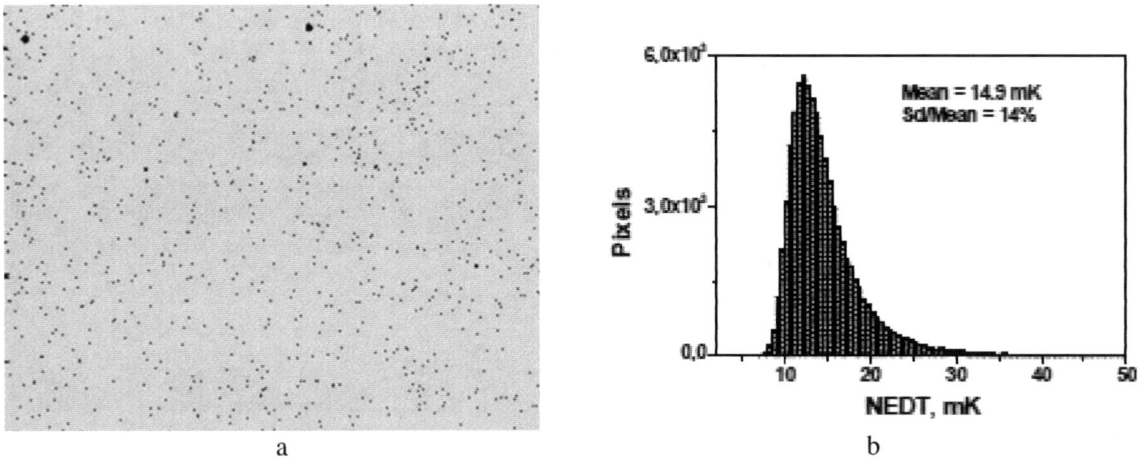

a b

Figure 10. The Tomogram of defect pixels (a) and histogram of NETD at 77K (b) of FPA 320× 256 ($\lambda_{1/2}$ = 5.2 μm).

Figure 11. Thermal image obtained from FPA 640×512 array based on MCT/Si(310)

2.6.2 LWIR

The LWIR photodiodes arrays 288×4 (pixel size 28×25 μm) were fabricated from the p-type MCT. The dark I-V characteristics are determined by different current mechanisms throughout n-p junction: diffusion current; generation-recombination current; band-to-band and trap-assisted tunneling current and surface leakage current. Experimental and calculated current-voltage characteristics for diodes fabricated from MCT (x = 0.231, ($\lambda_{1/2}$= 9.06 μm are presented in Fig. 12 (a).

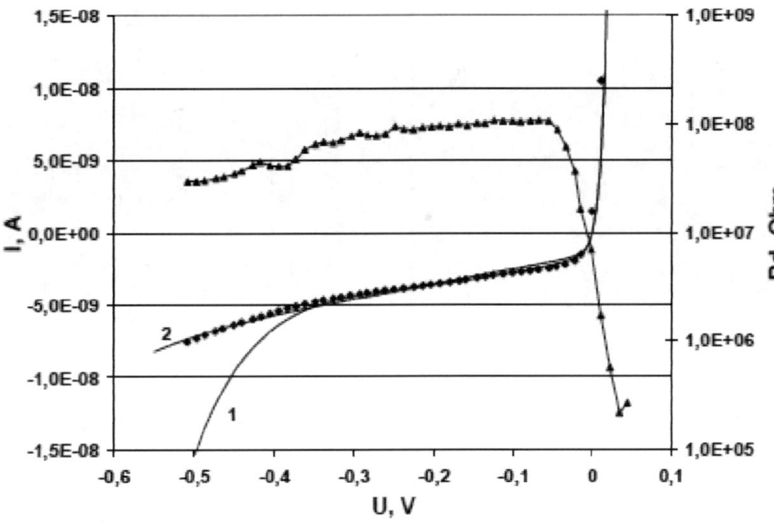

Figure 12. I-V and differential resistance-voltage characterisitics for a typical diode with ($\lambda_{1/2}$= 9.06 μm): Dots - experiment; lines – calculations.

The main contribution to the total amount of dark current in reverse bias less than 0.6 V is determined by generation in the depletion region and tunneling through traps. The contribution of other mechanisms is negligible. The dependence of the dark current on the reverse bias shown the predominance of the generation-recombination current. The R_0A value was calculated from the I–V curves, while the A value was evaluated from the measured photocurrent as described in [Gopal et al., 2001]. Fig. 13 shows R_0A product versus cut-off wavelength $\lambda_{1/2}$ in the photodiodes based on MCT/Si(310) at 77 K.

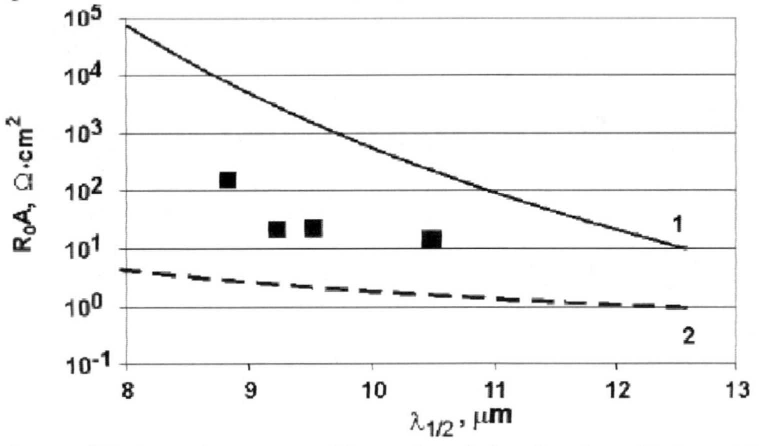

Figure 13. The dependence of R_0A product on cutoff wavelength $\lambda_{1/2}$ for photodiodes at 77 K based on MCT/Si(310).

As can be seen, the R0A product of these photodiodes in 8-12 μm is below the upper values calculated assuming the limitation by thermal generation (curve 1), while significantly exceeding the values determined for a regime limited by the background noise (curve 2) [25]. Thus, the obtained data are indicative of a high quality of the material, which is a necessary prerequisite for the development of multielement FPAs. The FPAs 288×4 parameters were measured at 77 K. Figure 14 shows tomogram of the specific detectivity (a) and voltage sensitivity (b).

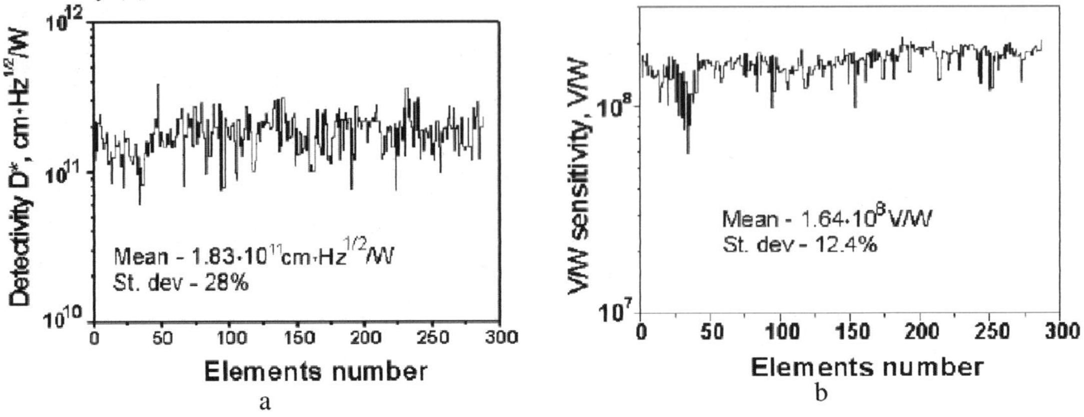

Figure 14. Topogram of the specific detectivity (a) and voltage sensitivity (b) of FPA 288×4 at 77K, $\lambda_{1/2}$= 9.5 μm.

As can be seen, all 288 channels are photosensitive and their characteristics are not inferior to those of the analogous FPA's based on MCT grown on lattice-matched CdZnTe substrates. The mean specific detectivity (D*) is 1.83×10^{11} cmHz$^{1/2}$W^{-1} at standard deviation of 28%, while the voltage sensitivity is 1.64×10^{8} V/W at standard deviation of 12.4%. When the linear IRFPAs are employed in technical imaging systems, additional requirements are imposed on the homogeneity of parameters of the elements. In long wavelength FPAs, channels with a specific detectivity below 5×10^{10} cmHz$^{1/2}$W^{-1} are conventionally classified as defect elements. According to this criterion, the presented 288 ×4 FPA based on HgCdTe/Si(310) heterostructure has no defect channels, since the minimum specific detectivity is about 6×10^{10} cmHz$^{1/2}$W^{-1}. So we showed that undoped hole-type HgCdTe layers with x =0.23 grown by MBE on Si(310) substrates ensure high photoelectric parameters limited by background radiation for LWIR FPAs 288×4.

3. HgTe QW MBE on GaAs(310)

3.1 Experimental procedure

Single and multiple HgTe-based QWs structures are shown schematically in Fig. 15 a. The buffer layers of ZnTe (0,3 μm) and sequence CdTe (5-7 μm) were grown on an atomic clean surface of (013)GaAs substrate prepared by chemical etching and thermal annealing in As flux in the an ultra vacuum chamber. Then 16-22 nm HgTe QWs with 24 nm $Hg_{1-x}Cd_xTe$ (x ~ 0,7) spacers were grown on the CdTe/ZnTe/GaAs at temperature 180-190 °C. The central part (10 nm) of spacer was doped *in situ* up to carrier concentration of approximately 10^{15} cm^{-3} using a conventional indium source. An ultrafast single wavelength (λ=0.6328) automatic ellipsometer LEF-755 (UFE) was used for layer buffer thickness and MCT composition and thickness control [26]. The evolution of ellipsometric parameters ψ and Δ during the growth of a single HgTe-based QW is shown in Fig. 15 b. It is represented as a sectionally smooth curve in the ψ-Δ plane. Smooth sections correspond to constant MCT composition layers growth. Their length determines layer thickness. The initial point O corresponds to the ellipsometric parameters of CdTe surface.

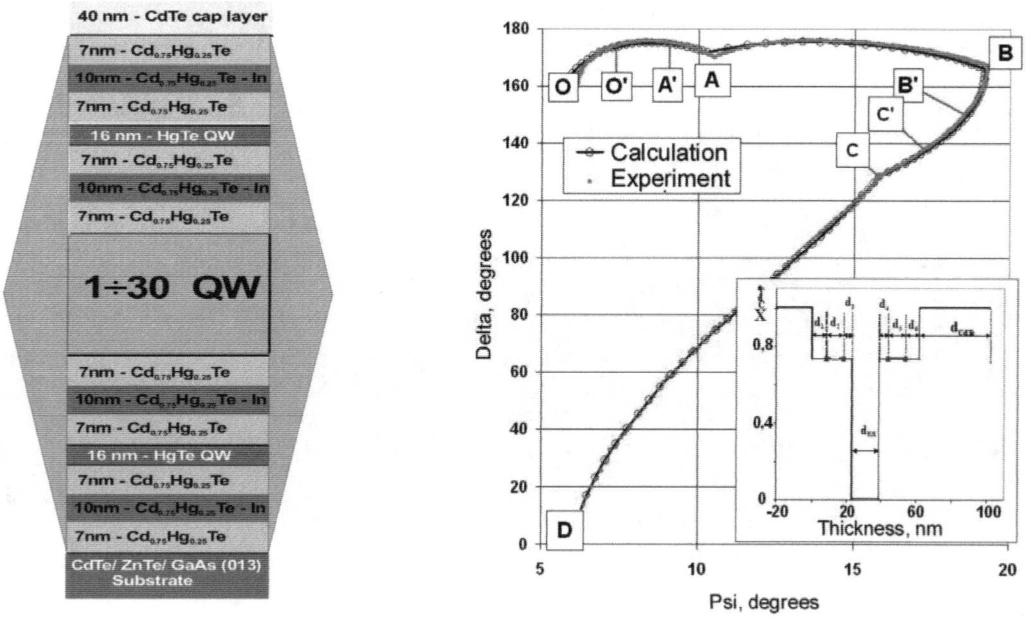

a b

Figure 15. The scheme of HgTe QW (a) and ellipsometric parameters evolution at single HgTe QW growth (b)

The fabrication of QW is beginning from the first spacer layer (x ~ 0.7) growth (curve O-A). The doping of central part of this layer is carried out by indium after opening in O' and closing in A' of the indium source shutter. Then the growth of wide-gap layer is continued up to point A. In the insertions, the sectors O-O', O'-A' and A'-A correspond to those undoped with thickness d_1, to In-doped with the thickness d_2 and to the undoped wide-gap layer with the thickness d_3 parts of spacer layer, respectively. After the closing of the cadmium source shutter, HgTe layer in thickness d_{QW} is growing and ellipsometric parameters ψ and Δ are changing between A-B points of dependence in ψ-Δ plane. The second spacer layer (see B-C curve) is growing in a manner which is analogous to the growth of the first spacer layer (O-A curve). The thicknesses d_4 , d_6 and d_5 correspond to undoped, In-doped and undoped parts of the second spacer layer. The grown single QW structure is covered by CdTe cap layer in thickness d_{CdTe} ~ 40 nm. The dots on the curve are experimental data which are measured with a 1s interval. The circles correspond to calculated values of the ellipsometric parameters through 1 nm in the thickness interval. The calculations were carried out using a one-layer model [27].

Multiple HgTe QWs were grown as periodic HgTe QW structures. Fig. 16 a shows the experimental ellipsometric trajectory measured during the growth of multiple HgTe QW. The curve breaks at each interface and, after approximately ten alterations, displays a nearly reproducible variation between points P_1 and P_2. This proves periodicity of growing structure parameters, layers composition and thickness. Fig. 16 b shows the thicknesses for all layers of the 30 multiple HgTe QW obtained by calculation from the experimental trajectory in Fig. 16 a. The average period of the structure is 37.7 nm (d_1 =21.1 and d_2 =16.6 nm). There is some dispersion in thicknesses, though the structure is periodical. It should be possible to improve the fabrication quality of the periodic structure by building feedback control implying automatic regulation (adjustment) of molecular sources relying on ellipsometric measurements.

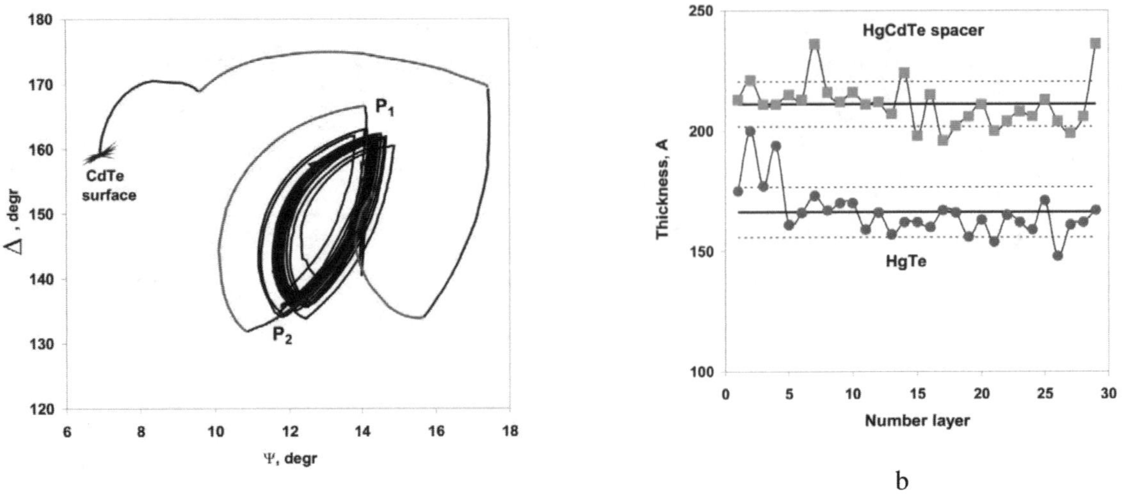

a b

Figure 16. The ellipsometric parameters evolution at 30 multiple HgTe QW growth (a) and thickness calculation from experimental data (b).

3.2 Electrophysical parameter.

Magnetotransport measurements of Hall bar HgTe QW were carried out in magnetic fields up to 17 T and temperatures from 50 mK to 3 K. Typical HgTe QW width 16 nm and 21 nm were investigated. Fig.17 shows the ρxx(B) and the Hall resistivity ρxy(B) in the temperature range (0.3-3) K and in magnetic fields up to 13 T.

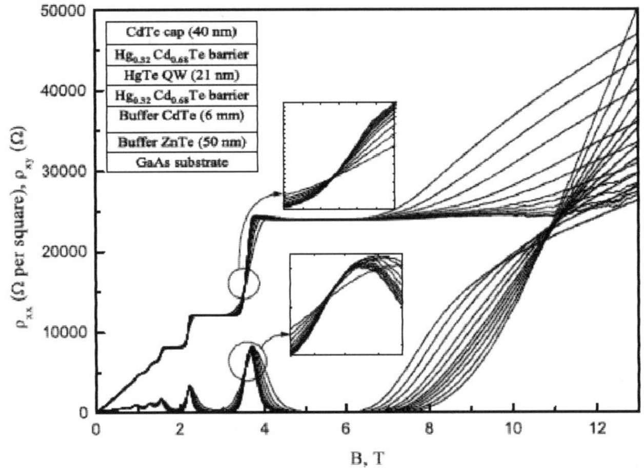

a

Figure 17. The longitudinal and the Hall resistance versus magnetic field in a 21 nm wide HgTe QW. The traces are taken at temperatures T= 0.3; 0.4; 0.5; 0.6; 0.7; 0.8; 1; 1.2; 1.6; 2 and 3 K.

One can see an ordinary QHE behavior beginning at B = 2 T with a wide plateau in the ρxy and a corresponding wide minimum in the ρxx. A behavior resembling a magnetic field induced QH liquid-insulator transition is observed at T < 1.6 K after the Fermi energy crosses the lowest Landau level. The transition is characterized by a critical magnetic field Bc = 10.9 T and a critical diagonal resistivity value ρxx = 0.9h=e^2. The quantum Hall effect investigations confirmed the high quality of HgTe QW. The observed magnetoresistances behavior is consistent with two-dimentional electron gas. Electron mobility at T = 1.6 K was (2.4-3.5) ×10 5 cm 2 /V·s for electron concentration N_s = (1.5-3) ×10^{11} cm^{-2} revealing the high quality of grown QW.

3.3 Detector application

We found a large sensitivity of HgTe QW to polarised radiation. The photocurrents were induced by direct band-to-band optical transitions or indirect intrasubband (Drude like) optical transitions in the lowest size-quantized subband applying mid-infrared (MIR) or terahertz (THz) radiation, respectively.

Figure 17 a shows the helicity dependence of the photocurrent measured at room temperature. The photocurrent was excited by THz radiation with wavelength λ= 148 μm and power P ≈ 5 kW from two perpendicular pairs of contacts. The solid lines are fits to phenomenological equations. Insets show corresponding experimental geometries. Along the top the polarization ellipses corresponding to key phase angles ' are sketched. The effect is observed in a wide spectral and temperature range, from 6 μm to 496 μm and from 4.2 K to 300 K.

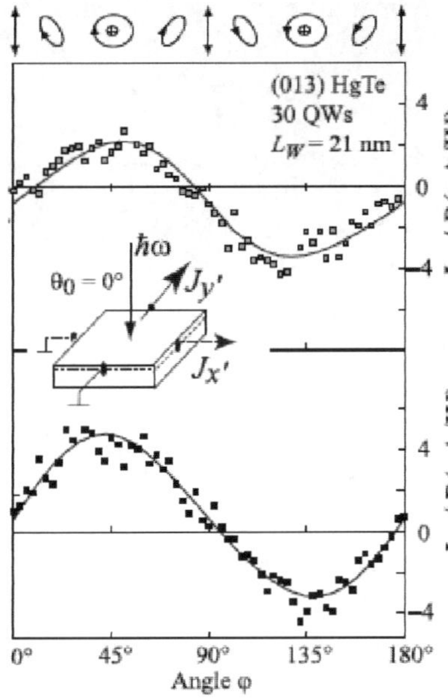

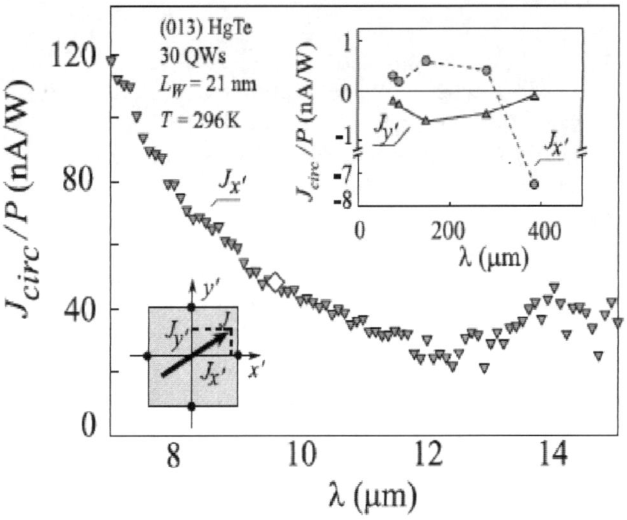

Figure 17. Photocurrent as a function of radiation helicity measured at room temperature in HgTe QWs grown on (013)-oriented GaAs substrates (a) and thickness calculation from experimental data (b).

Depending on the radiation wavelength and/or sample temperature the currents in both x′ and y′ directions can be well fitted simply by $J = \pm J_0 \times \sin 2\varphi \propto P_{circ}$ (as for the circumstances of Fig. 1) or by more complex dependence on the angle (see Fig. 3) given by $J = A\times\sin2\varphi + B\times\sin4\varphi + C\times\cos4\varphi + D$. (1) Here, A, B, C, and D are fitting parameters. The wavelength dependence of Jx′ and Jy′components are shown in Fig. 17 b. We attribute the observed time constant to the bandwidth of our electronic setup. A fast response is typical for photogalvanics where the signal decay time is determined by the momentum relaxation time being in our samples of the order of 0.3 ps at room temperature. As a large dynamic range is important for detection of laser radiation, we investigated the dependence of the sensitivity of the detection system on the radiation intensity applying cw and high-power pulsed radiation.We observed that the ellipticity detector at homogeneous irradiation of the whole sample remains linear up to 2 MW/cm^2 over more than nine orders of magnitude.

4. Conclusion

Investigations of growth processes of MCT MBE on Si(310) substrates for 3rd generation FPAs were carried out. It was shown that the optimization of processes of surface preparation and growth conditions allows to obtain MCT MBE on Si (310) without antiphase domains. Optimization of the growth process and the absence of antiphase boundaries allowed to reduce the density of morphological V-defects to ~ 1000 cm^{-2}. A technology for the device quality undoped p-type MCT was developed. It was demonstrated that HgCdTe/CdTe/ZnTe/Si(310) could be used to create reliable, resistant to thermal cycling FPAs for the spectral ranges 3-5 and 8-12 μm.

We demonstrated the growth of single and multiple HgTe quantum wells with precise control by single wavelength ellipsometry which allows us to measure the composition and thickness of each layer during the growth. High quality single and multiple (up to 30) HgTe QWs which were grown on (013) CdTe/ZnTe/GaAs substrates was confirmed by the presence of high mobility 2D electron gas. We demonstrated the application HgTe QW for detection of the polarization state of elliptically polarized radiation from the mid-infrared to the THz range. The sensitivity and linearity of the detection system developed here has been found to be sufficient to characterize the polarization of laser radiation from low-power cw-lasers to high-power laser pulses. The short relaxation time of free carriers in semiconductors at room temperature makes it possible to detect sub-nanosecond laser pulses.

5. Acknowledgments

Authors would like to acknowledge all colleagues for taking part in fabrication, measuring parameters and scientific investigation of MCT MBE and on basis of its PV FPA's. Especially thanks to S. Ganichev, S. Danilov, D. Kvon, E. Olshnetskiy for measuring HgTe QW and L.D.Burdina, O.I.Malyshev, D. Ikusov, L. Mironova for growing MCT HS.

6. References

[1] Kinch M A 2000 *J. Electron. Mater.* **29** 809
[2] Schulman J N , McGill TC 1979 *Appl. Phys. Lett.* **34** 663
[3] Reddy M, Peterson J M, Lofgreen D D , Franklin J A , Vang T, Smith E P G , Wehner J G A , Kasai I, Bangs J W and Jonson S M 2008 *J. Electron.Mater.* **37** 9 1274.
[4] Koestner R J and Schaake H F 1988 *J. Vac.Sci.Technol.* **A 6** 4 2834
[5] Ryu Y S, Song B S, Kang T W and Kim T W 2004 *J. Mater. Sci.* **39** 1147
[6] Carmody M, Pasko J G , Edwall D, Piquette E, Kangas M, Freeman S, Arias J, Jacobs R, Mason W, Stoltz A, Chen Y and Dhar N K 2008 **37** 9 1184
[7] Yakushev M V, Babenko A A, Sidorov Yu G 2009 *Neorganicheskiematerialy* **45** 1 15 (in Russian)
[8] Ganichev S D, Kiermaier J, Weber W, Danilov S N, Schuh D, Gerl Ch, Wegscheider W, Bougeard D, Abstreiter G and Prettl W 2007 *Appl.Phys. Lett.* **91** 091101
[9] Sidorov Yu G, Dvoretsky S A, Mihaylov N N, Yakushev M V, Varavin V S, Antsiferov A P 2000 *Opticheskiy zhournal* **67** 1 39 (in Russian)
[10] Kern W and Puotinen D A 1970 *RCA rev.* **31** 187
[11] Fenner D B, Biegelsen D K and Bringans R D 1989 *J.Appl. Phys.* **66** 419
[12] Blinov V V, Dvoretsky S A and Sidorov Yu.G. 1997 Patent of Russian Federation №2071985 Priority from 11.01.1993. Registration 20.01 1997. Bulletin №2 from 20.01.97. (in Russian)
[13] Blinov V V, Goryaev E P, Dvoretsky S A et al. 1997 Claim for invention № 95102853/25, priority from 01.03.95. Positive solution from 20.08. 1997. (in Russian)
[14] Bhan R K, Dhar V, Chaudhury P K et al. 1996 *Appl. Phys. Lett.* **68** 17 2453
[15] Buldygin A F, Vdovin A V, Studenikin S A, et al. 1996 *Avtometriya* **4** 73 (in Russian)
[16] Voitsehovsky A V, Denisov Yu A, Kohanenko A P, et al. 1996 *Avtometriya* **4** 51 (in Russian)
[17] Dvoretsky S., Varavin V, Mikhailov N, Sidorov Yu, ZakharyashT, V. Vasiliev V, Ovsyuk V, Chekanova G, Nikitin M, Lartsev I, Aseev A 2005 *Proc. SPIE* **5964** 75
[18] Wang C C and Me Farlane S H 1976 *Thin Solid Films* **31** 3 323
[19] Sidorov Yu G , Varavin V S , Dvoretsky S A et al. 1996 *J. Growth of Crystals* **20** 35
[20] Aoki T, Chang Y, Badano G, Zhao J, Grein C, Sivananthan S and Smith 2003 *J. Electron. Mater.* **32** 703
[21] Tashikawa M, Yamaguchi M 1990 *Appl. Phys. Lett.* **56** 5 484
[22] Sidorov Yu G, Dvoretsky S A, Mikhaylov N N and Varavin V S 2001 *Infrared focal planearrays*, Nauka, Novosibirsk (in Russian)
[23] Rheenen A D, Syversen H, Haakenaasen R, Steen H, Trosdahl-Iversen L and

Lorentzen T 2006 *Phys. Scr.* **T126** 101
[24] Kuleshov V F, Kuharenko Yu A, Fridrihov S A, et al. 1985 Spectroscopy and electron diffraction in the study of solid surfaces, Nauka, Moscow (in Russian)
[25] Rogalski A 2003 *Infrared Detectors* CRC Press
[26] Spesivtsev E V and Rykhlitski S V. Useful model " Ellipsometer " . Licence No 16314. Bulletin " Useful models. Industrial samples " 35 (20.12.2000) (In Russian)]
[27] Svets V A, Rykhlitski S V, Spesivtsev E V, Aulchenko N A, Mikhailov N N, Dvoretsky S A, Sidorov Yu G, Smirnov RN 2004 *Thin Solid Films* **455-456** 688

Journal of Physics: Conference Series **345** (2012) 012003 doi:10.1088/1742-6596/345/1/012003

NEW TECHNOLOGY OF ION-PLASMA MODIFICATION OF THE CONTACT SURFACES OF REED SWITCHES IN OSCILLATORY DISCHARGE

K A Arushanov, I A Zeltser[1], S M Karabanov, R M Maizels, and Y N Moos

Ryazan Metal Ceramics Instrumentation Plant Joint Stock Company, Novaya St. 51V, Ryazan, 390027, Russia.

[1]E-Mail: zeltseria@rmcip.ru

Abstract: The iron-nickel contact surfaces of reed switches after ion-induced modification have been studied using methods of Auger-electron spectroscopy, X-ray photoelectron spectroscopy, atomic force microscopy, and optical microscopy. It was demonstrated that the corrosion stability and erosion resistance of the modified contacts is associated with the features of surface topography as well as with a formation of nitride layers. Experimental grounds for a production possibility of reed switches with the modified contact surface instead of electroplating based on the precious metals are given.

Key words: reed switch, ion, plasma, nitriding, coating, modification, contact, nitride, Auger-electron spectroscopy, X-ray photoelectron spectroscopy, atomic force microscopy.

1. Introduction

One of the main technological problems of reed switch production is the deposition of special (corrosion- and erosion-resistant) coatings on the contact surfaces of permalloy springs. To deposit a coating, the electroplating technique is generally used and noble and precious metals (gold, palladium, rhodium, ruthenium, etc.) are employed as materials [1].

However, this technological approach is characterized by large electro- and material capacity, high prices of the equipment, low ecological level, and difficulty of the alloy deposition with the required chemical and phase composition and structure when forming thin nonporous or thick films having low internal stresses and high adhesion to the material of a contact piece.

It's known that one of the most promising methods for increase of corrosion and erosion resistance of metal surface is ion nitriding in pulsating plasma of glow discharge [2, 3].

The idea of usage of this method for producing of contact surfaces, that are alternative to electrodeposits based on precious metals, has found it's experimental verification for the first time in the work [4]. It was developed further in series of works [5-25].

The present paper is a follow-up of these works. The aim if this article is studying of influence of ion-plasma treatment parameters, switching regimes on properties of nickel-iron contact surfaces and creation of a new technology of reed switch modification with usage of nitriding and reactive cathode sputtering in oscillatory discharge.

2. Materials and Methods

2.1.Samples

The samples were reed switches MKA-14108 [1] structurally made on base of off-the-shelf devices MKA-14103 [1] (**Figure 1**). The special feature of design of such reed switches in comparison with the off-the-shelf devices consists in lack of any special coatings on permalloy contacts.

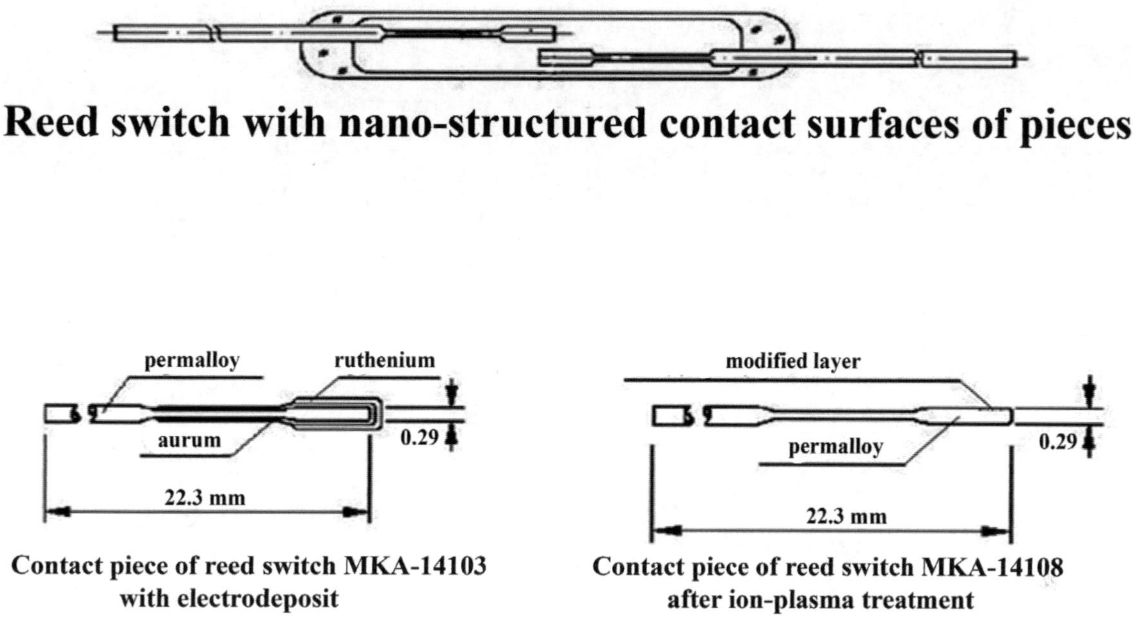

Figure 1. Reed switch design.

2.2. Technology

The processing route for production of reed switches MKA-14108 is shown in **Figure 2**, and the primary processes – in **Figure 3**. The specificity thereof as compared with the typical flow layout [1] consists in replacement of the operation "Plating" by the operation "Ion-Plasma Treatment" ("IPT").

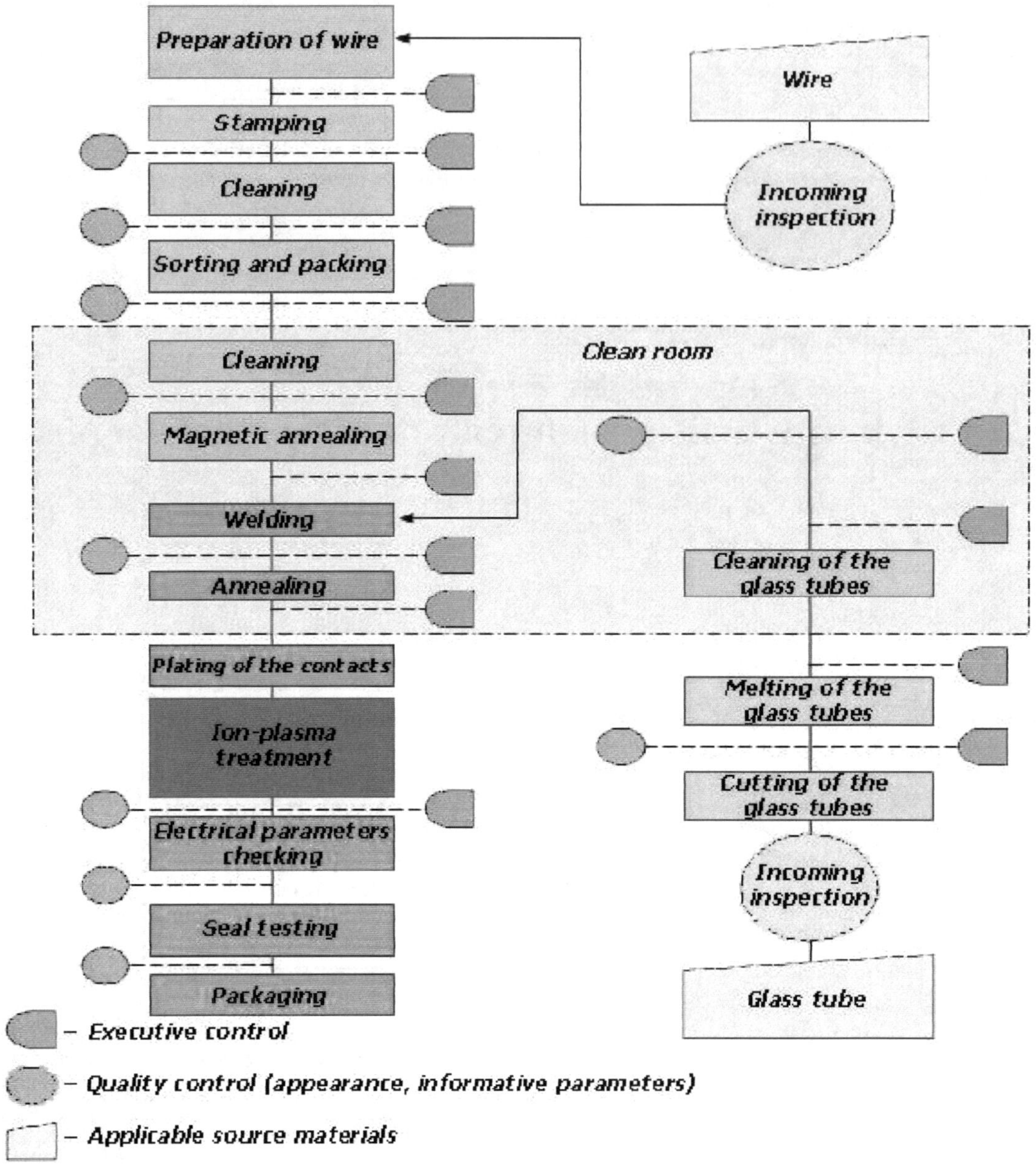

Figure 2. Processing route for reed switches MKA-14108 with the modified contact surface (instead of electroplating).

Contact springs were pressed from Permalloy Dilaton wire, degreased and subjected to hydrogen annealing. The wire was stretched from vacuum-melt Permalloy (52% Ni, 48% Fe). Spectrally pure (99.999 %) nitrogen with the pressure in the envelope of reed switch in the value of $33\cdot 10^3 - 40\cdot 10^3$ Pa was used as gas filling when sealing.

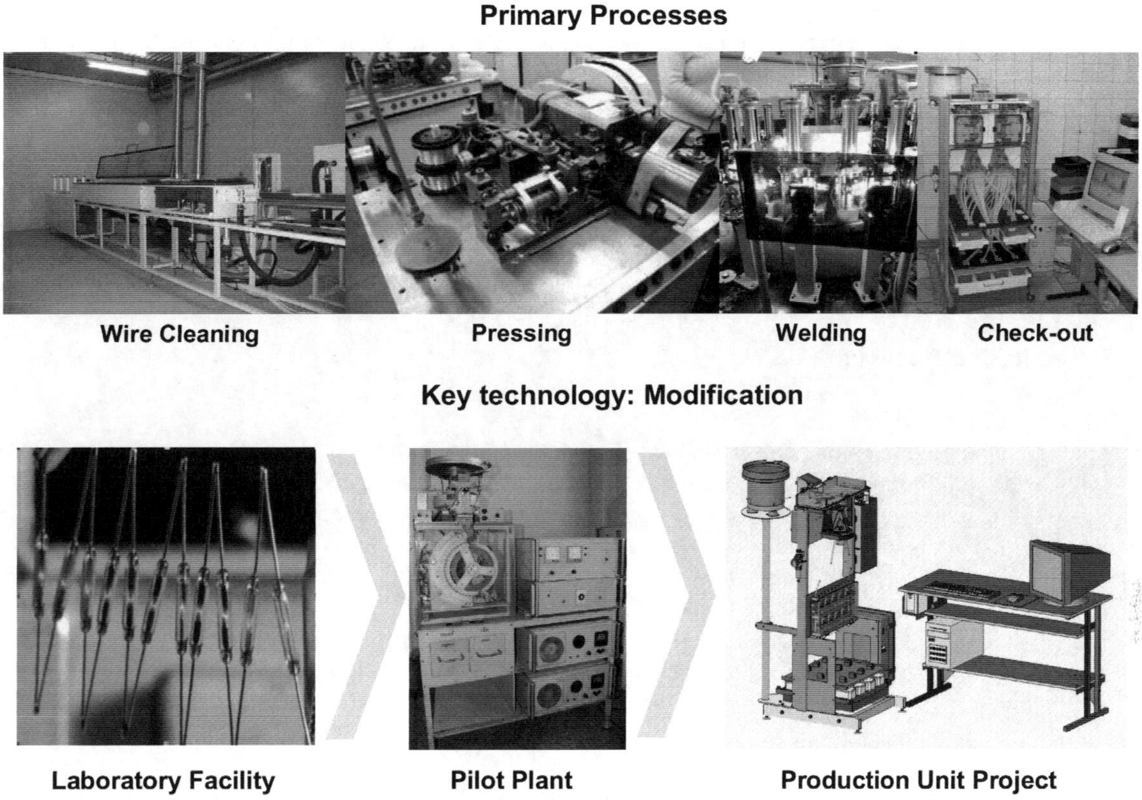

Figure 3. Technology.

2.3. Ion-Plasma Treatment

Ion-plasma treatment (IPT) of the contact surfaces was carried out by high-voltage pulsed discharges which were initiated on the open contacts (with a gap d = 27–30 μm) by means of a specially developed device [5, 7] (**Figure 3**).

Two modes of IPT characterized by the electrophysical conditions of discharge, and as a consequence, - by a heating degree of the substrate, were used. The first IPT mode is the mode of ion nitriding, providing the diffusion saturation of near-surface layers of contacts in the region of overlap by nitrogen atoms, formation on the treated parts of nitride layer with the prescribed contact properties. Properties of the hardenable surface, corrosion- erosion resistance and electroconductivity are optimized due to the necessary combination of nitride and diffusion layers which grow into a base material. Depending on a chemical composition, the nitride layer is either γ'- phase (Fe_4N) or ε-phase ($Fe_{2\text{-}3}N$). The second mode is the mode of reactive cathode sputtering. During this process, a film comprising iron and nickel nitrides is sputtered onto the surface of contacts. Like ion-nitrided layers, this coating also meets the modern requirements for the contact surfaces of reed switches. Duration of a single IPT in both modes was 30 sec.

The distinctive feature of the ion nitriding and reactive cathode sputtering of reed switches is that in contrast to the conventional technologies [2, 3, 26], the saturation of near-surface layers of contacts by nitrogen and sputtering of contact coatings is carried out not in nitrogen-containing vacuum (p = 150-1000 Pa), but at rather high pressure ($p = 33\cdot 10^3 - 40\cdot10^3$ Pa) of nitrogen. By this it's not necessary to apply the special vacuum equipment for ion nitriding of near-surface layers and forming of coatings by methods of reactive cathode sputtering [2, 26]. Nitriding and sputtering processes are performed direct in the reed switch. The glass envelope functions as a working chamber, and contact pieces serve as anode and cathode alternately.

2.4. Research technique

The samples after treatments by high-voltage discharges and the samples without treatments were put to the switching tests. Resistance tests, breakdown voltage measurements and switching reliability tests of reed switches were performed with the help of the special-purpose equipment [1] according to methods presented in [1].

About the state of modified surface it was judged by the results of Auger-electron spectroscopy (AES), X-ray photoelectron spectroscopy (XPS), atomic-force microscopy (AFM) and optical microscopy (OM). Gas composition and pressure were controlled by means of a magnetic mass-spectrometer. The findings were matched with the measured results of a number of characteristics, particularly, of breakdown voltage and transient contact resistance.

3. Results and Discussion

3.1. Ion-Nitrided Layers

Analysis of topography of contact surface (before and after treatment thereof) was performed in air with the help of atomic-force microscope (by probe nanolaboratory (PNL) NTEGRA manufactured by NT-MDT Company, Zelenograd, Moscow, Russia). For optical researches of the working surfaces of contacts of reed switches, the metallurgical microscope MMP-4 with a computer system of image visualization was used. Si cantilevers of NSG10/W_2C type, with a hard conducting coating of W_2C 30 nm thick, with work function $e\varphi_p = 4.902$ eV were used.

The contact surface of reed switch after IPT according to the first mode (ion nitriding) may be divided into three areas, typical by surface topography, which are revealed on images of contact pieces, obtained in optical (**Figure 4**) and atomic-force (**Figure 5**) microscopes.

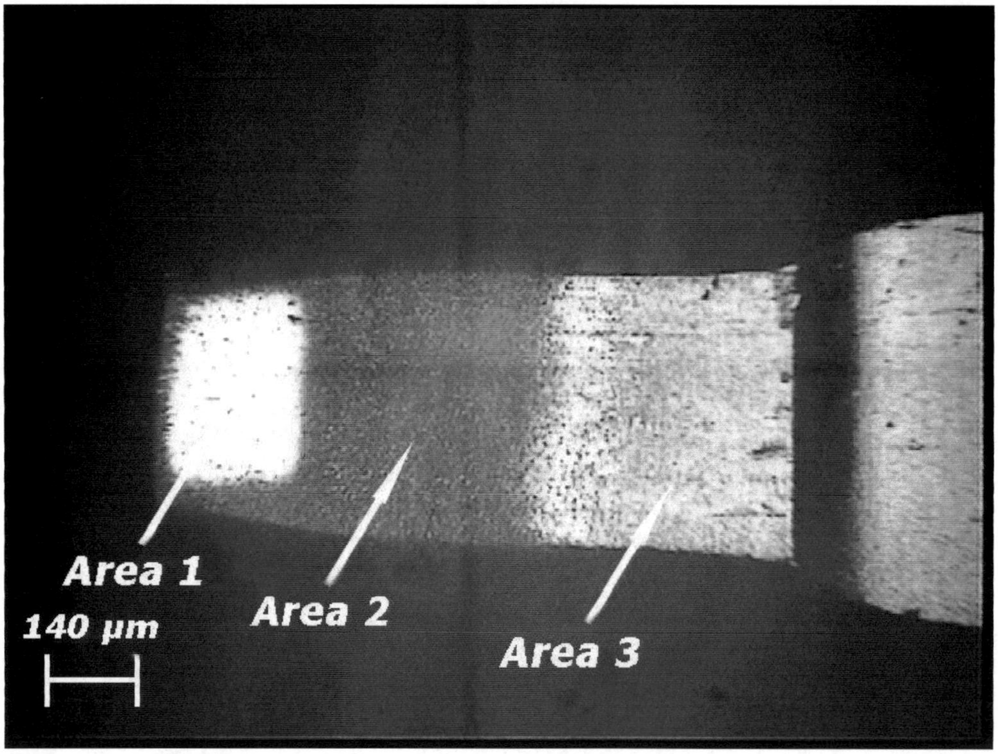

Figure 4. Contact piece of the reed switch after the hundred-fold ion-plasma treatment.

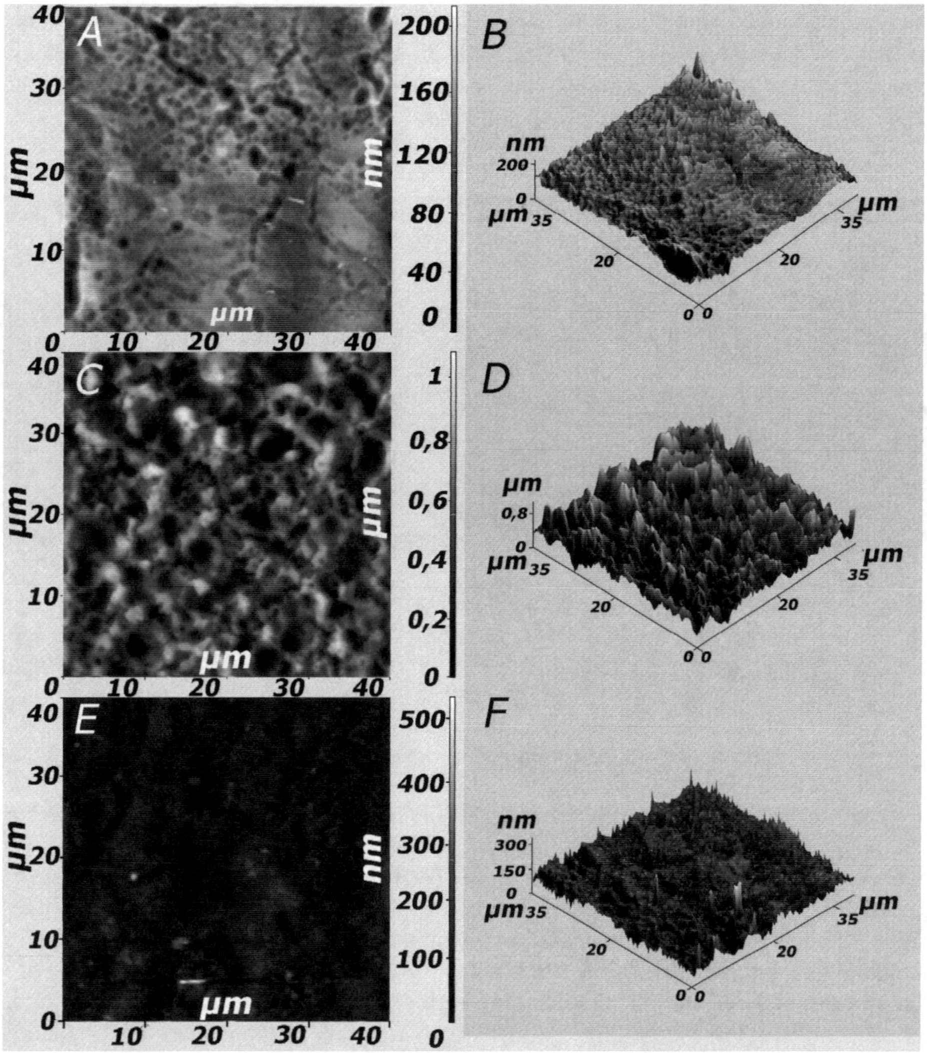

Figure 5. AFM-image of the contact surface of reed switch after 100-fold IPT: **(A)** and **(B)** -2D- and 3D-image of area 1; **(C)** and **(D)** - 2D- and 3D-image of area 2; **(E)** and **(F)** - 2D- and 3D-image of area 3.

Area 1 (**Figures 4, 5a,b**) is a working area (contact overlap area), where gas discharge is ignited when voltage is supplied on the open contacts of reed switch. As a result of reactive cathode sputtering, the nitrides (FeN and NiN) are formed in gas phase of overlap area, the part of which is deposited on the contact surface in area 1, where under ion bombardment nitriding of surface is performed according to Kelbel mechanism [26].

Reactions which determine the ion nitriding process (the first IPT mode) in conditions of the glow discharge according Kelbel mechanism are shown in **Figure 6,** and Figure 6b shows reactions determining the sputtering process of contact coatings by method of the cathode sputtering.

When accelerating in the area of cathode drop of the potential, the nitrogen ions bombard the target – cathode. At that, the majority of ion energy (about 90 %) is expended on the target heating, and the rest – on the electron emission, ion implantation and sputtering of atoms (Fe, Ni, C, O) located on the target surface. As a result of reactive cathode sputtering, in the gas phase, iron and nickel nitrides are formed which are deposited on the contacts and glass envelope of the reed switch. When metal gets to the surface at the temperature of 300-500° C, the molecules of metal nitrides MeN (Fe, Ni) of permalloy dissociate and form low-order nitrides Me_2N, Me_3N, Me_4N. Nitrogen escaped in this case is diffused partially into the target and partially is vaporized into plasma.

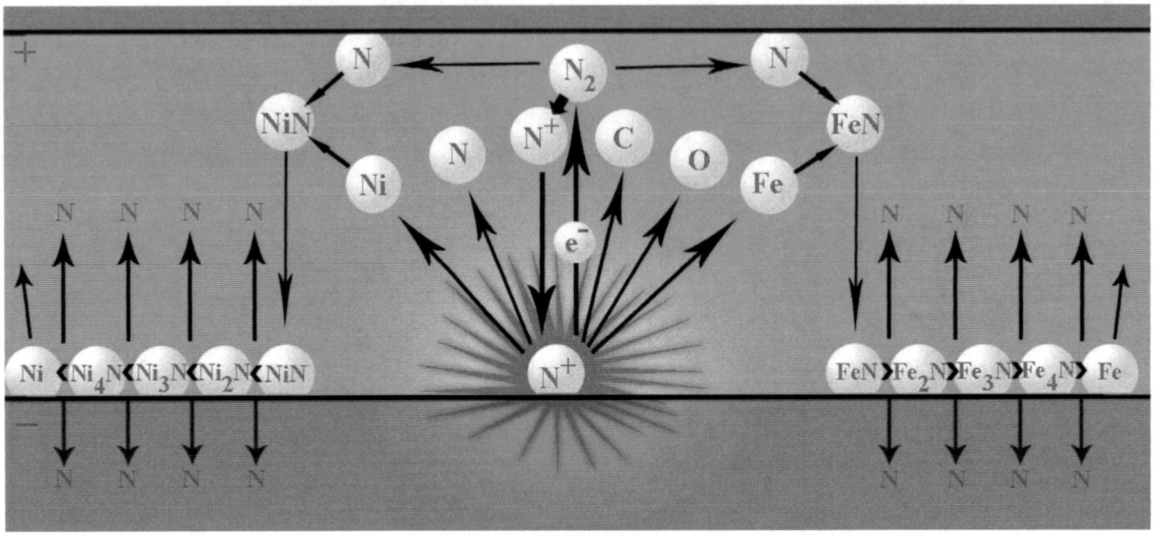

Figure 6. Ion nitriding.

As a result of diffusion, a part of FeN and NiN molecules will be deposited on the surface of contacts, in area 2 adjacent to the area 1 (**Figures 4, 5 c, d**). Area 3, after estimating in accordance to optical (**Figure 4**) and atomic-force (**Figure 5 e, f**) images should not contain nitrides or contains it in very small quantity. The surface of area 1 is heated as a result of ionic bombardment, and of areas 2 and 3 - in consequence of heat exchange with the areas 1 and 2, respectively. Therefore, because of low-grade heating, dissociation of FeN and NiN molecules, and, consequently, the ion nitriding of surfaces of areas 2 and 3 is unlikely. On the surface of contacts in area 2 (**Figure 4**) and on the inner surface of glass envelope, located near to the region of overlap, reactive cathode sputtering products form a black coating (**Figure 7**). The same coating can be formed in area 1, if, due to low ion energy (or current density), the temperature threshold of dissociation of FeN and NiN molecules will not be overpassed. It's possible that during the diffusion a small amount of cathode sputtering products comes to the reverse sides of reed switch contacts.

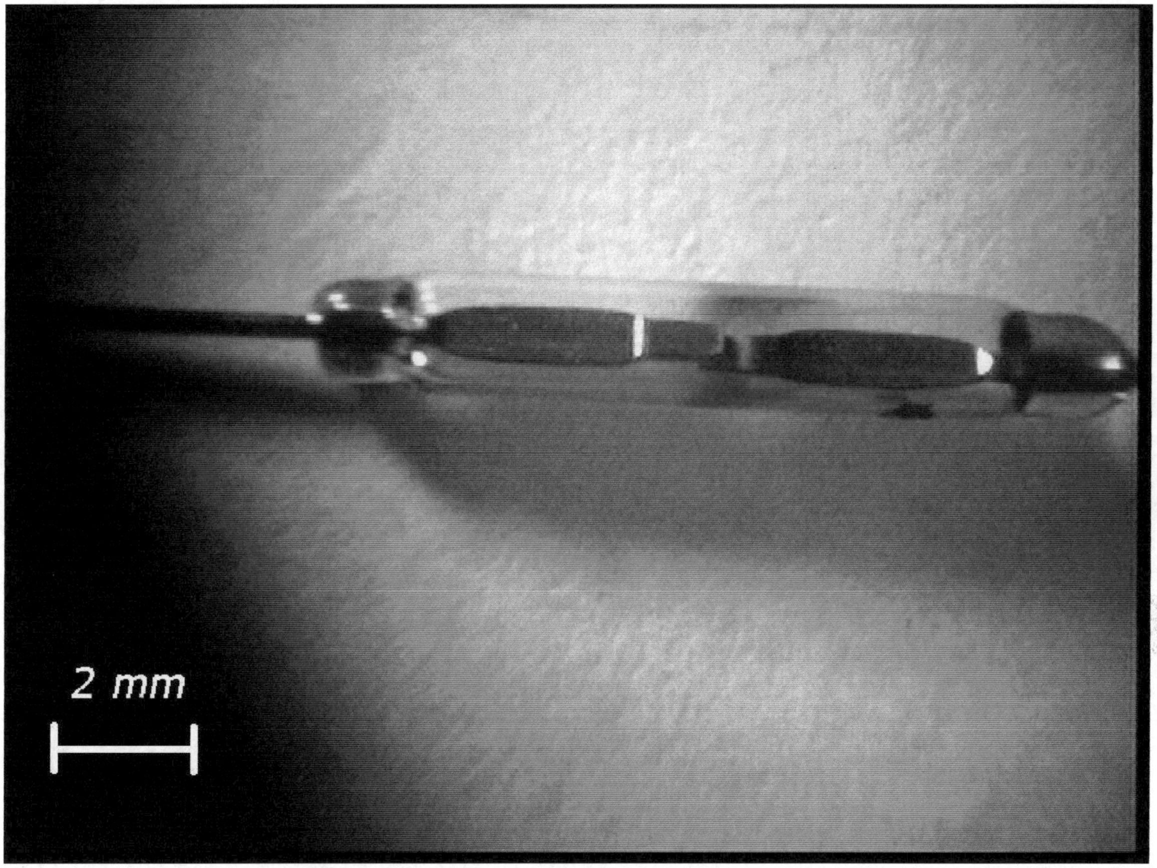

Figure 7. Reed switch after a hundred-fold ion-plasma treatment.

3.1.1.Auger- and AFM-Investigations

Elemental composition analysis of the surface of contacts (before and after treatment) was carried out using Auger electron spectrometer [27]. The device is equipped with the cylindrical mirror electron energy analyzer (resolution of 0.25 %) with an integrated electron gun having a beam current up to 1 μA if diameter of the beam is 100 μm. To clean the surface of contact pieces of reed switches and for layer-by-layer Auger analysis, an ion gun with a differential pumping of the working gas (Ar) and ion current density up to 3.5 mA/cm2 was used. All measurements were conducted in vacuum 2×10^{-7} Pa to eliminate effects of electron-stimulated adsorption of molecules of residual gases on the investigated surfaces.

Conductive nano-asperities

The received Auger profile of phase distribution in depth together with an indication of nitrogen concentration in stoichiometric compounds (Fig. 8) as well as images of nanostructured contact 50 nm asperities (Fig. 9) on a new heterogeneous Fe_3N (30 nm) and Fe_4N (40 nm) base are evidence of formation of the nanostructured surface contact layer.

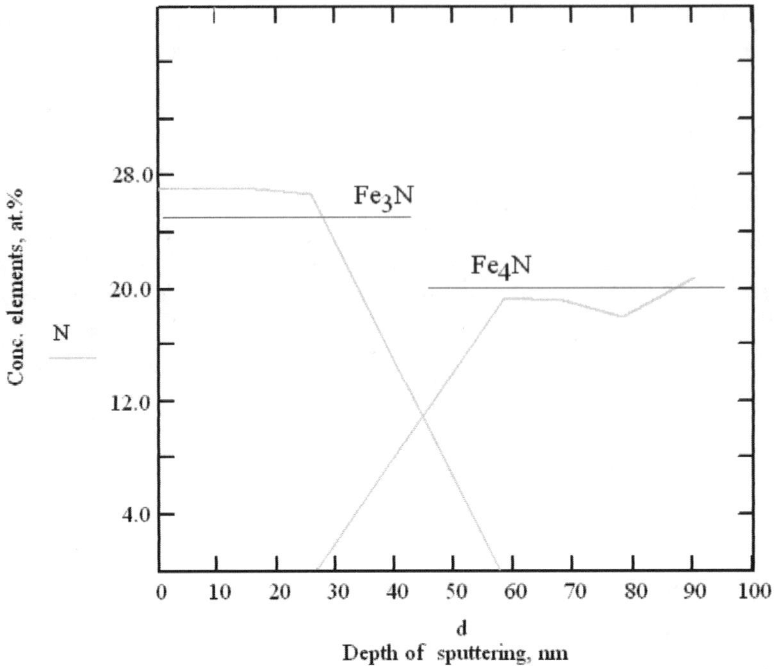

Figure 8. Restored phase distribution depth profile with an indication of nitrogen concentration in stoichiometric compounds of Fe_3N and Fe_4N. Single treatment (duration 30 s).

The surface nanostructure (nano-relief with the conductive nano-asperities) provides realization of the higher reliability principle – a multilevel echelon protection of contacts from erosion, corrosion and mechanical damages, improves the noise immunity from foreign particles, and as the final result, solves cardinally a quality problem of reed switches, what substantially reduces a failure number while in operation.

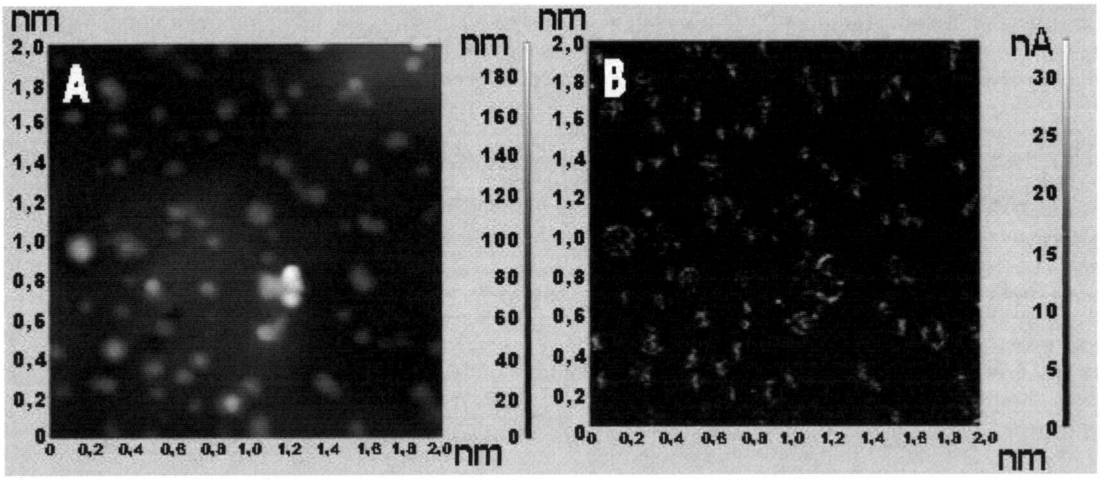

Figure. 9. AFM-image of nitrated surface area of contact spring of reed switch: **(A)** – constant actuating force method; **(B)** – spreading resistance imaging method. Single treatment.

Surface pores and cones.

Effects of ion-plasma fluxes on the surface can change its energy state, and that is showed by the relief evolution, for example, heterogeneous etching and reveal of the polycrystalline structure of the solid body. In addition, relief changes reveal ion-induced voltages, processes of recrystallization, compositional change of the near-surface layers and change of the dislocation mobility.

Ion-plasma fluxes initiate a formation of the surface stresses, diffusion activation, change of the dislocation structure, phase state. The development of these processes usually leads to the relief modifications. It is also necessary to take into account a result from the interaction of all processes of the forming relief indicated above with the processes of sputtering, ion-stimulated segregation, desorption, etc.

In the consequence of ion bombardment, the surface pores are formed in area 1 on the surface of contact pieces (**Figures 5a, b, 11**). The concentration thereof, as is evident from **Figures 5 a, b, 11,** grows with increase of duration of IPT.

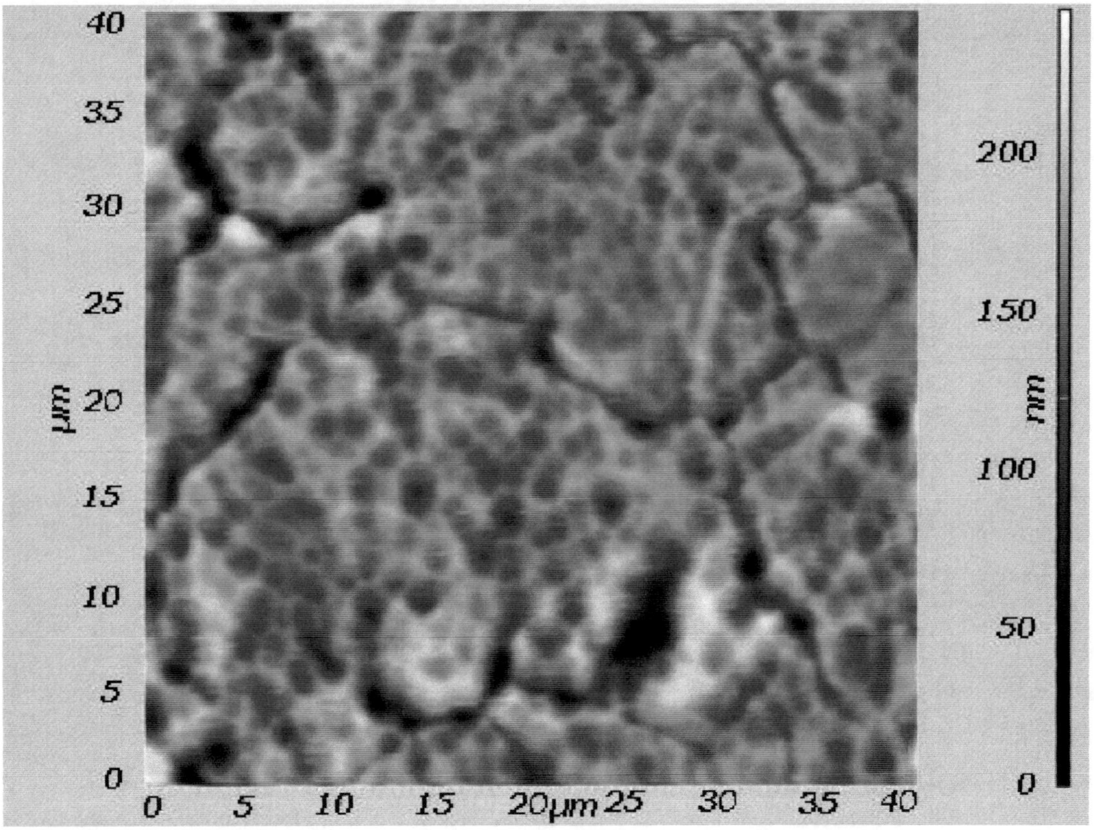

Figure 11. 2D AFM-image of area 1 of the contact surface of reed switch after 200-fold IPT. On the certain stage of development, the conical asperities start to grow on the bottom of pores (Figure 12 a, b).

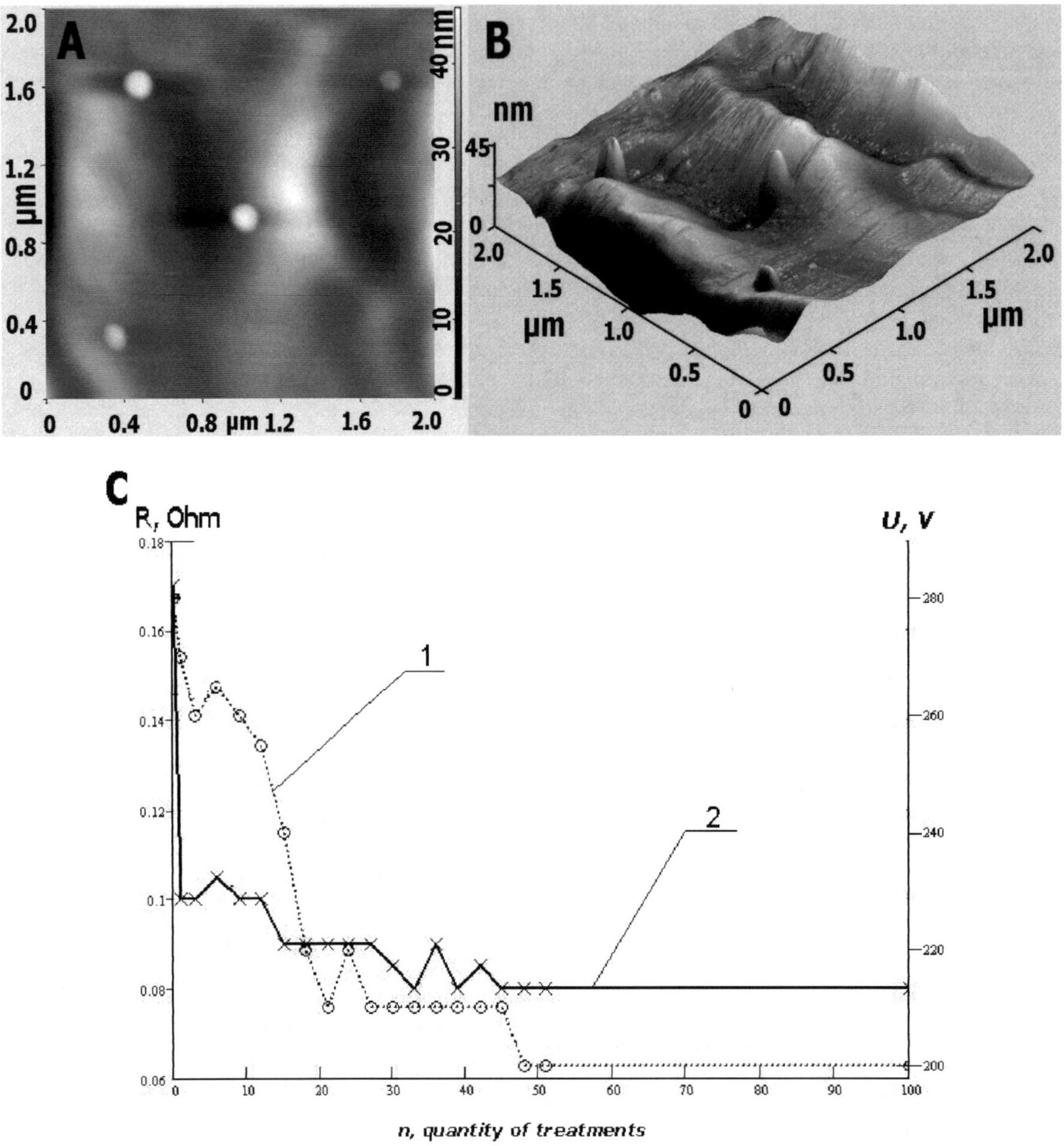

Figure 12. 2D **(A)** and 3D **(B)** AFM – image of area 1 of contact surface of a reed switch after 100-fold IPT; **(C)** dependences of resistance R (curve 2), breakdown voltage Uп (curve 1) of reed switch MKA-14108 median direction on quantity of treatments n. The quantity of reed switches in a lot is 100 pcs.

The conditions of appearance and mechanism of development of surface pores and cones on the solid surface treated by ions are considered in the paper [28]. On the author's opinion the main reason of their formation are processes caused by ion-induced voltages and connected with atom movement in the surface layer. It is conditionaled on ion-enhanced diffusion, dislocation displacement and recrystallization. Pores formation must arise from supersaturation of surface with interstitial atoms and filling of microscopic vesicles of lattice with it is the result of ion-stimulated diffusion and recombinational processes in the near-surface layers. Yield of this formations to the surface in the field of ion-induced voltages leads to creation of open pores that at first have annular shape with gradual increase of slope of sidewalls. Pore development is completed with the growth of cone ledges on their bottom.

At ion bombardment redeposition of atoms from pore slopes to its bottom leads to formation of areas with lower voltage level and consequently to formation of cones on pore bottom on account of local diffusion sink. Meanwhile cone formations lower voltage level on the bottom, material flow from pore to the surface becomes slower, as a result the pore decreases and then disperses [28].

Surfaces with such relief (**Figure 12**) can possess reduced yield of sputtering and secondary electron emission. They reflects ion flows in a less degree, are characterized by selectivity of optical properties, high field-emission parameters. The structure of cones can change catalytic, corrosion and emission properties of reed switch contact surface. As follows from experimentally obtained dependence of breakdown voltage on quantity of treatments (**Figure 12 c**) stabilization of cone surface structure comes after 30 (thirty) IPT.

3.1.2. XPS and AFM-researches

Analysis of elemental and chemical composition of contact surface (before and after treatment, switching) is realized on scanning X-ray photoemissive microprobe PHI Quantera SMX (made by Physical Electronics, Inc., USA-Japan).

All measurements were carried out in vacuum 1×10^{-8} Pa for removal of effect of electron-stimulated adsorption of residual gas molecules on the surfaces under analysis. Sputtered cleaning was not carried out. Surface sensing was carried out in the overlap area of contact-pieces by X-ray beam of about 7 μm diameter.

Emission from aluminum anode Al Kα 1486.6 eV was used. Measurements of resistance, breakdown voltage of reed switches and reliability tests (in regimes 30 V - 0.5 A - $1.25\cdot10^6$ operations at temperature 155 ºC - $1.25\cdot10^6$ operation in normal conditions (n.c.) and 50 mV – 5 μA - 10^8 operation in n.c.) were carried out with the help of special-purpose equipment [1] in methods, represented in [1].

There are obtained review XPS – spectra (**Figure 13**) and photoelectron spectra of lines C 1s, O 1s, Ni 2p, Fe 2p, N 1s, Si 2p (**Figure 14**).

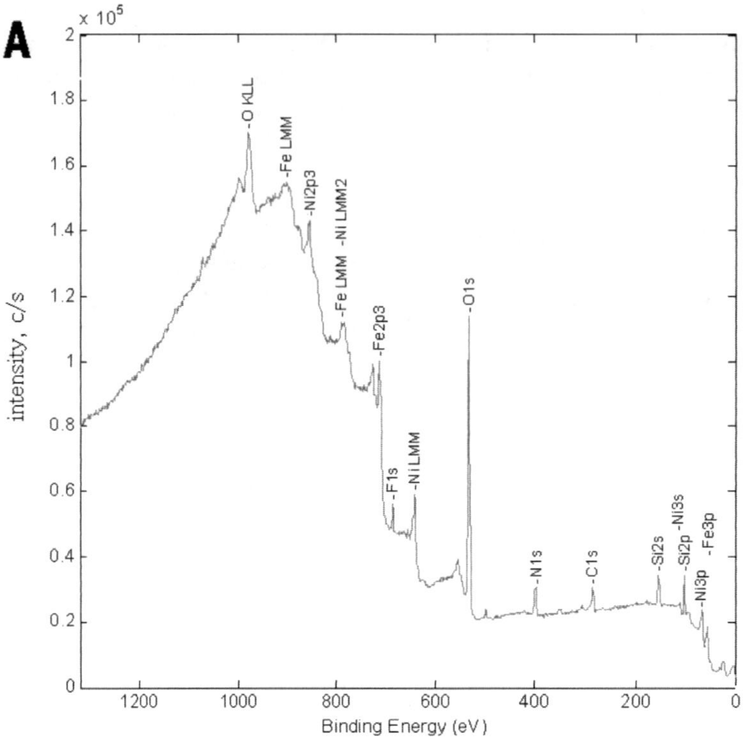
A
x 10^5
intensity, c/s
Binding Energy (eV)
-O KLL
-Fe LMM
-Ni2p3
-Ni LMM2
-Fe LMM
-Fe2p3
-O1s
-F1s
-Ni LMM
-N1s
-C1s
-Si2s
-Si2p
-Ni3s
-Ni3p
-Fe3p

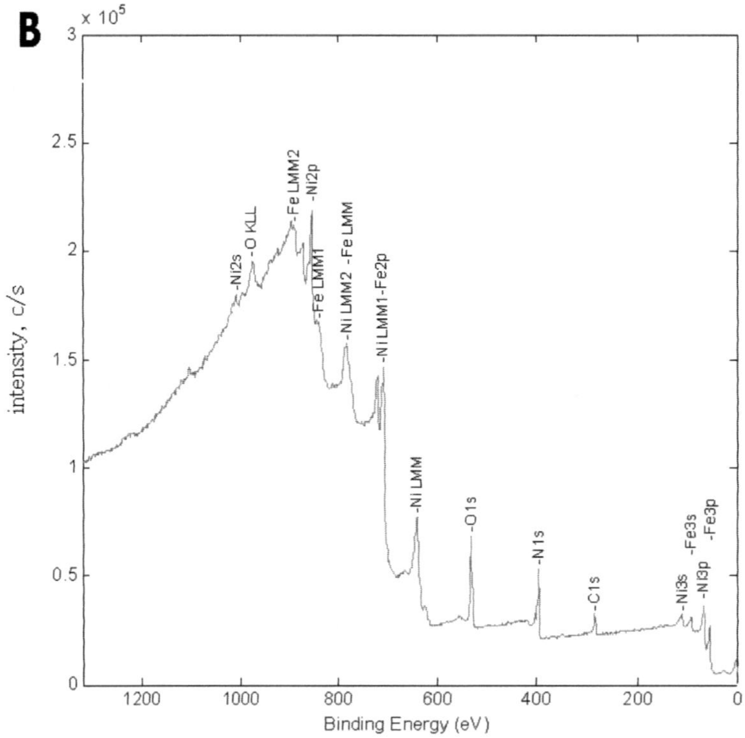
B
x 10^5
intensity, c/s
Binding Energy (eV)
-Ni2s
-O KLL
-Fe LMM2
-Ni2p
-Fe LMM1
-Ni LMM2
-Fe LMM
-Ni LMM1
-Fe2p
-Ni LMM
-O1s
-N1s
-C1s
-Ni3s
-Fe3s
-Ni3p
-Fe3p

C

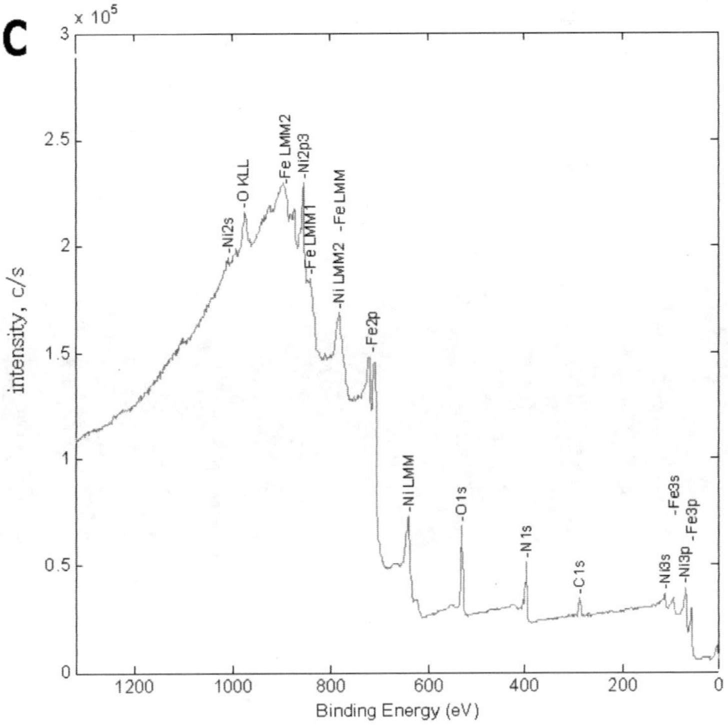
x 10^5
intensity, c/s
Binding Energy (eV)
-Ni2s
-O KLL
-Fe LMM2
-Ni2p3
-Fe LMM1
-Ni LMM2
-Fe LMM
-Fe2p
-Ni LMM
-O1s
-N1s
-C1s
-Ni3s
-Fe3s
-Ni3p
-Fe3p

D

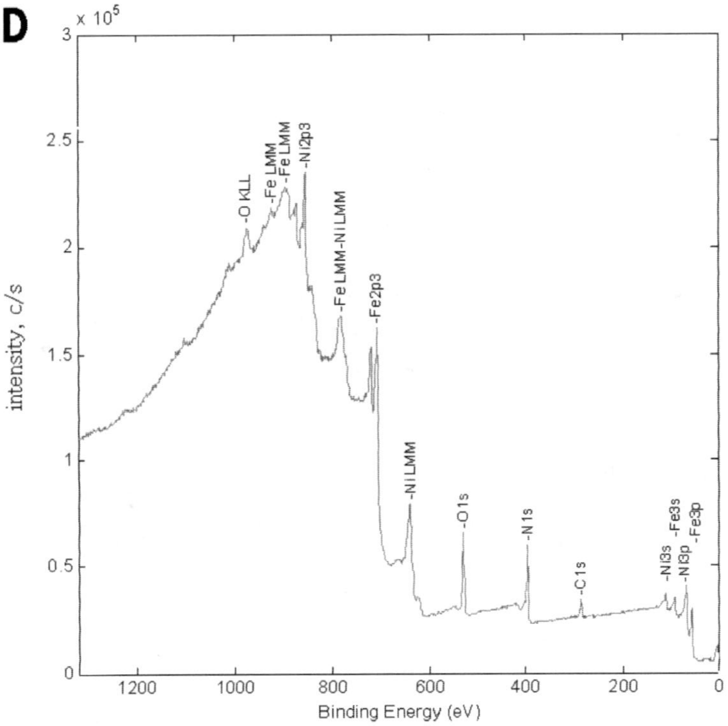
x 10^5
intensity, c/s
Binding Energy (eV)
-O KLL
-Fe LMM
-Fe LMM
-Ni2p3
-Fe LMM-Ni LMM
-Fe2p3
-Ni LMM
-O1s
-N1s
-C1s
-Ni3s
-Ni3p
-Fe3s
-Fe3p

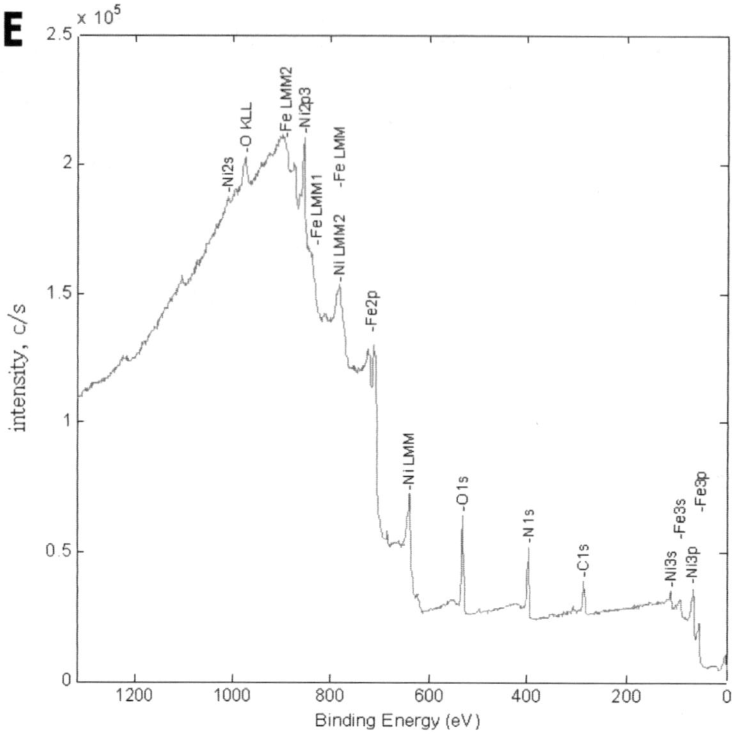

Figure 13. Review XPS-spectra: **(A)-** reed switch 5, **(B)-** reed switch 14, **(C)-** reed switch 35, **(D)-** reed switch 43, **(E)-** reed switch 52.

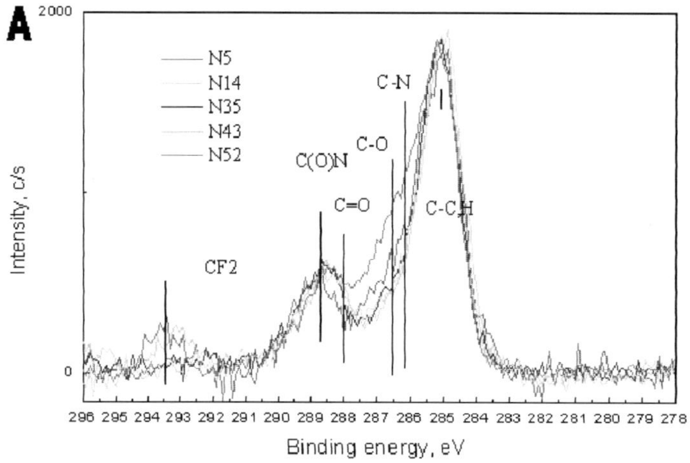

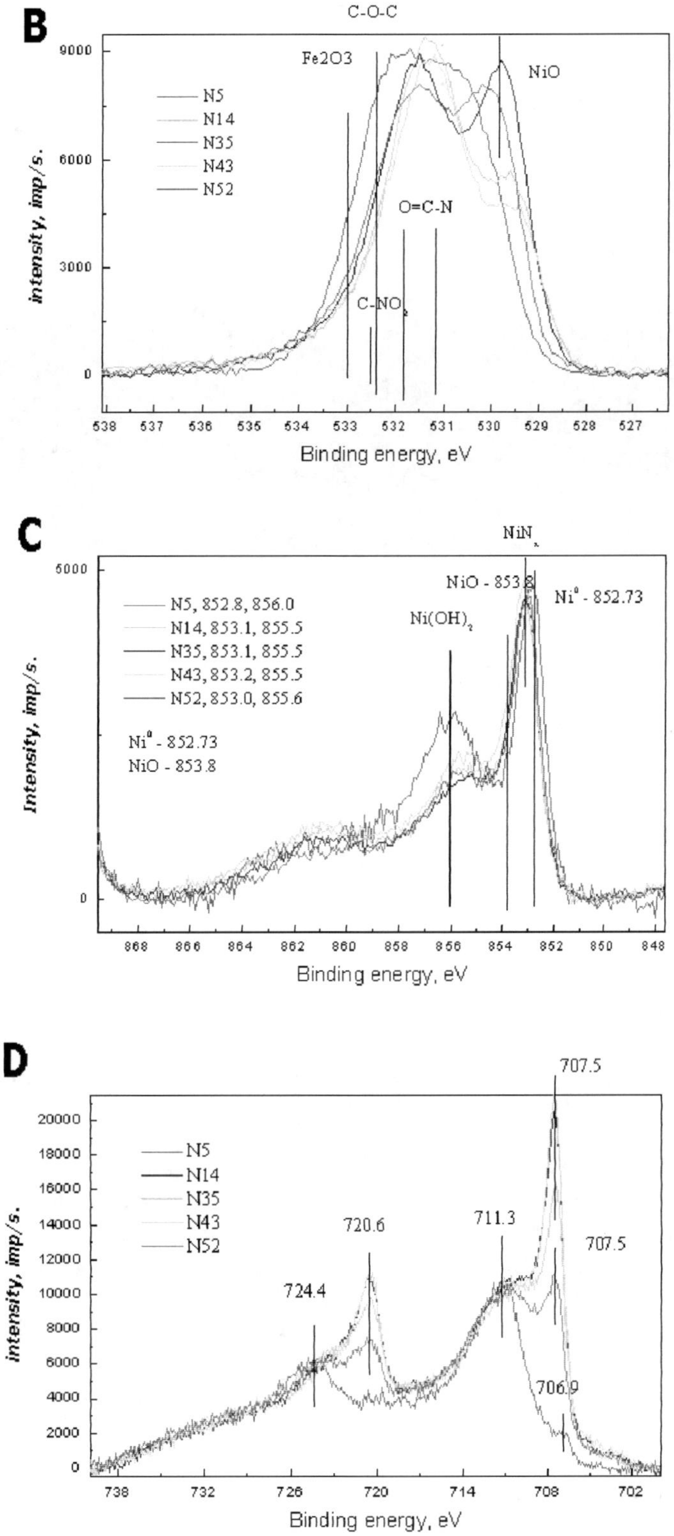
B
C-O-C
Fe2O3
NiO
N5
N14
N35
N43
N52
O=C-N
C-NO2
intensity, imp/s.
Binding energy, eV
C
NiNx
NiO - 853.8
Ni0 - 852.73
Ni(OH)2
N5, 852.8, 856.0
N14, 853.1, 855.5
N35, 853.1, 855.5
N43, 853.2, 855.5
N52, 853.0, 855.6
Ni0 - 852.73
NiO - 853.8
Intensity, imp/s.
Binding energy, eV
D
707.5
N5
N14
N35
N43
N52
720.6
711.3
707.5
724.4
706.9
intensity, imp/s.
Binding energy, eV

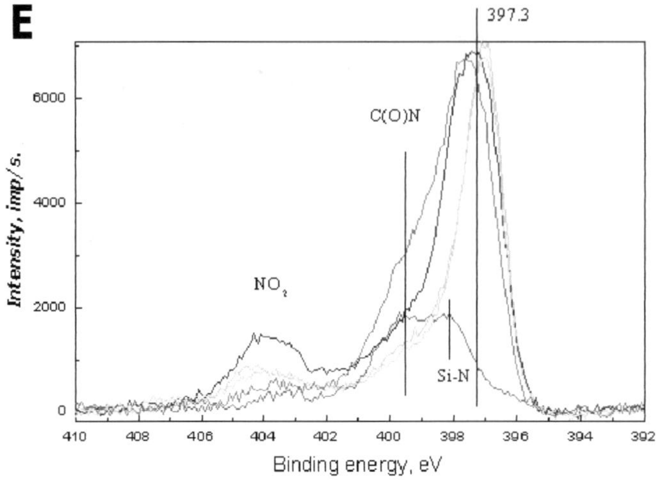

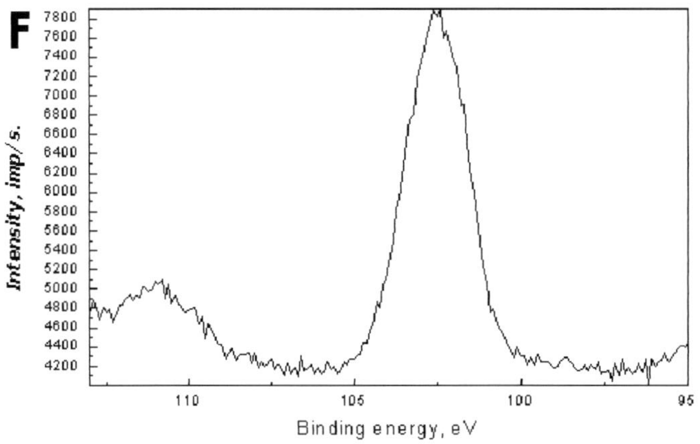

Figure 14. Photoelectron spectra **(A)**- C 1s, **(B)**- O 1s, **(C)**- Ni 2p, **(D)**- Fe 2p, **(E)**- N 1s, **(F)**- Si 2p.

The results of processing of review XPS-spectra of contact surfaces and reed switch resistance and breakdown voltage measurements are given in **Table 1** [14, 21].

Welding

In the review photoelectron spectrum taken in the overlap area of contact pieces of reed switch №5 (**Figure 13a**, **Table 1**) after welding, except of iron and nickel lines, lines of oxygen, carbon, silicon, fluorine and nitrogen are present. XPS-spectra of lines O 1s, Ni 2p, Fe 2p (**Figures 14 b, c, d**) indicate that in the virgin sample Fe and Ni are in oxidized state. Spectrum C 1s (**Figure 14 a**) denotes the presence of different carbonaceous groups on the contact surface. The position of the main peak in spectrum C 1s (285.0 eV) denotes good conductivity and small thickness of surface contamination layer. Photoelectron spectra N 1s (**Figure 14 e**) give states N in test samples. The main peak with energy of about 397.3 eV corresponds to nitride state. It is seen that in the virgin sample (№ 5) Si_3N_4 is present, that also follows from binding energy of the peak Si 2p (102.5 eV) **(Figure 14 f)**.
So we can suggest that the growth of reed switch resistance after welding is connected with oxidation of contact surface, nanodrops of glass and other products of thermal decomposition of glass and remains of technological environment (for example remains of hydrofluoric acid HF, used for glass envelope manufacturing), adsorption of oil vapor in the air with the further formation of polymer films.

Table 1. Regimes of IPT and switching, reed switch resistance R, breakdown voltage U, element concentration on contact surfaces of samples № 5, 43, 35, 14, 52.

№ of a sample	IPT regime	Switching regime	R, Ohm	U, V	C1s, at. %	N1s, at. %	O1s, at. %	Fe2p, at. %	Ni2p, at. %	F1s, at. %	Si2p, at. %
35	200 V - 30 s - 100 tr.	Without switching	0.08	210	13.5	29.2	27.6	17.8	11.9	-	-
43	200 V - 30 s - 30 tr.	Without switching	0.09	200	13.4	31.6	22.4	18.5	14.1	-	-
52	200 V - 30 s - 30 tr.	30 V – 0.5 A - $1.25 \cdot 10^6$ operat. at 155 °C - $1.25 \cdot 10^6$ operat. in n.c.	0.09	250	22.4	27.6	26.0	13.9	10.1	-	-
14	200 V - 30 s - 30 tr.	50 mV - 5 μA - 10^8 operat. in n.c.	0.09	220	14.5	30.5	24.7	16.8	13.5	-	-
5	Non-IPO	Without switching	0.22	290	9.3	8.7	53.5	8.3	4.6	3.3	12.5

Ion-plasma nitriding

After 30 treatments by high-voltage pulsed discharges resistance of reed switches MKA-14108 reduced and became to correspond to resistance of off-the-shelf devices MKA 14103 **(Table 1)**.

In the review XPS-spectrum taken from the reed switch after 30-fold IPT (sample № 43) (**Figure 13 d)**, as on the spectrum from untreated in gas discharge contact surface of sample №5 **(Figure 13 a)**, lines of iron, nickel, oxygen, carbon and nitrogen are observed. However after IPT the situation has greatly changed. The lines of silicon, fluorine are absent, intensity of oxygen line decreased, and lines of iron, nickel, carbon and nitrogen became more intensive. Observed in practice change of surface structure, reduction of resistance and breakdown voltage after IPT is directly connected with sputtering by nitrogen ions of polymer films and spurious coatings (from silicon oxide, ferric and nickel oxide) that improves red switch resistance and also with nitriding and gasabsorbtion processes.

In the photoelectron spectrum Ni 2p (**Figure 14 c)** the shift of the main peak to the side of high binding energy denotes the formation of connection with N. The presence of NiO in the first place follows from spectrum O 1s (**Figure 14 b)**. The formation of oxynitride phase is possible. The state with energy 707.5 eV in photoelectron spectrum Fe 2p (**Figure 14 d)**, as in the case with Ni should be considered as connection with N. The main peak with energy about 397.3 eV in photoelectron spectrum N 1s (**Figure 14 e)** also corresponds to nitride state of surface of reed switch № 43.

After 100-fold treatment by high-voltage pulsed discharges the resistance of reed switches MKA-14108 median-direction matches with the resistance of off-the-shelf devices MKA 14103.

In the review XPS-spectrum taken from reed switch № 35 after its 100-fold IPT (**Figure 13 c)**, as in the spectrum of sample № 43 after 30-fold IPT (**Figure 13 d)**, lines of iron, nickel, oxygen, carbon and nitrogen are observed. In the spectrum it is seen that after 100-fold IPT the situation has not considerably changed. The lines of silicon, fluorine are also absent, the intensity of carbon and oxygen lines has a little bit increased (C by 0.1 at. %, O by 5.2 at. %), and intensity of lines of iron, nickel, nitrogen has on the contrary reduced (Fe by 0.7 at. %, Ni by 2.2 at. %, N by 2.4 at. %).

Switching

At comparison of XPS-spectra of samples № 43, 14 and 52 taken before and after reliability tests (**Figures 13 b, d, e)** the same tendency if found as at hundred-fold treatment. After tests in microregime (50 mV - 5 μA - 10^8 operations in n.c.) intensity of carbon and oxygen lines increased (C by 1.1 at. %, O by 2.3 at. %), and intensity of iron, nickel, nitrogen lines on the contrary decreased (Fe by 1.6 at. %, Ni by 0.7 at. %, N by 2.4 at. %). In the power regime (30 V - 0.5 A -$1.25 \cdot 10^6$ operations at 155 °C - $1.25 \cdot 10^6$ operations in n.c.) this tendency is more obvious. The intensity of carbon and oxygen lines has considerably increased (C by 9 at. %, O by 3.6 at. %), and intensity of iron, nickel, nitrogen lines on the contrary has considerably decreased (Fe by 4.6 at. %, Ni by 4.0 at. %, N by 4 at. %).

It is connected with that in the power regime plasma arc of opening appears [1]. Because of arc thermal effect in contact microspots surface heating higher than boiling temperature occurs, that leads to intensive metal vaporization [1]. As a result, as it is seen from AFM-images (**Figure 15**) cone-shaped craters (sphalerite formations) of about 2000 nm diameter and about 120 nm depth are formed on the surface of contact pieces.

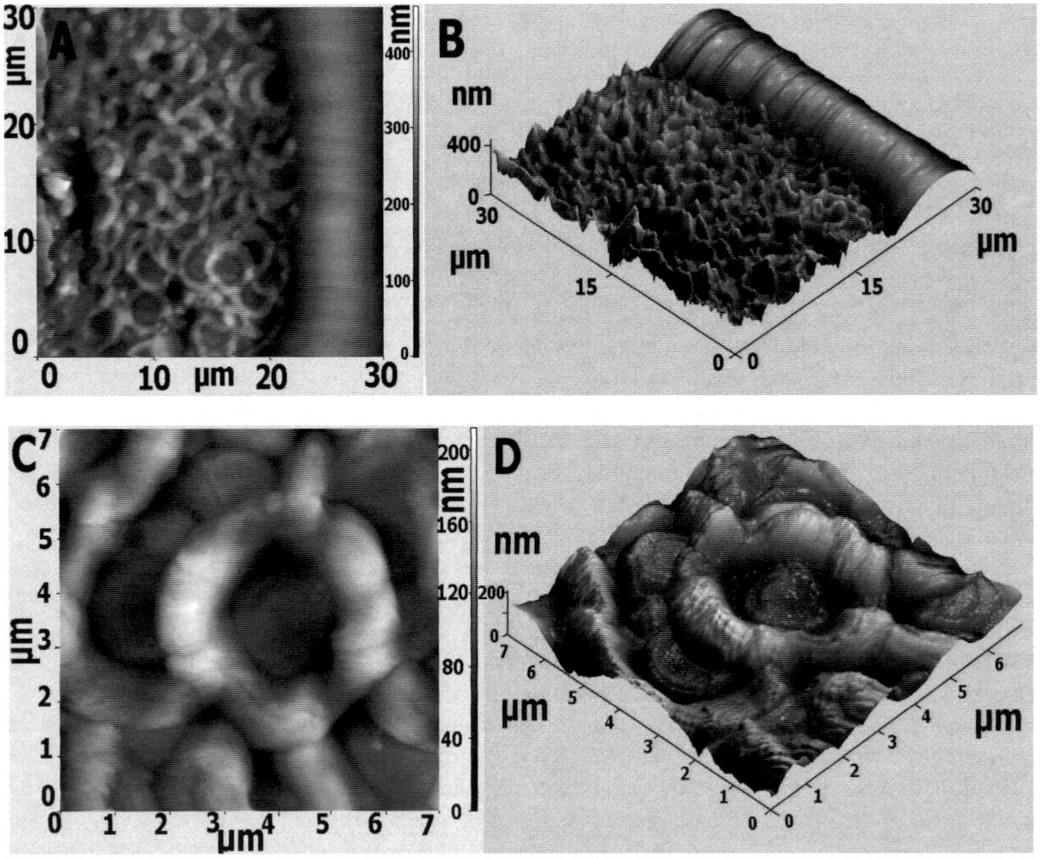

Figure 15. AFM – image of contact surface of sample № 52 (after IPT and switching in regime: 30 V – 0.5 A -$1.25 \cdot 10^6$operat. at 155 °C -$1.25 \cdot 10^6$operat.in n.c.): **(A), (B)** – 30 x 30 μm; **(C), (D)**, - 7 x 7 μm.

Simultaneously diffusion carry-over of carbon and oxygen to the surface with the formation of carbon dioxide according to the data of mass spectrometric analysis [18, 22]. On account of arc thermal effect in contact microspots decomposition of nitrides occurs with nitrogen release. Meanwhile reed switch breakdown voltage increases by 50 V. This growth is connected with gas pressure growth in reed switch, because the product (pd) for contact gap of reed switch corresponds to the right branch of Paschen [18], (**Figure 16**).

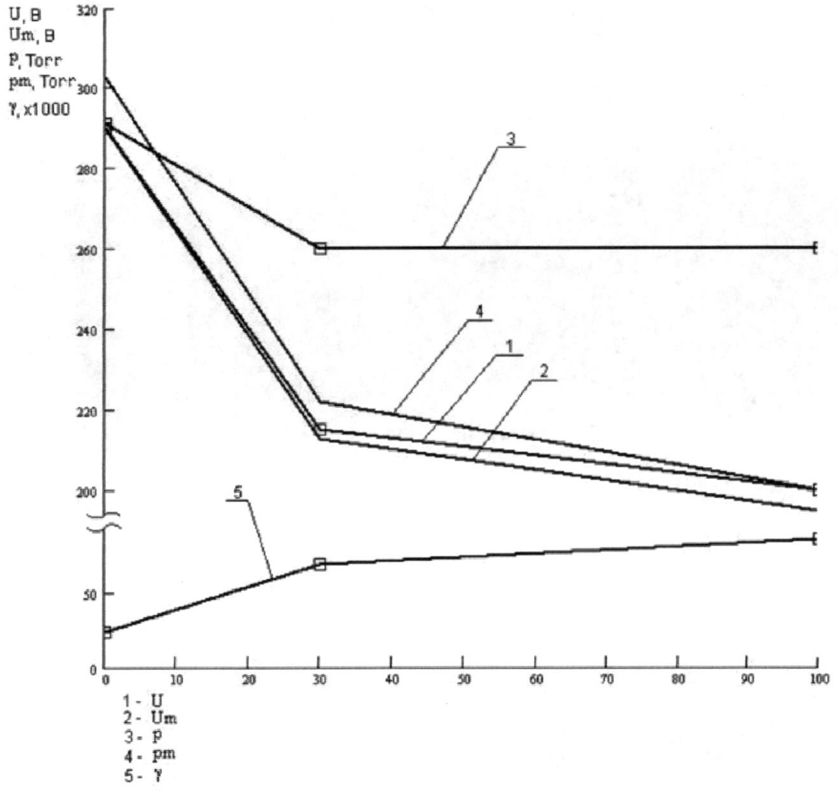

Figure 16. Dependences of breakdown voltage U -1, gas filling pressure P - 3, coordinate of minimum of Paschen curve U_m - 2, p_m - 4, cumulative rate of secondary electron emission γ - 5 of reed switch MKA-14108 median-direction on quantity of treatments n.

However in spite of partial destruction (to a lesser degree – microregime or father – power regime) of near-surface nitride layers by switching, failures of the samples under test were not observed.

3.2. The coatings obtained by reactive cathode sputtering method

Sputtering

The surface of contact piece after IPT according to the second regime (of reactive cathode sputtering) can be divided for convenience by images of contact piece surfaces, taken with the help of optical (**Figure 17**) and atomic-force (**Figure 18**) microscopes in two areas of distinctive topography.

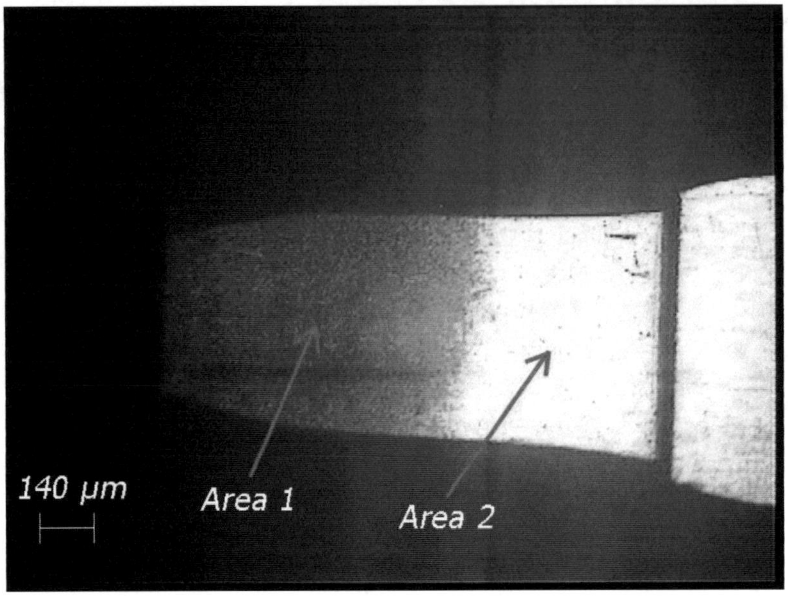

Figure 17. Reed switch contact-piece after 30-fold ion-plasma treatment (reactive cathode sputtering mode).

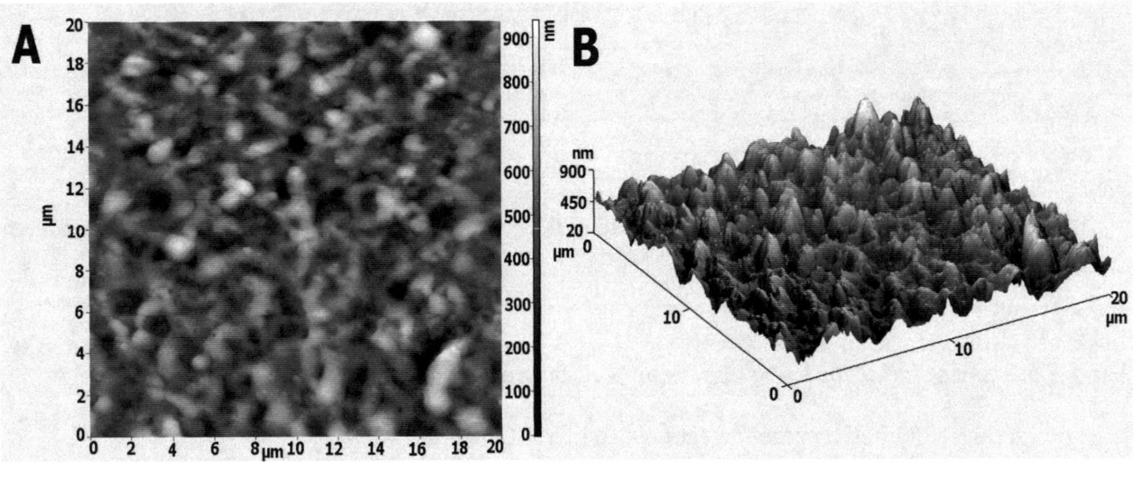

Figure 18. AFM-image of reed switch contact surface after 30-fold ion-plasma treatment (reactive cathode sputtering mode): **(A)** and **(B)** -2D and 3D image of area 1.

Area 1 (**Figures 17, 18a, b**) is a working area (contact-pieces overlapping area) where, in fact, a gas discharge is allowed with the voltage supply to the opened reed switch contacts. As a result of the reactive cathode sputtering, nitrides (FeN and NiN) the most part thereof is deposited on the contacts surface in area 1 are formed in the gas phase of the overlapping area.

Figure 19 demonstrates the reactions determining the process of the contact plating deposition by the reactive cathode sputtering method. Nitrogen ions accelerating in the area of the potential cathode loss bombard the target – cathode. Thereby, the most part of ion energy (about 90%) is consumed for the target heating, and the rest part – for the electron emission, ion implantation, and atom sputtering (Fe, Ni, C, O) located on the target surface. As a result of the reactive cathode sputtering, iron and nickel nitrides which thereafter are deposited on the reed switch contact-pieces and glass envelope are formed in the gas phase. With the deposition on the contact surface in area 1 (with temperature less than 400-500° C) molecules of metals MeN (Fe, Ni) nitrides of permalloy do not dissociate as in case of ion nitriding but form a black-colored film (**Figure 17**). The coating thickness, according to the data of metallographic investigations, was (after 30-fold IPT) about 400 μm.

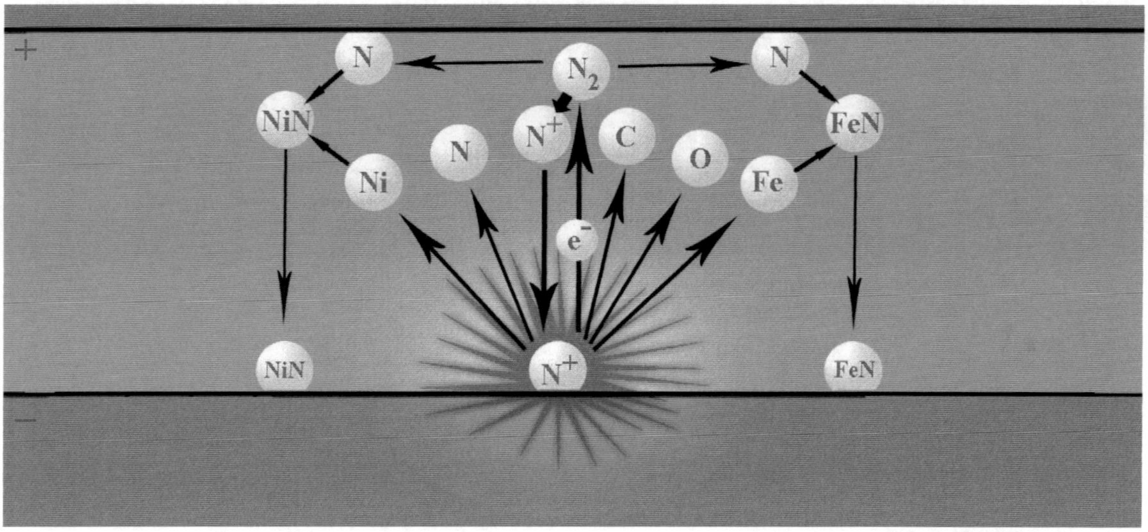

Figure 19. Coating deposition by reactive cathode sputtering method.

As a result of diffusion, a part of FeN and NiN molecules will be deposited on the contacts surface in area 2 neighboring with area 1 (**Figures 17, 18 c, d**) and the inner surface of the envelope, located near the overlapping area (**Figure 7**). A small number of the products of cathode sputtering also gets on the reverse sides of the reed switch contact-pieces.

Switching tests

It is known [1] that the basic effect of the corrosion, erosion and contaminations impact on the electrical contacts lies in the conductivity trouble therefore the contacts resistance to these processes can be characterized by a number of closures, whereby the conduction changes tragically. The contacts erosion resistance can be characterized by a number of closures whereby the contacts conduction trouble or the contacts non-breaking occurs.

Experimental prototypes of the reed switches after high-discharge treatments in the reactive cathode sputtering mode (the second mode of IPT) and without such were under comparative switching tests as per the technique developed earlier in the paper [25] for MKA-14108 reed switches with ion- nitrided contact surfaces. The tests were conducted for the active load and in the idling mode (without load). It allowed optimizing the modes of the contact spring surface modification and studying physical processes occurring on all stages of the reed switch treatment and operation.

Comparative switching tests of MKA-14108 pilot reed switches and MKA 14103 off-the-shelf devices showed that MKA-14108 pilot reed switch operation time, pretreated by high-voltage pulsed discharges in the reactive cathode sputtering mode, meets the requirements specified for the durability of MKA 14103 off-the-shelf devices (**Table 2**).

Table 2. Results of comparative switching tests.

Switching tests mode	Reed switch type	Without failures, %	Rmax Before tests, Ohm	Rmax After tests, Ohm
50 mV - 5 mkA, 100 Hz, 10^6 operations	MKA-14103	100	0.1	0.1
	MKA-14108, without treatment	21	0.25	10
	MKA14108, with treatment	100	0.08	0.09
5 V - 10 mA, 100 Hz, 10^6 operations	MKA-14103	100	0.17	0.11
	MKA14108, without treatment	27	0.35	22.7
	MKA14108, with treatment	100	0.07	0.08
20 V - 0.5 A, 50 Hz, $5 \cdot 10^6$ operations	MKA-14103	38	0.16	40
	MKA14108, without treatment	47	0.31	0.11
	MKA14108, with treatment	100	0.09	0.10
24 V - 400 mA, 50 Hz, $5 \cdot 10^5$ operations	MKA-14103	100	0.13	0.5
	MKA14108, without treatment	100	0.29	0.11
	MKA14108, with treatment	100	0.07	0.01
36 V - 15 mA, 50 Hz, $5 \cdot 10^6$ operations	MKA-14103	25	0.2	8.7
	MKA14108, without treatment	53	0.49	0.34
	MKA14108, with treatment	100	0.07	0.09
50 V - 50 mA, 50 Hz, $5 \cdot 10^5$ operations	MKA-14103	100	0.14	0.22
	MKA14108, without treatment	100	0.38	0.25
	MKA14108, with treatment	100	0.07	0.08
100 V -100 mA, 50 Hz, $5 \cdot 10^5$ operations	MKA-14103	2	0.17	0.19
	MKA14108, without treatment	100	0.27	0.23
	MKA14108, with treatment	100	0.08	0.09

For the treatment mode optimization and identification the tests of MKA-14108 reed switch pilot samples before and after treatment without load (in idling mode) were conducted. The switching number changed step-by-step from 0 to 107 collisions. R was measured on each stage of the reed switch testing (**Figure 20**).

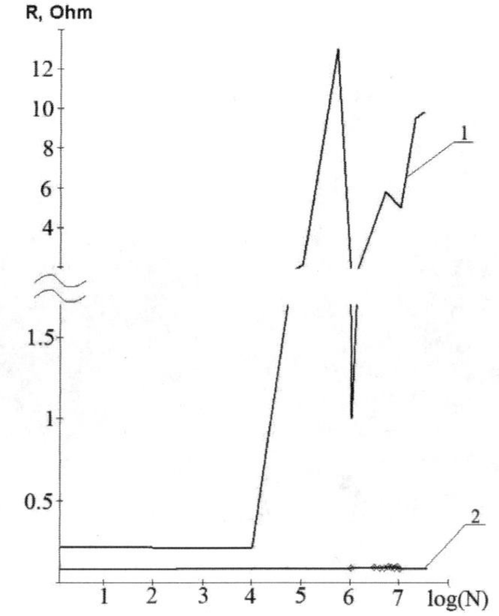

Figure 20. Dependence of resistance R of MKA14108 reed switch along the median upon the number of operations N in dry circuit (1 – untreated contact-pieces, 2 - contact-pieces after IPT in the reactive cathode sputtering mode). Reed switch batch quantity is 100 pcs.

Moreover, the pilot samples were tested for the reliability with an active load as per the combined mode (**Figure 21**).

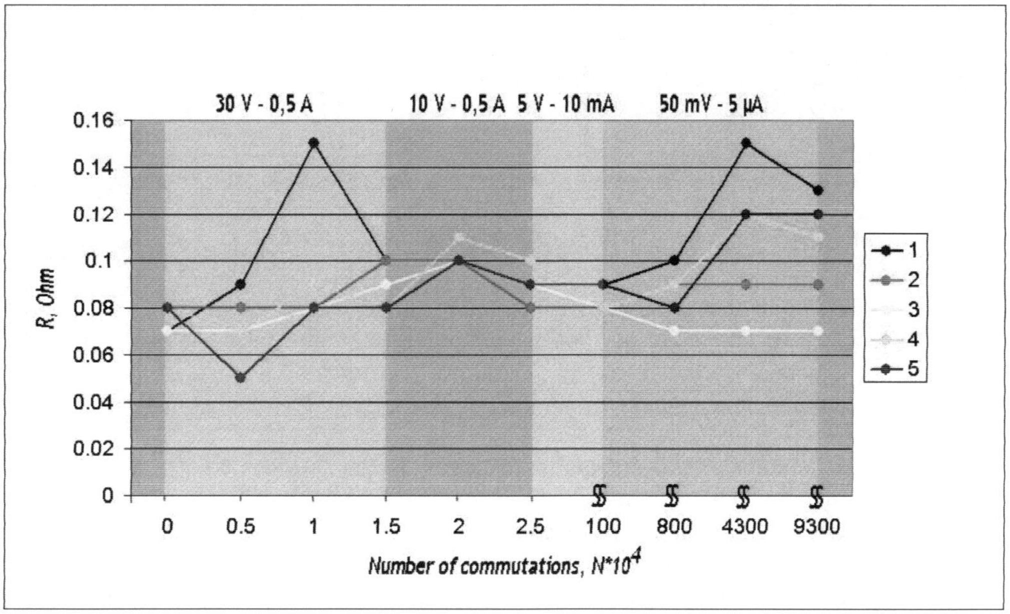

Figure 21. Dependence of resistance R of MKA14108 reed switches (5 pcs.) upon a number of operations N in the combined mode.

AFM-image of the contact surface, obtained by the reactive cathode sputtering method, before and after switching are shown in **Figures 18** and **22**.

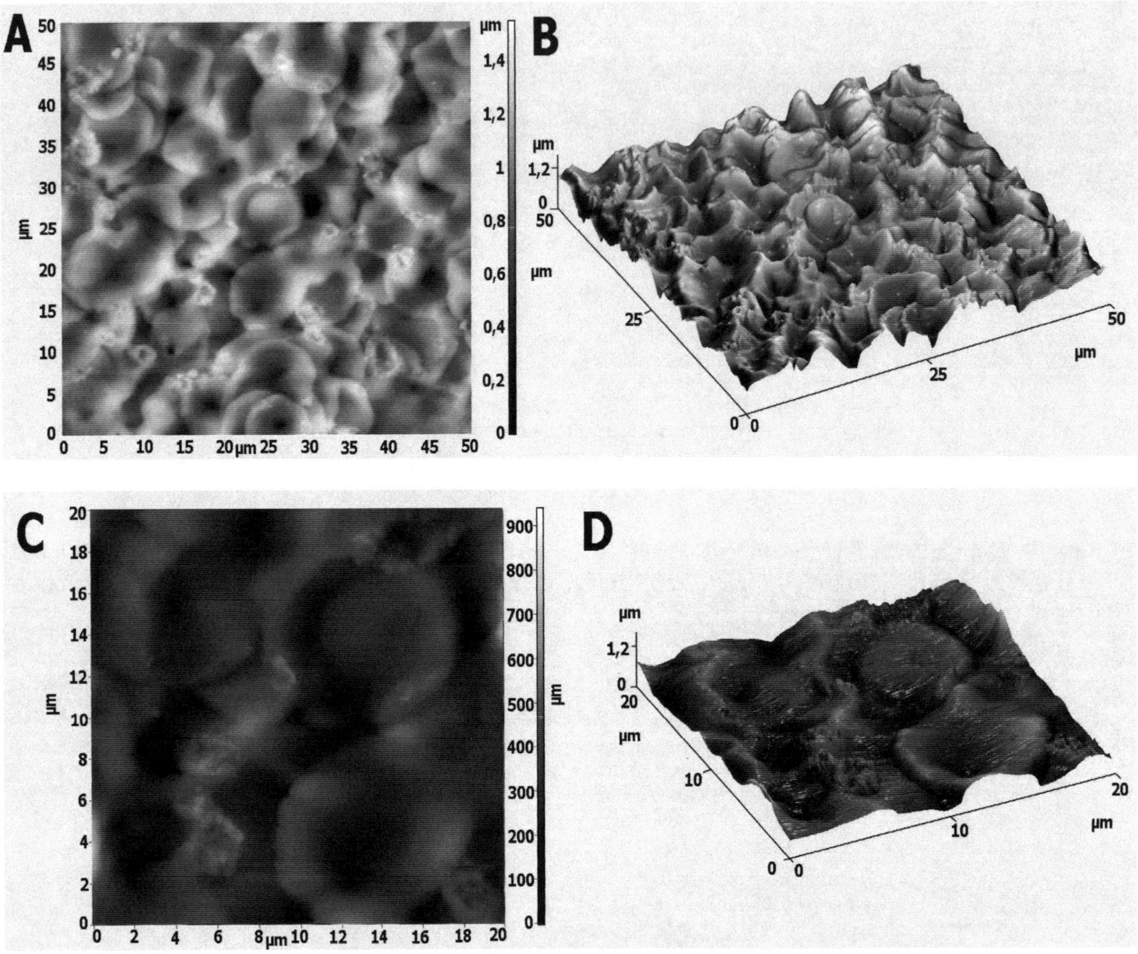

Figure 22. 2D and 3D AFM – image of the contact surface of MKA-14108 reed switch (after IPT in the reactive cathode sputtering mode and following switching in the combined mode: **(A), (B)** – 50 x 50 µm; **(C), (D)**, - 20 x 20 µm.

Due to the heating effect of the breaking arc [1] occurring in the range of 0 – $2.5 \cdot 10^4$ operations (**Figure 21**), the surface heating over the boiling temperature occurs in the contact micro-spots. It leads to the intensive metals evaporation. As a result, AFM-images show (**Figure 22**) that the cone-shaped craters of about 8000 µm in diameter and about 100 µm in depth are formed on the surface of the contact-pieces.

Considering the experimental curves delineated in **Figures 20, 21,** one can make a conclusion about a full compliance of the deposited coatings to the requirements [1] for reliability, resistance value and stability, specified to the coatings of stock-produced MKA-14103 reed switches.

Therefore, switching reliability tests for the reed switch with the contact coatings fabricated by the reactive cathode sputtering method showed a positive result. They confirmed a high film adhesion to the substrate material. Apparently, the adhesion bonds strength is stipulated by an additional activation of the surface by plasma operation.

Journal of Physics: Conference Series **345** (2012) 012003 doi:10.1088/1742-6596/345/1/012003

4. Conclusion

A principally new technological process of the magnetically operated contacts (MC) contacting surfaces modification (MK) when after the contact springs sealing within the glass enveloped fulfilled with nitrogen through MCs being opened, they pass current pulses which cause forming micro-nanosized layers with the specified contact properties in the surficial area of MCs, was developed.

As a result of the implemented investigations it was managed to provide such electrophysical discharge conditions under which in the surficial area and on the surface of the permalloy reed switch contacts, corrosion- and erosion-resistant nitride layers with a high electrical conduction are formed by the methods of ion nitriding or reactive cathode sputtering, that allows refusing electroplating application made of precious metals.

Therefore, the obtained results denote the viability of ion nitriding and reactive cathode sputtering methods application for the reed switch production.

Reference

[1] Karabanov S.M.; Maizels R.M.; Shoffa V.N. *Magnetically operated contacts (reed switches) and units based thereof*; (Publishing house “Intellect”: Dolgoprudny, Russia, 2011).

[2] Arzamasov B.N.; Bratukhin A.G.; Eliseev Y.S.; Panayoti T.A. *Ion chemicothermal treatment of alloys in gas medium*; (Publishing house “MSTU named after Bauman N.E.”: Moscow, Russia, 1999).

[3] Lyaschenko B.A.; Mironenko V.I.; Radko O.V.; Bobyr S.A. *Features of 30XГCA steel nitriding in pulsating discharge*; (Published in VCNU, Cherkassy, Ukraine, 2007; pp. 107-110); (in Russian).

[4] Zeltser I.A.; Karabanov S.M.; Maizels R.M.; Moos E.N.; Sablin V.A. Modifications of sealed magnetically operated permalloy contacts surface by pulsed discharges. In the collected papers of the second international theoretical and practical conference *Magnetically operated contacts (reed switches) and units based thereof*; (Publishing house “Poligraph”: Ryazan, Russia, 2009; pp. 174-177); (http://www.rmcip.ru – in Russain).

[5] Zeltser I.A.; Karabanov S.M.; Maizels R.M.; Sablin V.A. Investigation and development of modification methods of hermetically sealed magnetically operated contacts surface. In the collected papers of the second international theoretical and practical conference *Magnetically operated contacts (reed switches) and units based thereof*; (Publishing house “Poligraph”: Ryazan, Russia, 2009; pp. 184-207); (http://www.rmcip.ru – in Russain).

[6] Zeltser I.A.; Karabanov S.M.; Kuznetsov A.A.; Maizels R.M.; Sablin V.A.; Chernyak E.Y. Investigation of ion-plasma modification of iron-nickel sealed magnetically operated contacts by Auger spectroscopy. In the collected papers of the Second International Theoretical and Practical Conference *Magnetically operated contacts (reed switches) and units based thereof*; (Publishing house “Poligraph”: Ryazan, Russia, 2009; pp. 178-183); (http://www.rmcip.ru – in Russain).

[7] Karpov A.S.; Maizels R.M.; Shishkina L.V.; Shkutenko L.N. Plant for automatic ion-plasma treatment of reed switches. In the collected papers of the second international theoretical and practical conference *Magnetically operated contacts (reed switches) and units based thereof*; (Publishing house “Poligraph”: Ryazan, Russia, 2009; pp. 169-173); (http://www.rmcip.ru – in Russain).

[8] Kuznetsov A.A.; Vasilyev E.V.; Zeltser I.A.; Chernyak E.Y. Analysis of elemental and chemical composition of reed switch contact zones by Auger spectroscopy. In the collected papers of the Second International Theoretical and Practical Conference *Magnetically operated contacts (reed switches) and units based thereof*; (Publishing house “Poligraph”: Ryazan, Russia, 2009; pp. 84-87); (http://www.rmcip.ru – in Russain).

[9] Arushanov K.A.; Zeltser I.A. Instrument and technological aspects of reed switch with nanostructured contact surfaces creation. (Vestnik RSURE, 2009, 3, pp. 93-98); (in Russian).

[10] Zeltser I.A.; Karabanov S.M.; Maizels R.M.; Moos E.N.; Sablin V.A. Modification of sealed magnetically operated iron-nickel contacts surface. In reports theses of X International Workshop *Structural fundamentals of materials modification by methods of non-traditional technologies*, (MHT-X; Obninsk, Russia, 2009; pp. 58-59); (in Russian).

[11] Zeltser I.A.; Karabanov S.M.; Maizels R.M.; Moos E.N.; Sablin V.A. Nanostructure

Modificated Contact Surface. In *Abstract Book of International Conference "Information and Structure in Nanoworld"*; (Saint-Petersburg, Russia, 1-3 July, 2009; p. 83).

[12] RF Patent No.2391733. Magnetically operated sealed contact. / Karabanov S.M., Zeltser I.A., Maizels R.M., Trunin E.B.

[13] RF Patent No.2393570. Method of reed switch with nitrided contact-pieces production. / Karabanov S.M., Maizels R.M., Arushanov K.A., Zeltser I.A., Provotorov V.S.

[14] Arushanov K.A.; Zeltser I.A.; Karabanov S.M.; Maslakov K.I.; Naumkin A.V. Investigation of plasma modification of magnetically operated contacts by methods of X-ray photoelectron spectroscopy and atomic force microscopy. In materials of 7 International Theoretical and Practical Conference *Recent developments of the European science, Volume. 36. Chemistry and chemical technologies. Physics*; (Sofia, Bulgary, 2011; pp. 56 – 61); (in Russian).

[15] Arushanov K.A.; Zeltser I.A.; Karabanov S.M.; Maizels R.M.; Moos E.N. New method of reed switch contact surfaces modification. In materials of 7 International Theoretical and Practical Conference *Recent developments of the European science, Volume. 36. Chemistry and chemical technologies. Physics*; (Sofia, Bulgary, 2011; pp. 52–55); (in Russian).

[16] Arushanov K.A.; Zeltser I.A.; Karabanov S.M.; Maizels R.M.; Moos E.N. Nanorelief formation of contact surfaces. In XI International Workshop *Structural fundamentals of materials modification* (MHT-XI) Reports theses; (Institute of Atomic Energy of Scientific Research Nuclear University of Moscow Engineering and Physical Institute, Obninsk, Russia, 2011; pp. 20-22).

[17] Arushanov K.A.; Zeltser I.A.; Karabanov S.M.; Maizels R.M.; Moos E.N. Ion-induced modification of contact surfaces. In materials of XX International Conference *Ion interaction with surface*, VIM-2011; (Zvenigorod, Russia, 25-29 of August 2011; pp. 206-209).

[18] Arushanov K.A.; Zeltser I.A.; Karabanov S.M.; Trunin E.B. Diffusion nitrogen saturation of surficial layers of reed switch contacts in pulsating plasma. In 3 International Theoretical and Practical Conference *Magnetically operated contacts (reed switches) and units based thereof*, reports theses; (Ryazan, Russia, 28-30 of September 2011; pp. 31-38).

[19] Avachev A.P.; Arushanov K.A.; Zeltser I.A.; Karabanov S.M.; Konobeev V.A. Investigation of plasma modification of magnetically operated contacts surface by methods of atomic force and optical microscopy. In 3 International Theoretical and Practical Conference *Magnetically operated contacts (reed switches) and units based thereof*, reports theses; (Ryazan, Russia, 28-30 of September 2011; pp. 47-53).

[20] Arushanov K.A.; Zeltser I.A.; Karabanov S.M.; Maizels R.M.; Moos E.N. New method of magnetically operated contacts modification. In 3 International Theoretical and Practical Conference *Magnetically operated contacts (reed switches) and units based thereof*, reports theses; (Ryazan, Russia, 28-30 of September 2011; pp. 54-60).

[21] Arushanov K.A.; Zeltser I.A.; Karabanov S.M.; Maslakov K.I.; Naumkin A.V. Investigation of ion-plasma modification of iron-nickel magnetically operated contacts surface by X-ray photoelectron spectroscopy. In 3 International Theoretical and Practical Conference *Magnetically operated contacts (reed switches) and units based thereof*, reports theses; (Ryazan, Russia, 28-30 of September 2011; pp. 39-46).

[22] Zhuravlev S.A.; Zeltser I.A.; Maizels R.M.; Polyakov A.S.; Chernyak E.Y. Analysis of chemical gas composition inside reed switches. In 3 International Theoretical and Practical Conference *Magnetically operated contacts (reed switches) and units based thereof*, reports theses; (Ryazan, Russia, 28-30 of September 2011; pp. 68-69).

[23] Arushanov A.K.; Zeltser I.A.; Kuznetsov A.A.; Chernyak E.Y. Application of Auger spectroscopy for investigation of plasma modification of magnetically operated contacts. In 3 International Theoretical and Practical Conference *Magnetically operated contacts (reed switches) and units based thereof*, reports theses; (Ryazan, Russia, 28-30 of September 2011; pp. 61-67).

[24] Drozdov M.N.; Zeltser I.A.; Karabanov S.M.; Teodoru O.M.N.D. Application of mass-spectrometry of secondary ions for investigation of ion-plasma modification. In 3 International Theoretical and Practical Conference *Magnetically operated contacts (reed switches) and units based*

thereof, reports theses; (Ryazan, Russia, 28-30 of September 2011; pp. 129-133).
[25] Karabanov S.M.; Zeltser I.A.; Maizels R.M.; Moos E.N.; Arushanov K.A., Creation of Principally New Generation of Switching Technique Elements (Reed Switches) with Nanostructured Contact Surfaces. *Journal of Physics: Conference Series*, 2011, V. 291, No.01 2020, pp. 1-17.
[26] Chatterjee – Fischer R.; Eysell F. – W. *Nitriding and carbonitriding*; Editted by Supov A.V.; (Publishing house "Metallurgy": Moscow, Russia, 1990; p. 280).
[27] Kuznetsov A.A.; Abramova S.Yu.; Potapova T.E.; Protopopov O.D. *Journ. of Electron Spectr. and Rel. Phenom* 1994, V. 68, pp. 407-412; (in Russian).
[28] Begrambekov L.B.. Modification of solid bodies surface with ion and plasma impacts; (Moscow Engineering and Physical Institute, Moscow, Russia, 2001).

Supracrystalline nanoribbons for nanoelectronics

P.A. Arefeva[1], R.A. Brazhe[2]

[1,2]Ulyanovs State Technical University

[1]e-mail: lifmuz@rambler.ru

Abstract.

In the last two decades various forms of carbon nanostructures have attracted a great deal of interest due to their novel fundamental properties and possible applications in electronics. With the developments in production methods, these materials have emerged as one of the most promising materials for non-silicon electronics. Among carbon nanostructures graphene is a flat monolayer of carbon atoms tightly packed into a 2-D honeycomb lattice and is considered as the basic building block of other graphitic materials in all dimensions [1].

In the past, 2-D materials were thought to be thermodynamically unstable and could not exist. In 2004, experimental discovery of graphene showed that these materials can exist in liquid suspensions or on non- crystalline substrates. Because of this fact, theoretical studies on graphite are much more than experimental ones.

The possibility of more complicated than graphene two-dimensional periodic structures existence has been theoretically proved. There are not separate atoms or ions but its symmetric complexes at the nodes at their crystalline lattice [2]. They were called as supracrystals.

1. Materials and methods

Since graphene is a semimetal, it's impossible to save from carriers in it when gate voltage applied, and therefore there will always be a high leakage current in graphene structures. To overcome this undesirable effect it is suggested to use narrow strips of graphene, because of the size called nanoribbons, where a formation of the band gap is possible because of the quantum size effect, and the width of the band gap is inversely proportional to transverse size of the ribbons. When defects appear on the border the nanoribbons change their state from the metallic to the semiconducting state.

Theoretically proved the possibility of the existence of more complex than graphene, 2D-carbon nets of sp2-hybridized carbon atoms forming the structure consisting of 3- and 12-polygons, 4-, 6- and 12-polygons. They can be considered as two-dimensional crystals formed by square or hexagonal cells with junctions consisted of proper atomic (or ionic) polygons but not separate atoms or ions. Such structures were called 2D-supracrystals [2].

The letters X and Y in brackets determine the chemical element symbol in indications of supracells in 2D- supracrystals (Fig. 1). The indexes after these letters mean: the first index defines the type of supracell, the next indexes describe the types of the cells enclosed. Firstly, the number of sides of the junction cells, then the number of sides of the surrounding cells (if they exist).

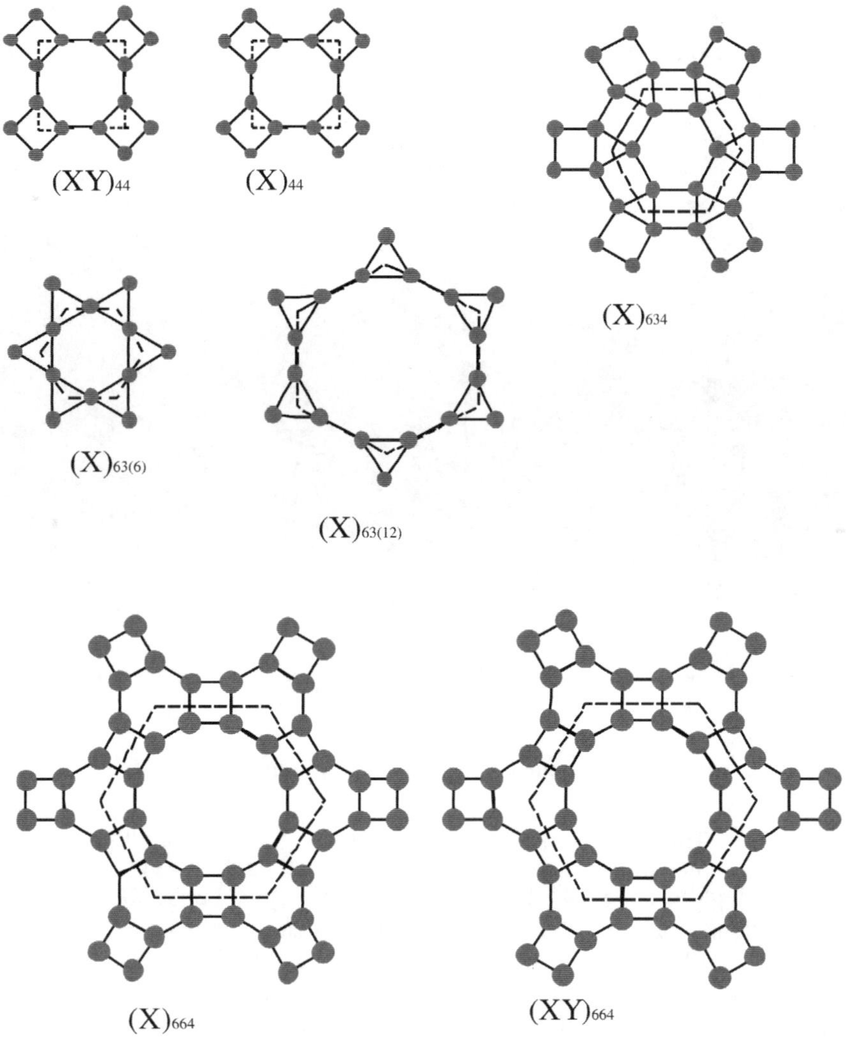

Figure 1. Supracells of the 2D-supracrystals and their indications.

2. The results of modeling and calculations.

With the help of a software package NanoEngineer supracrystalline nanoribbons were modeled and tested for stability which are obtained by various methods of cutting. Various types of nanoribbons $(C)_{44}$ are shown in Fig. 2.

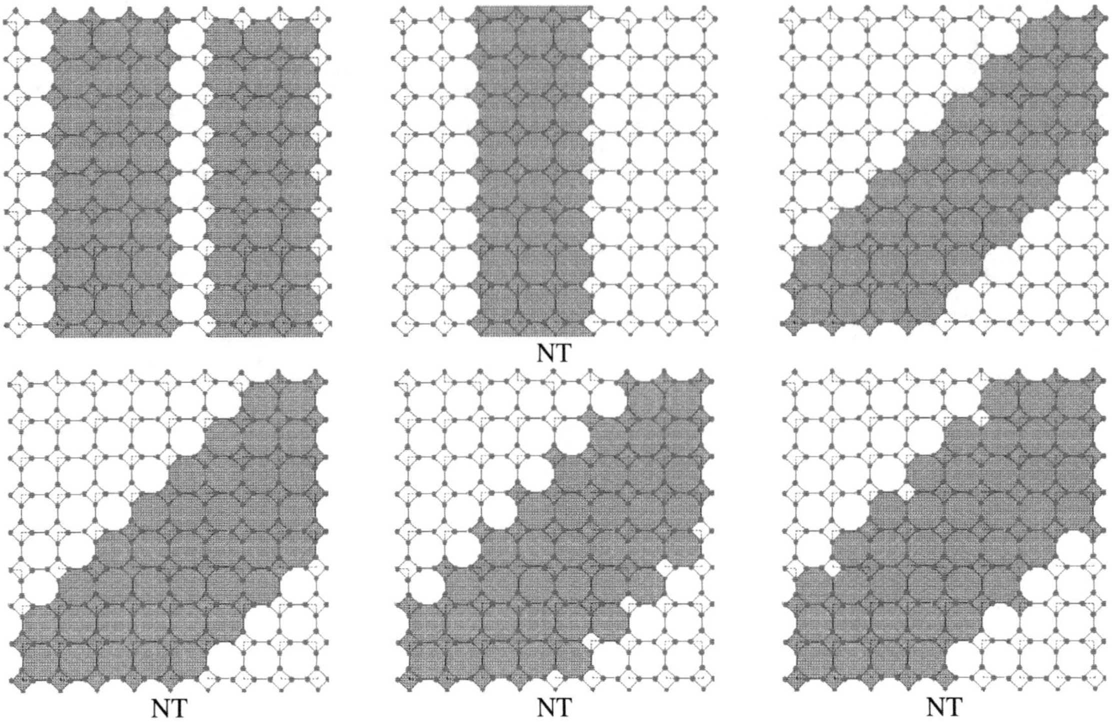

Figure. 2. Supracrystalline sheets and various tapes of nanoribbons $(X)_{44}$. The ribbons that can be rolled into nanotubes are marked by the letters NT [3].

The method of molecular dynamics was chosen to give a theoretical appreciation of band gap for graphene and supracrystal nanoribbons, as it requires the least computer resources.

The fundamental principles of molecular dynamics and particle dynamics are based on twin charges of Lennard-Jones interactions. The idea of a twin charge means that the interaction of two particles depends on their relative positions and does not depend on the positions of any other particles. All Lennard-Jones charges describe the interactions where repulsions are typical for smaller distances and attractions for greater distances. The Lennard-Jones charges allow to describe properly many physical phenomena and properties of substances on a high-quality level.

The graphs are showing the results of calculations of the band gap of the «zig-zag» (Fig.5) and «armchair» (Fig.6) graphene nanoribbons and supracrystal $(C)_{44}$ (Fig.3) and $(Si)_{44}$ nanoribbons, depending on the width of the nanoribbons.

Journal of Physics: Conference Series **345** (2012) 012004 doi:10.1088/1742-6596/345/1/012004

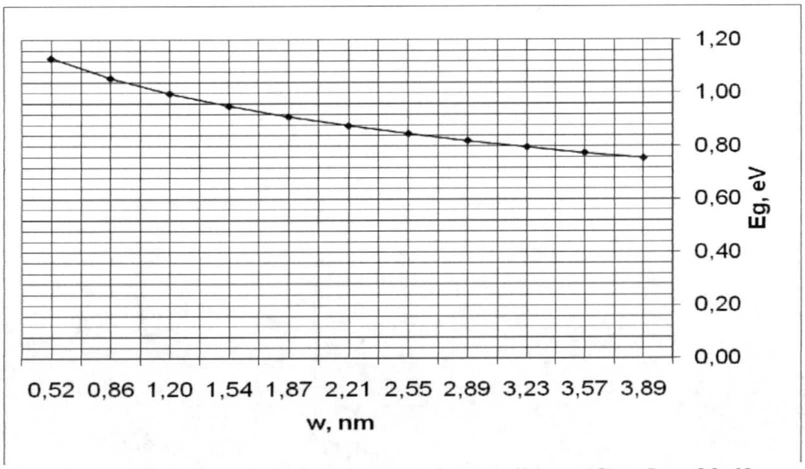

Figure.3. The dependence of the band gap supracrystal nanoribbon $(C)_{44}$ L = 23,68 nm from the width of the ribbon.

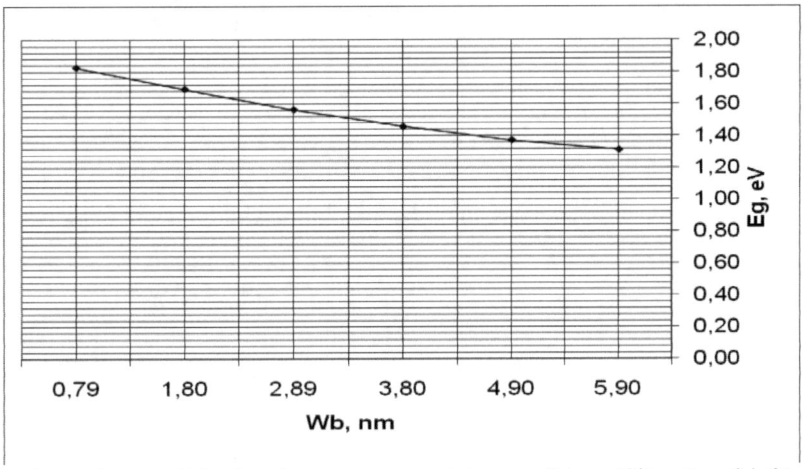

Figure.4. The dependence of the band gap supracrystal nanoribbon $(Si)_{44}$ L = 24,67 nm from the width of the ribbon.

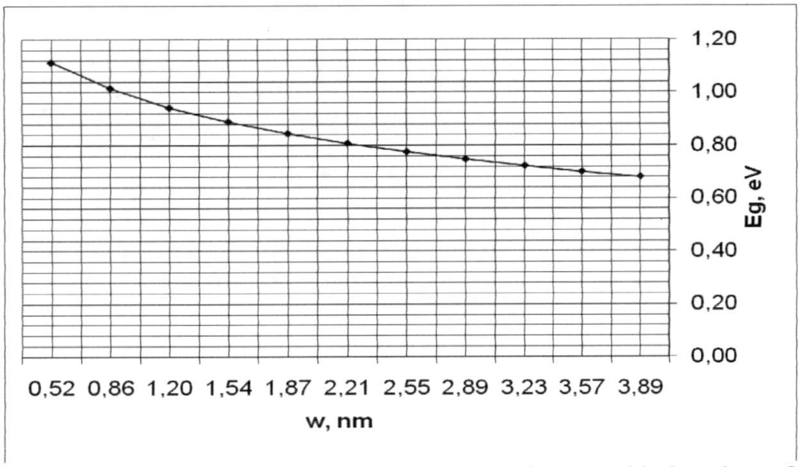

Figure.5. The dependence of the band gap of grephene nanoribbons with the edges of the "zig-zag» type L = 21,73 nm from the width of the ribbon.

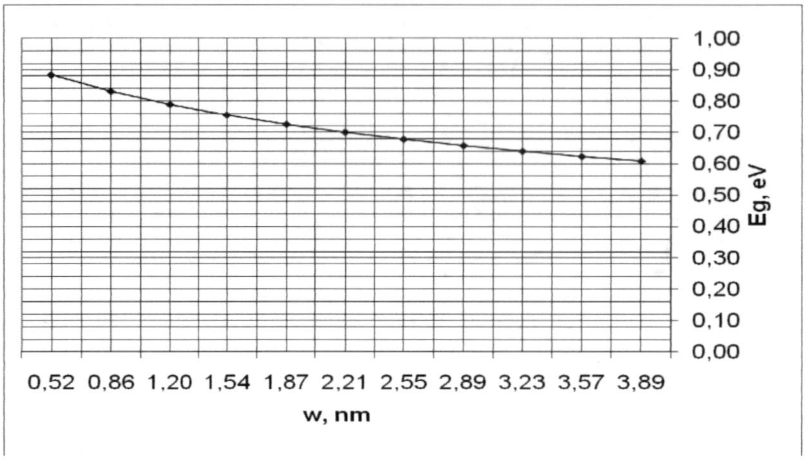

Figure.6. The dependence of the band gap of graphene nanoribbons with the edges of the "armchair» type L = 28,27 nm from the width of the ribbon.

3. Conclusions

The variations of the boundary conditions, chemical composition and chirality permit to obtain the supracrystalline nanoribbons with different electrical characteristics which are of interest for practical applications. In this work we present the results of simulation and calculations of energy stability and electrical characteristics (band gap) for supracrystalline nanoribbons of different symmetry made on the base of C, Si, S, B-N. We have demonstrated, that such nanoribbons may be successfully used in nanoelectronics.

References

1. Novoselov K. S. et al. Two-dimensional atomic crystals // PNAS. 2005. V. 102. P. 10451.
2. Brazhe R.A., Karenin A.A. Simulation of the supracrystal's physical properties // Izvestia VUZOV. Povolzhskii region. Fiziko-matematicheskie nauki. 2011. 17 (1).
3. Arefeva P.A., Brazhe R.A., Karenin A.A. Supracrystalline nanoribbons // Radioelektronnaya technika. Ulyanovsk, 2010. P. 141-147.

Journal of Physics: Conference Series **345** (2012) 012005 doi:10.1088/1742-6596/345/1/012005

Direct measurements of three-dimensional localization of light near the nanostructures with sub-wavelength spatial resolution

A V Demin[1,3], A A Ezhov[2], G G Levin[1] and V I Panov[2]

1 All-Russian research institute of optico-physical measurements, 46 Ozernaya st., Moscow, 119361, Russia

2 Faculty of Physics, Lomonosov Moscow State University, 1-2 Leninskiye Gory, Moscow, GSP-1, 119991, Russia

demin@vniiofi.ru

Abstract. Here we report on the development of a scanning near field microscope to map the sub-wavelength localization of light. The equipment can detect the intensity and polarization of an alternating electromagnetic field of optical frequency, which is localized in the close proximity to nanoobjects and nanostructures.

1. Introduction

The development of nano-photonics is defined by the growing demand in new solutions in the areas of storage and communication of optical data [1].

Scanning near-field optical microscopy (SNOM) is the method allowing one to study optical properties of nanostructured media by direct observation of the three dimensional (3D) local electromagnetic field intensity with spatial resolution better than the diffraction limit.

2. Materials and methods

Special test grating consist of vanadium rhombs on quartz substrate was measured by means of aperture type SNOM probe based on tapered single mode optical fiber, which was covered by aluminum film. The diameter of aperture was about 100 nm. Probe scanned laterally above the sample surface illuminated by normally incident light of electroluminescent diode, which was equipped with polarizer. A configuration of transmitted light with the probe as sub-wavelength collector was used.

3. Results

3D mapping of light intensity and polarization at sub-wavelength scale play an important role in the understanding of an interaction between the light and the matter at nanometer scale [2, 3]. Recently an influence of the structure anisotropy to local intensity of light was shown by means of SNOM [1]. During the measurements carried out with the help of developed equipment, we have shown that in the case of the polarization direction parallel to the long axis of rhomb there were no significant

[3] To whom any correspondence should be addressed.

Journal of Physics: Conference Series **345** (2012) 012005 doi:10.1088/1742-6596/345/1/012005

interference fringes. Clearly resolved fringes were seen if the polarization direction was orthogonal to the long axis.

The results which demonstrate that the distribution of intensity of the orthogonally polarized near field components is sensitive to the polarization of the incident light are shown on figures 1-3.

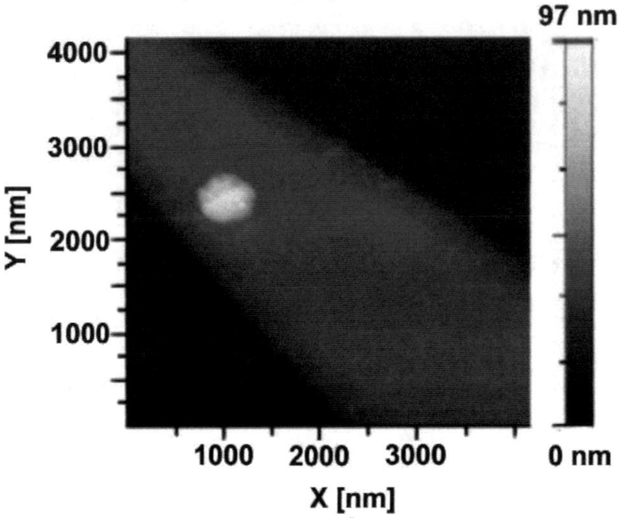

Figure 1. Topography of the vanadium rhomb on the quartz substrate with the single particle of contamination.

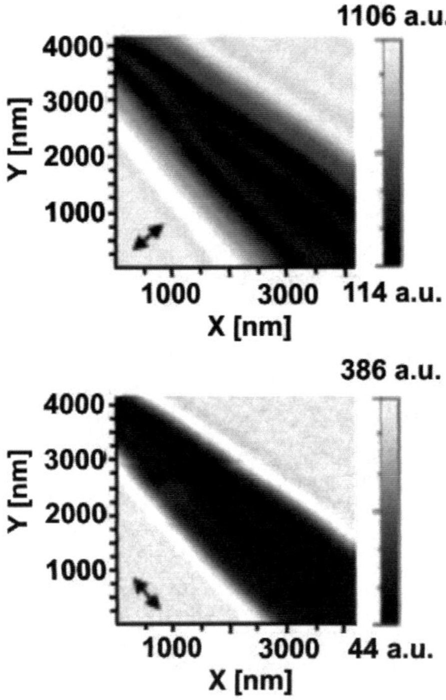

Figure 2. Two dimensional (2D) maps of electromagnetic field intensity with orthogonal directions of light polarization, which measured above sample surface shown on figure 1. The direction of linear light polarization is indicated by arrows.

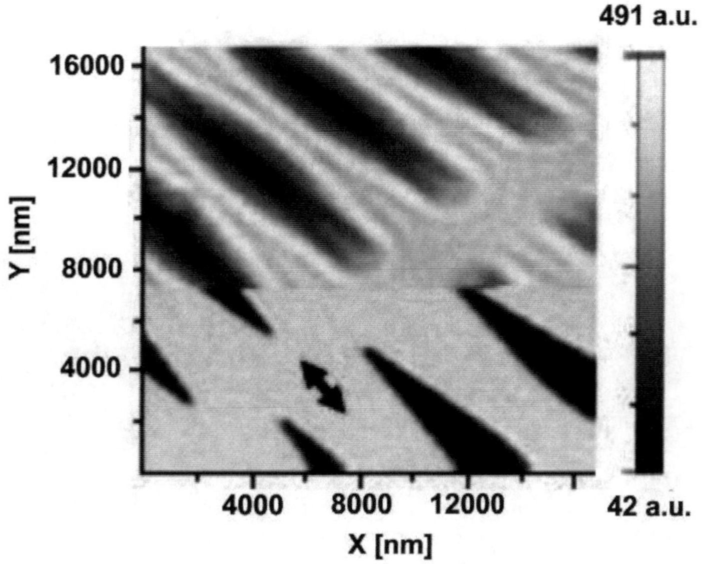

Figure 3. 2D maps of light intensity obtained by SNOM at different distance between the probe and the sample surface: few nanometers (bottom) and at 1.5 microns above the sample surface (upper). Direction of linear light polarization is indicated by arrow.

4. Acknowledgements
The work was performed within the bounds of state contract "Generation of the metrological complex and guideline base for the measurements of the optical constants of nanostructured media and metamaterials in nanophotonics" in the frame of state targeted program "Development of the nanoindustry infrastructure in Russian Federation in 2008-2011".

The authors thank B.B. Tsema, M.R. Sherbakov and Dr. A.M. Dubrovkin for their kind assistance during the experiments and Prof. A.A. Fedyanin for helpful discussions.

References
[1] Shcherbakov M R, Tsema B B, Ezhov A A, Panov V I and Fedyanin A A 2011 *JETP Lett.* **93** 720 (2011)
[2] Gramotnev D K and Bozhevolnui S I. 2010 *Nat. Photonics* **4**, 83
[3] Bozhevolnui S I, Volkov V S, Devaux E, Laluet J-Y and Ebbesen T W 2006 *Nature* **440**, 508

Journal of Physics: Conference Series **345** (2012) 012006 doi:10.1088/1742-6596/345/1/012006

Quantum regime of resonant-tunnelling diode oscillations

I A . Elant'ev, I U Bezotosny, M P Bezhko, V F Elesin

National Research Nuclear University MEPh, Moscow, 31 Kashirskoe road, Russia

E-mail: ivan.elantev@physicist.net

Abstract. Resonant-tunneling diode (RTD) is one of the first quantum devices. In contrast to devices settled on quantum dots and quantum threads, which works only at low temperatures, RTD can works in a wide range of temperatures from liquid-helium temperature till 100 °C. Also, RTD has one more feature, which makes this device attractive for use in integrated circuits – high operating frequency and, as a consequence, fast response. In this work we investigated free-running lasing of RTD in quantum regime.

Quantum regime of free-running lasing on RTD was predicted in paper [1]. To operate in this regime we need to shift the energy of the incident electrons relative to resonant level. For the frequency 1 THz shift value is 40 meV. Our pattern has width of the negative differential resistance (NDR) region about 200 mV, so if suppose that resonant level position corresponds to the peak current quantum regime must be occur in the NDR region. To get more exact values of parameters we measured generation spectra at different bias voltages. Harmonic amplitudes data subject to bias voltage was obtained. Also we measured dependence main frequency of the bias voltage.

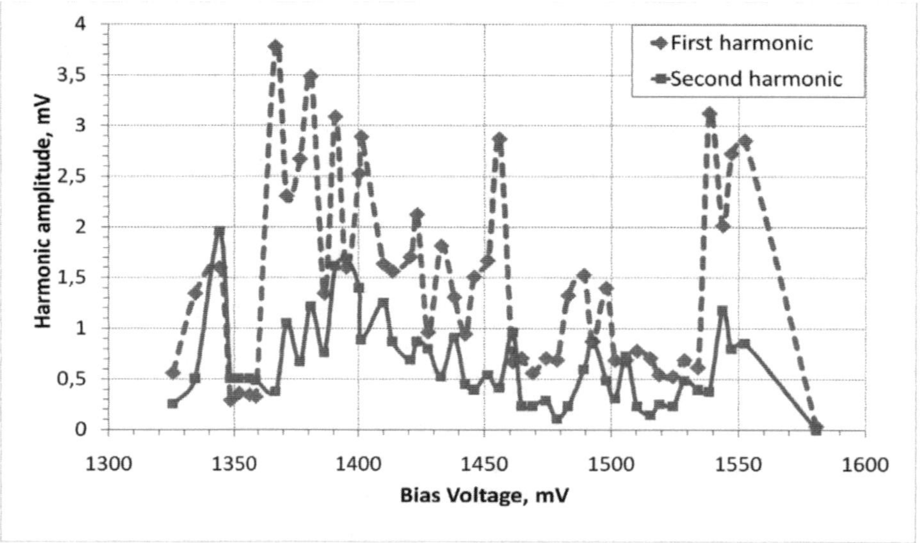

Figure 1. Dependence first and second harmonics of the bias voltage.

Dependences of harmonic amplitudes is very complicated, there are many maximums and minimums (figure 1). It may be explained by existence of several generation mechanisms.

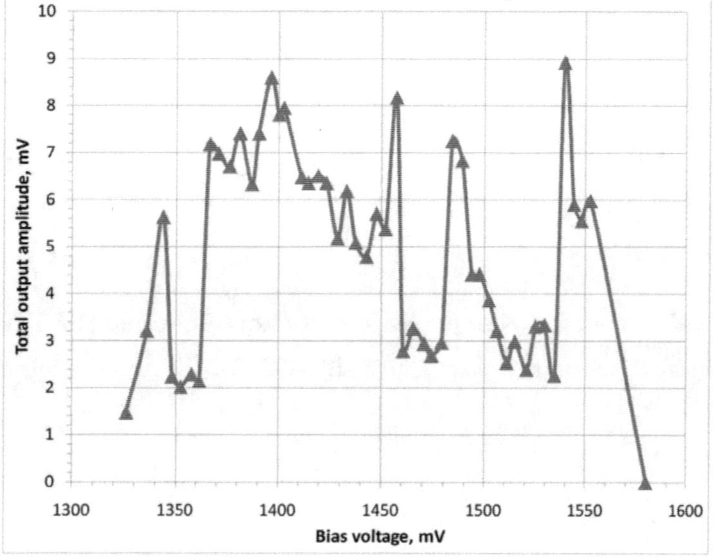

Figure 2.Dependence total output amplitude of bias voltage

Harmonics spectrum was integrated by bias voltages, so we obtain total value of output signal subject to bias voltage (figure 2). One can see that there are four regions with maximal amplitude. Two of them are on the borders of NDR region, and two other are in the NDR region. That character of the generation attracted only to RTD. This feature can be explained by existing several mechanisms of the generation.

References

[1] Kim G, Koh K M, Kim C H 12.2001 Journal of the Korean Physical Society Vol. 39 pp. S279-S282.

IV Nanotechnology International Forum (Rusnanotech 2011) IOP Publishing
Journal of Physics: Conference Series **345** (2012) 012007 doi:10.1088/1742-6596/345/1/012007

2D supracrystals as a promising materials for planar nanoacoustoelectronics

A I Kochaev, A A Karenin, R M Meftakhutdinov and R A Brazhe

Ulyanovsk State Technical University, 432027, Ulyanovsk, Russia

E-mail: a.kochaev@gmail.com

Abstract. The elastic and acoustic properties of 2D carbon supracrystals are simulated within the scope of modified density functional theory. The central and non-central force constants of carbon atoms and the corresponding elastic moduli, have been calculated in order to determine the velocities of elastic wave propagation in 2D supracrystalline nanoallotropes. It is shown that in sp^2 nanoallotropes and graphene these velocities are approximately of the same value, and in sp^3 nanoallotropes they are much less than in graphene.

1. Introduction

The supracrystals are defined by us as the supra atomic periodic structures where the atoms in nodes of a structure are replaced by their symmetric complexes (Fig.1).

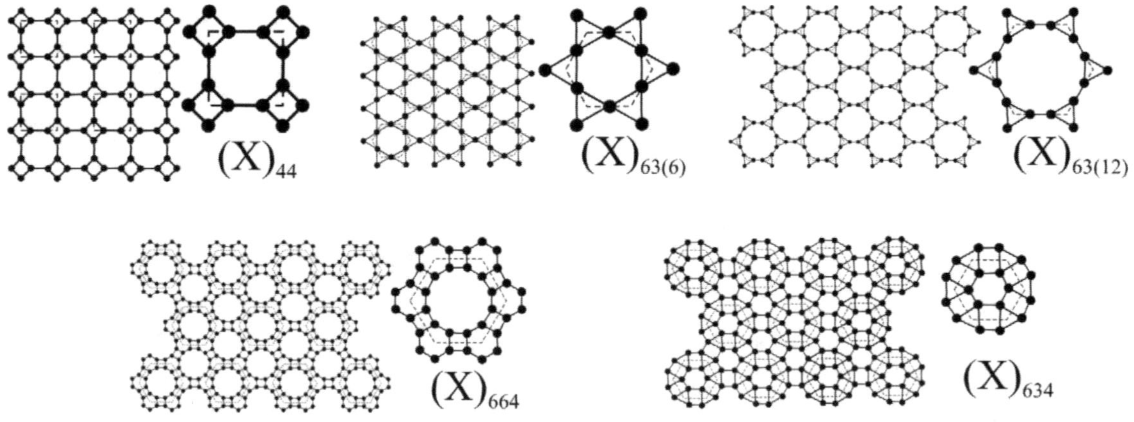

Fig. 1. 2D supracrystalline structures and the form of the corresponding supracrystalline cells. Here X is the symbol of the chemical element, and the order of the indexes behind the parentheses is following: the first index determines the type of supracell, the next indexes describe the cell's of enclosing. At first, the numbers of sides of the nodal cell are shown, than the same are shown for the surrounding cells (if there are). Numbers in parentheses indicate the form of a polygon in the cell's center.

The possibility of such structures existence is shown in [1]. Some of the structures and similar to them have been investigated in [2–11]. By a random structure searching Pickard and Needs [5] found many structures, one of which was the chiral framework structure (CFS). Note, that there are a certain relationship between CFS and $(X)_{664}$ structure. One of the framework structures consist of boron and nitrogen atoms at first synthesized in [8, 9] is our $(X)_{44}$ structure. Here we summarized these previous results in order to show the common character of 2D supracrystalline structures.

In this paper, the estimations for electric and energetic properties of 2D supracrystals consist of carbon, silicon, boron, nitrogen etc. atoms are also adduced. Our paper is devoted to the mechanical and acoustic properties of supracrystals. Note, that mechanical properties of nanoscale objects are studied worse than their electrical ones. Nevertheless, one can find the papers devoted to the mechanical properties of carbon systems [12–16]. In [16] the elastic constants of graphene are obtained using the Harrison bond-orbital method. In our terms the graphene is a special case of 2D supracrystalline carbon sp^2 nanoallotropes. Therefore, we have a right to use the checked but slightly modified scheme [15, 16] to study the mechanical and acoustic properties of carbon 2D supracrystals.

Moreover, using this scheme one can find the agreement between the numerical estimations of the diamond phonon frequencies made in [14] and the experimental data on Raman scattering [17–19] and lattice (phonon) thermal conductivity [20].

2. Calculation of the constants of central and non-central interaction for the carbon atoms in 2D supracrystals

The idea of approach is as follows: the constants of the central and non-central interactions are to express from the system energy per one atom which, in its turn, is to express from the metallic energy and the covalent energy [15, 16].

With the notations adopted in [15] the covalent energy

$$V_2 = \eta_2 \frac{\hbar^2}{m_0 l^2}, \tag{1}$$

where η_2 = -3,26 for sp^2 hybridization and -3,22 for sp^3 hybridization of an electron wave function, $\hbar$ is the reduced Planck constant, m_0 is the free electron mass, l is the bond length. In [15] the metallic energy was calculated from the energy of s- and p-states but in the case of supracrystals we really do not have any simple way to find them. Therefore, the particularity of approach used here is that metallic energy was calculated from energy per one atom [15]:

$$E_{atom} = -\frac{n_2}{3}|V_2|\left[1 + \frac{10}{3}\left(\frac{V_1}{V_2}\right)^2\right], \tag{2}$$

where n_2 is a number of nearest atoms in 2D structures. Knowing V_2 and E_{atom} one can find V_1 from (2) as

$$|V_1| = \sqrt{\frac{3}{10n_2}}\left[|V_2|\left(3|E_{atom}| - n_2|V_2|\right)\right]^{\frac{1}{2}}. \tag{3}$$

Using the software package Abinit-5.8.4 [21] within the scope of the density functional theory [22] in Hartree–Fock approximation the values of l and E_{atom} were calculated for the supracrystals considered. To generate k-points in Brillouin zone the Monkhorst–Pack algorithm [22] was used. As a mathematical approximation the Broyden method of iterations was used. The cutoff energy was chosen for $k = 6$.

Defining l, E_{atom} and calculating $|V_1|$, $|V_2|$ from Eqs. (1), (3) we can find the constants of the central (α) and the non-central (β) interactions [15] as

$$\alpha = \frac{2}{l^2}|V_2|\left[1-\frac{10}{3}\left(\frac{V_1}{V_2}\right)^2\right], \tag{4}$$

$$\beta = \frac{\lambda}{3}\alpha. \tag{5}$$

Here λ is the coefficient determined through the matrix elements of the covalent energy operator between the corresponding atomic wave function of s- and p-states [14, 15, 23]. For 2D structures of graphene type in [14, 15] is adopted that $\lambda = 0.66$. In the case of 2D supracrystals we used the value of $\lambda = 0.69$ from [23]. The results of calculations are summarized in the Table 1.

Table 1. Parameters and the results of the calculations for graphene and carbon 2D supracrystals

Parameter	$(C)_6$ graphene	$(C)_{44}$	$(C)_{63(6)}$	$(C)_{63(12)}$	$(C)_{664}$	$(C)_{634}$
Hybridization	sp^2	sp^2	sp^3	sp^2	sp^2	sp^3
n_2	3	3	4	3	3	4
l, Å	1.42	1.43	1.86	1.71	1.56	2.17
$\|E_{atom}\|$, eV	13.5	12.9	14.2	14.7	11.3	13.3
$\|V_2\|$, eV	12.3	12.1	7.10	8.50	10.2	5.20
$\|V_1\|$, eV	2.08	1.70	2.75	3.98	1.83	2.73
α, N/m	177	177	3.27	25.1	120	3.47
β, N/m	38.9	40.8	0.75	5.80	27.6	0.80

Note: The results of calculations for graphene $(C)_6$ are cited from [15].

3. Calculation of the elastic moduli for carbon 2D supracrystals

The expressions for elastic moduli of 2D crystalline structure of 6mm symmetry class (graphene-like system) versus the constants of the central and non-central interactions have been obtained in [16] and are of the form

$$c_{11} = \frac{1}{\sqrt{3}}\left(4\alpha+\beta+18\frac{\alpha\beta}{4\alpha+\beta}\right),$$
$$c_{12} = \frac{1}{\sqrt{3}}\left(4\alpha+\beta-18\frac{\alpha\beta}{4\alpha+\beta}\right). \tag{6}$$

Obtain the corresponding expressions for $(X)_{44}$ supracrystalline structure of 4mm symmetry class (Fig. 2).

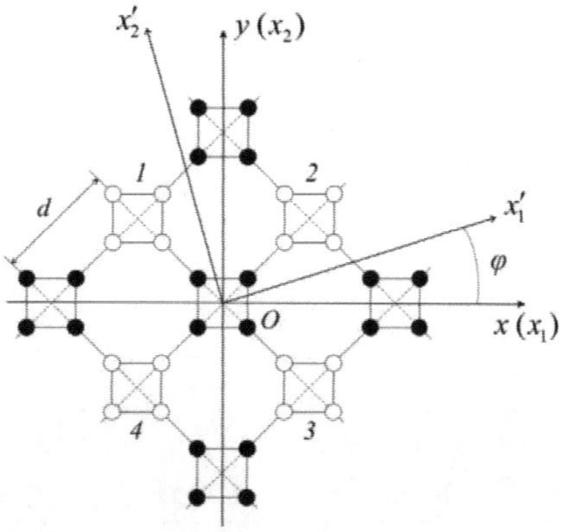

Fig. 2. Disposing of the structure $(X)_{44}$ with respect to the crystalline axes x_1 and x_2

Let the origin of coordinates is placed in "zeroth" atomic complex. Then the expressions for the energy of central W_C and non-central W_{NC} interactions take the form [16]

$$W_C = \frac{\alpha}{d^2}\sum_{i=1}^{4}\left(\mathbf{R}_{0i}^2 - \mathbf{r}_{0i}^2\right)^2,$$
$$W_{NC} = \frac{\beta}{d^2}\sum_{i,j>1}^{4}\left(\mathbf{R}_{0i}\mathbf{R}_{0j} - \mathbf{r}_{0i}\mathbf{r}_{0j}\right)^2. \tag{7}$$

Here d is the length of supracell rib (it may be expressed via the bond length), where $\mathbf{R}_{0i} = \mathbf{r}_{0i} + \delta\mathbf{r}_{0i}$, $\delta\mathbf{r}_{0i} = u_{0i}\mathbf{i} + v_{0i}\mathbf{j}$ is the displacement of the i-th atomic complex under the lattice deformation. The coordinates of displacement radius-vector end u_{0i} and v_{0i} along x_1 and x_2 axes accordingly are of the form

$$\begin{aligned}
u_{01} &= u' - \frac{\sqrt{2}}{2}de_{xx} + \frac{\sqrt{2}}{4}de_{xy},\\
u_{02} &= u' + \frac{\sqrt{2}}{2}de_{xx} + \frac{\sqrt{2}}{4}de_{xy},\\
u_{03} &= u' + \frac{\sqrt{2}}{2}de_{xx} - \frac{\sqrt{2}}{4}de_{xy},\\
u_{04} &= u' - \frac{\sqrt{2}}{2}de_{xx} - \frac{\sqrt{2}}{4}de_{xy},\\
v_{01} &= v' + \frac{\sqrt{2}}{2}de_{yy} - \frac{\sqrt{2}}{4}de_{xy},\\
v_{02} &= v' - \frac{\sqrt{2}}{2}de_{yy} + \frac{\sqrt{2}}{4}de_{xy},\\
v_{03} &= v' - \frac{\sqrt{2}}{2}de_{yy} + \frac{\sqrt{2}}{4}de_{xy},\\
v_{04} &= v' - \frac{\sqrt{2}}{2}de_{yy} - \frac{\sqrt{2}}{4}de_{xy},
\end{aligned} \tag{8}$$

where u' and v' are the inner displacements, e_{xx}, e_{yy}, e_{xy} are the components of the strain tensor.

Taking into account (8) we'll expend Eq. (7) in powers of u_{0i} and v_{0i} neglecting terms above the second order. Then we'll minimize the elastic energy $W = W_C + W_{NC}$ with respect to inner displacements considering that $\partial W/\partial u' = \partial W/\partial v' = 0$. It allows us to get the expression for the elastic energy density $w = W/S$ as a function of α, β, e_{xx}, e_{yy}, e_{xy} where $S = d^2$ is an area per one supracell. This expression may be compared with the corresponding expression for two-dimensional structure of 4mm symmetry class [24]:

$$w = \frac{1}{2}\lambda_{xxxx}\left(e_{xx}^2 + e_{xy}^2\right) + \lambda_{xxyy}e_{xx}e_{xy} + 2\lambda_{xyxy}e_{xy}^2. \tag{9}$$

So far as in (9) $\lambda_{xxxx} = c_{11}$, $\lambda_{xxyy} = c_{12}$, $\lambda_{xyxy} = c_{33}$ [25], the expressions for non-zero components of elastic moduli tensor for $(X)_{44}$ type of 2D supracrystal can be written as

$$c_{11} = \frac{4(2\alpha + 3\beta)}{\left(1+\sqrt{2}\right)^2}, \quad c_{12} = \frac{4(2\alpha - \beta)}{\left(1+\sqrt{2}\right)^2}, \quad c_{33} = \frac{2\alpha + \beta}{\left(1+\sqrt{2}\right)^2}. \tag{10}$$

4. Elastic waves in 2D supracrystals: the velocities of propagation and pure modes

Consider the peculiarities of elastic waves propagation in 2D supracrystals regarding them as an infinite two-dimensional continuum where only in-plane displacements of the particles are possible. Love waves and other types of shell waves are excluded from the consideration here.

In general, along an arbitrary direction x_1' (Fig. 2) only one quasi-longitudinal and two quasi-transverse elastic waves can propagate. These waves satisfy to Green-Christoffel equation [26]

$$\rho_2 v^2 u_\alpha = \lambda_{\alpha\beta\gamma\delta} a_{1\beta} a_{1\delta} u_\gamma, \tag{11}$$

where ρ_2 is the two-dimensional (in this case) medium density, v is phase velocity of the wave, u_α and u_γ are the components of the particles displacement, $\lambda_{\alpha\beta\gamma\delta}$ is the tensor of elastic moduli, $a_{1\beta}$ and $a_{1\delta}$ are the elements of the direction cosines matrix-column:

$$\begin{pmatrix} a_{11} \\ a_{12} \end{pmatrix} = \begin{pmatrix} \cos\varphi \\ \sin\varphi \end{pmatrix}. \tag{12}$$

The velocities of quasi-longitudinal and quasi-transverse elastic waves are the roots of a corresponding characteristic equation

$$\left|\lambda_{\alpha\beta\gamma\delta} a_{1\beta} a_{1\delta} - \rho_2 v^2\right| = 0 \tag{13}$$

and depends on the elastic moduli, density of the crystal and the direction cosines. A similar problem was solved successfully for three-dimensional crystals. Moreover, there are computer programs which allow to construct 3D surfaces of phase velocities for the elastic waves propagating in these crystals [27–29]. They also may be used for 2D lines of phase velocities construction in 2D crystals. Propagation directions for pure longitudinal and pure transverse elastic waves correspond to the extreme values of the phase velocities and are perpendicular to the velocities lines at the extremes points [30].

The analytical methods for propagation directions calculation of pure modes for elastic waves were suggested in [31–36]. Their application to the effective elastic moduli in the case of pure longitudinal and pure transverse waves in 2D crystal yields to the following expressions:

– for the 4mm symmetry class

$$\begin{aligned} \lambda'_{1111} &= (\sin^4\varphi + \cos^4\varphi)c_{11} + 2\sin^2\varphi\cos^2\varphi(c_{12} + 2c_{33}), \\ \lambda'_{2121} &= (\cos^4\varphi + \sin^4\varphi)c_{33} + 2\sin^2\varphi\cos^2\varphi(c_{11} - c_{12} - c_{33}), \\ \varphi &= n\frac{\pi}{4}, \quad n = 0, 1, 2, \ldots, 7. \end{aligned} \tag{14}$$

Journal of Physics: Conference Series **345** (2012) 012007 doi:10.1088/1742-6596/345/1/012007

– for the 6mm symmetry class

$$\lambda'_{1111} = c_{11}, \quad \lambda'_{2121} = \frac{1}{2}(c_{11} - c_{12}). \tag{15}$$

The propagation velocities of pure longitudinal and pure transverse elastic waves accordingly may be defined from the expressions

$$v_L = \sqrt{\lambda'_{1111} s}, \quad v_T = \sqrt{\lambda'_{2121} s}, \tag{16}$$

where $s = \rho_2^{-1}$ is a specific surface of 2D crystal. Its values for carbon 2D supracrystals, in comparison with graphene, are presented in Table 2. In the equations written there N_A is the Avogadro number, $\mu = 0.012$ kg/mol is molar mass for carbon, l is bond length.

Table 2. Specific surface for carbon 2D structures

	Specific surface	
Symbol of structure	Formula	s, 10^6 m^2/kg
$(C)_6$ graphene	$s = \frac{3\sqrt{3}}{2}\frac{N_A}{\mu} l^2$	2.63
$(C)_{44}$	$s = \frac{1}{2}\left(1+\sqrt{2}\right)^2 \frac{N_A}{\mu} l^2$	2.99
$(C)_{63(6)}$	$s = \frac{4\sqrt{3}}{3}\frac{N_A}{\mu} l^2$	4.01
$(C)_{63(12)}$	$s = \frac{\sqrt{3}}{6}\left(2+\sqrt{3}\right)^2 \frac{N_A}{\mu} l^2$	5.79
$(C)_{664}$	$s = \frac{\sqrt{3}}{12}\left(3+\sqrt{3}\right)^2 \frac{N_A}{\mu} l^2$	3.94
$(C)_{634}$	$s = \frac{\sqrt{3}}{6}\left(1+\sqrt{3}\right)^2 \frac{N_A}{\mu} l^2$	5.09

The phase velocities for elastic waves in 2D supracrystal of type $(C)_{44}$ and graphene constructed with the computer program used are shown in Fig. 3. Our program is based on the solution of Green-Christoffel equation. In this figure you can see that in $(C)_{44}$ structure belonging to 4mm symmetry class there are four directions (over each 45°) in which pure modes of elastic waves can propagate. Graphene as well as others 2D supracrystals belonging to 6mm symmetry class is acoustically isotropic two-dimensional medium.

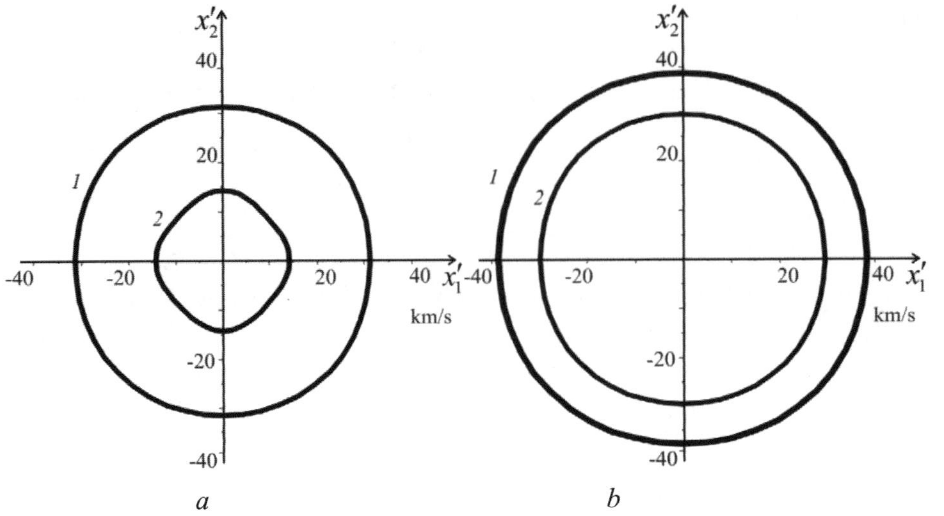

Fig. 3.The phase velocities lines for longitudinal (*1*) and transverse (*2*) elastic waves in 2D supracrystals $(C)_{44}$ (*a*) and graphene (*b*)

The results of calculations by formulas (14)–(16) for longitudinal and transverse velocities (v_L and v_T, accordingly) in carbon 2D structures are represented in Table 3. The lower and upper values of the velocities range correspond to pure longitudinal and pure transverse waves propagating at the angles $\varphi_1 = 0$ and $\varphi_2 = 45°$ to the x_1' axis.

Table 3. Characteristics of the elastic waves in carbon 2D structures

Parameter	$(C)_6$ graphene	$(C)_{44}$	$(C)_{63(6)}$	$(C)_{63(12)}$	$(C)_{664}$	$(C)_{634}$
c_{11}, N/m	533	328	9.84	75.7	361	10.5
c_{12}, N/m	331	215	6.15	47.1	226	6.52
c_{33}, N/m		68				
v_L, km/s	37.4	31.3–31.9	6.30	20.9	37.7	7.30
v_T, km/s	29.5	13.0–14.3	5.00	16.5	29.8	5.80

5. Conclusions

The calculated Young modulus E for graphene in this paper is 267 N/m. In [37, 38] and [39] was obtained 235 N/m and 312 N/m, respectively. Thus, the method of the nanostructures study is successful and its application to supracrystals is justified.

The force constants which would describe the interaction of carbon atoms in 2D supracrystals depend on their structure and may differ each other by a factor of tens. We see that sp^3 nanoallotropes unlike sp^2 nanoallotropes have the least values of the force constants so far so all four their valence electrons are bonded in the plane of 2D supracrystals by strong σ-bonds. Accordingly, the covalent energy in sp^3 nanoallotropes is higher (less by modulus) then in sp^2 nanoallotropes although their metallic energy is the same of order, and formulas (4), (5) give that the constants α and β are less, than in sp^2 allotropes.

As it follows from Fig. 3 and Table 3, the velocities of elastic waves propagation in graphene are nearly twice higher their values for the bulk elastic waves in a diamond [40]. The velocities of elastic waves in 2D supracrystals of $(C)_{44}$ and $(C)_{664}$ types are close to them. But due to small value of c_{33} compared with c_{11} and c_{12} the velocity of pure transverse wave in $(C)_{44}$ structure are essentially less than in graphene and $(C)_{664}$ structure. The velocities of elastic waves in $(C)_{63(12)}$ structure are characterized by some less values. As for 2D carbon sp^3 nanoallotropes, the velocities of elastic waves propagation in these structures are less than in sp^2 nanoallotropes in several times due their much worse elastic characteristics.

The calculated values of force constants, elastic moduli and propagation velocities for elastic waves in carbon 2D supracrystals should be considered as estimations. For more accurate values the experimental data on lattice parameters, bond lengths and elastic moduli of corresponding structures are required. Nevertheless, these results do seem quite useful. We believe that the possibility to change within wide limits the structural, electric and mechanical characteristics of 2D nanoscale materials consist of the same basic chemical element can stimulate the work on their synthesis.

Acknowledgment
This work was supported by the Ulyanovsk State Government.

References

[1] Brazhe R A and Karenin A A 2011 *Izvestiya vysshih uchebnyh zavedenyi. Fiziko-matematicheskie nauki (Volga region)* **2(18)** 105
[2] Novoselov K S, Geim A K, Morozov S V et al. 2004 *Science* **306** 666
[3] Terrones H, Terrones M, Hernández E et al. 2000 *Phys. Rev. Lett.* **84** 1716
[4] Kara A, Leandri C, Davila M E et al. 2008 *ArXiv* 08.11.2611
[5] Pickard C J, Needs R J *Phys. Rev. B* 2010 **81** 014106.
[6] Lisenkov S V, Vinogradov G A, Astakhova T Yu et al. 2005 *JEPT Lett.* **81** 346
[7] Lisenkov S V, Vinogradov G A, Astakhova T Yu et al. 2006 *Phys. Sol. State* **48** 192
[8] Golberg D, Bando Y, Bourgeois L, Kurashima K et al. 2000 *Appl. Phys. Lett.* **77** 1979
[9] Biró L P, Márk G I, Horváth Z E et al. 2004 *Carbon* **42** 2561
[10] Shevchenko E V, Talapin D V, Kotov N A et al. 2006 *Nature* **495** 55
[11] Lisiecki L, Albouy P and Pileni M 2003 *Adv. Mat.* **15** 712
[12] Keating P N *Phys. Rev.* 1966 **145** 637
[13] Davydov S Yu, Lebedev A A and Smirnova N Yu *Phys. Sol. St.* 2009 **51** 481
[14] Davydov S Yu 2009 *Phys. Sol. St.* **51** 2166
[15] Davydov S Yu 2010 *Phys. Sol. St.* **52** 184
[16] Davydov S Yu 2010 *Phys. Sol. St.* **52** 810
[17] Ferrari A C, Meyer J C, Scardaci V et al. 2006 *Phys. Rev. Lett.* **97** 187401
[18] Aouissi M, Hamdi I and Meskin N 2006 *Phys. Rev. B* **74** 054302
[19] Calizo I, Bejenari I, Rahman M, Liu G et al. 2009 *J. Appl. Phys.* **106** 043509
[20] Nika D L, Ghosh S, Pokatilov E P et al. 2009 *J. Appl. Phys.* **94** 203103
[21] URL http://www.abinit.org
[22] Honenberg P and Kohn W 1964 *Phys. Rev.* **136** B864
[23] Harrison W A 1989 *Electronic structure and the properties of solids* (Torronto: Dover)
[24] Landau L D and Lifshitz E M 1970 *Theory of Elasticity (Course of Theoretical Physics, vol. 7), third ed.* (Oxford: Butterworth-Heinemann)
[25] Nye J F 1957 *Physical Properties of Crystals* (Oxford: University Press)
[26] Christoffel E B 1877 *Ann. di matematica pura ed applicata* **8** 193
[27] Duarte M, Piedrahita C, Salinas T et al. 2004 *Earth Sciences Research Journal* **8** 63
[28] Laboratory for scientific visual analysis URL: http://www.sv.vt.edu.

[29] Brazhe R A and Kochaev A I 2011 *Acta Mechanica* **222** 193
[30] Ting T C T 2006 *Acta Mechanica* **185** 147
[31] Borgnis F E 1955 *Phys. Rev.* **98** 1000
[32] Brugger K 1965 *J. Appl. Phys.* **36** 759
[33] Chang Z P 1968 *J. Appl. Phys.* **19** 5669
[34] Brazhe R A, Grigoriev M A and Nayanov V I 1975 *Fyzika Tverdogo Tela (in Russian)* **17** 886
[35] Kochaev A I and Brazhe R A 2011 *Acta Mechanica* **220** 199
[36] Brazhe R A and Kochaev A I 2010 *Radio-electronic equipment: 40 Interuniversity Collection of Scientific Papers* (Ulyanovsk: USTU)
[37] Flannery C M, Whitfield M D and Jackman R B 2003 *Semicond. Sci. Technol.* **18** S86.
[38] Zhou Z and Huang R 2008 *J. Mech. Phys. Solids* **56** 1609
[39] Arroyo M and Belytshko T 2004 *Phys. Rev. B* **69** 115415
[40] Cadelano E, Palla P L, Giordano S et al. 2009 *Phys Rev. Lett.* **102** 235502

Silicon nanowire structures as high – sensitive pH - sensors

S O Belostotskaya[1], O V Chuyko[1], A E Kuznetsov[1], E V Kuznetsov[1], E N Rybachek[1]

E-mail: kae@tcen.ru

Abstract. Sensitive elements for pH-sensors created on silicon nanostructures were researched. Silicon nanostructures have been used as ion-sensitive field effect transistor (ISFET) for the measurement of solution pH. Silicon nanostructures have been fabricated by "top-down" approach and have been studied as pH sensitive elements. Nanowires have the higher sensitivity. It was shown, that sensitive element, which is made of "one-dimensional" silicon nanostructure have bigger pH-sensitivity as compared with "two-dimensional" structure. Integrated element formed from two p- and n-type nanowire ISFET ("inverter") can be used as high sensitivity sensor for local relative change [H+] concentration in very small volume.

1. Introduction

In recent years, system-on-chip technology has been increasingly used to create complex, integrated systems on a single chip. The development of ISFET sensors, a new generation of analytical devices, is one of today's most promising directions of investigation in the field of analytical biology and chemistry.

2. ISFET sensors technology

The technologies of pH-sensitive elements on the basis of silicon nanostructure include a combination of two basic stages: formation of silicon structures with nanodimensions and sequence of operations CMOS of process.

SOI and polysilicon layer have been used for formation of nanostructures. Suspended nanowire structures were formed by methods of photolithography and selective etching. Diameter of 1μm-length nanowires was varied from 20-100 nanometers. On the surface of Si nanostructures low temperature ultrathin SiO_2 (thickness ~ 20Å) was formed by plasma oxidation

3. Results and discussion

Two kinds of ISFET: "two-dimensional" nanostructures (silicon films) and "one-dimensional" nanostructures (nanowires) were investigated as pH sensors (figures 1, 2). Both devices respond appropriately to pH changes.

[1] SMC «Technological Center», MIET, Moscow, Russia.

Published under licence by IOP Publishing Ltd

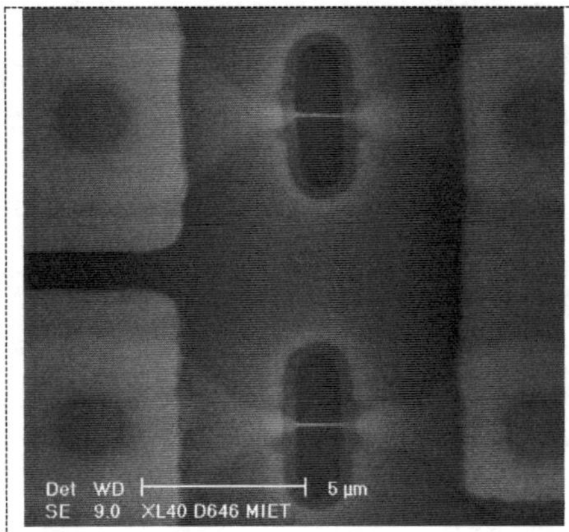

Figure 1. SEM image of nanowires.

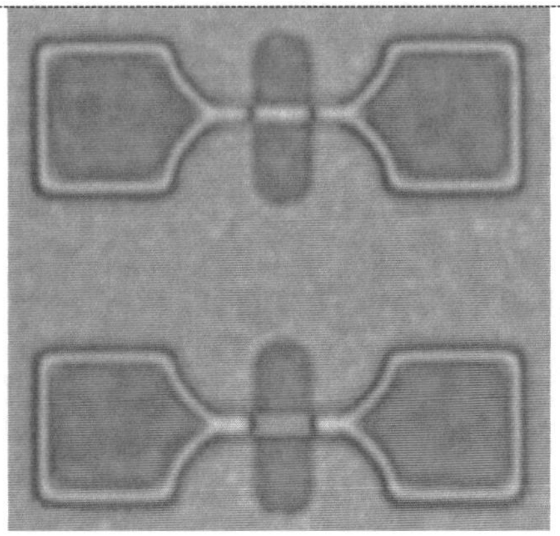

Figure 2. Optical image of nanowires.

The Agilent B1500A semiconductor parameter analyzer was used to obtain the pH from 3 to 9 buffer solutions, where the temperature was room temperature (25°C). Threshold voltages were applied using a gold reference electrode. The buffer solutions were used phosphate buffered saline (PBS) 1M.

Threshold voltage of ISFETs (V_t) dependences of pH value are presented on figures 3, 4.

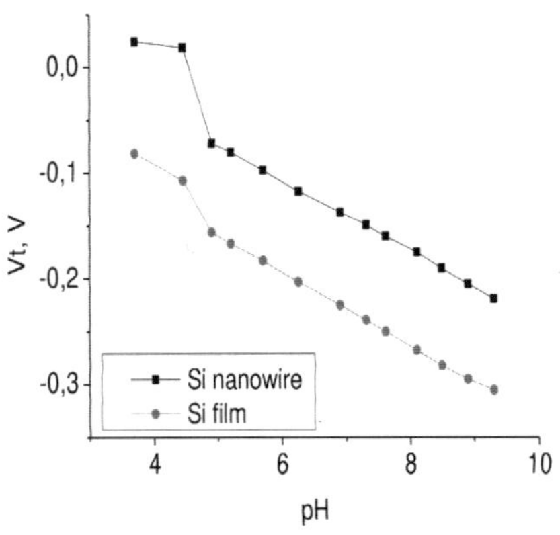

Figure 3. Characteristic of the threshold voltage vs pH values.

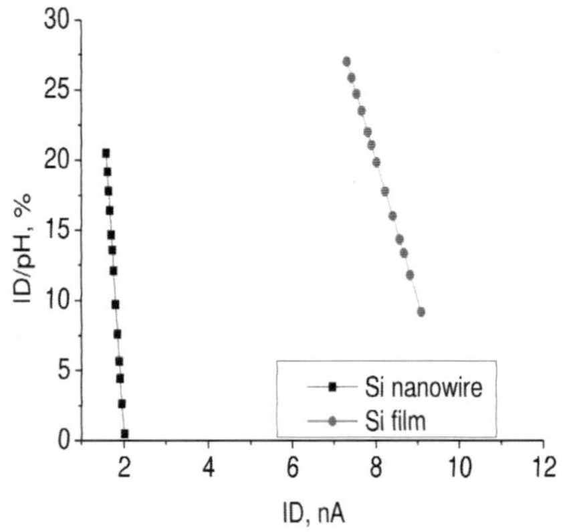

Figure 4. Characteristic of nanostructure ISFET pH-sensitivity.

ISFET sensors constructed on the basis of "one-dimensional" nanostructures (nanowire) and "two-dimensional" nanostructures (silicon film) have a linear pH sensitivity – 41,8 mV/pH and 39 mV/pH respectively in a concentration range between pH 3 and pH 9. With nanowires greater sensitivity in comparison with films on voltage (7%) and on current (25%) was observed.

For the purpose of creation of the sensors for relative change of value pH with high sensitivity, us have been developed integrated elements formed from two n - and p - type nanowire ISFET ("inverter"), made on structures SOI and polysilicon (figures 5, 6).

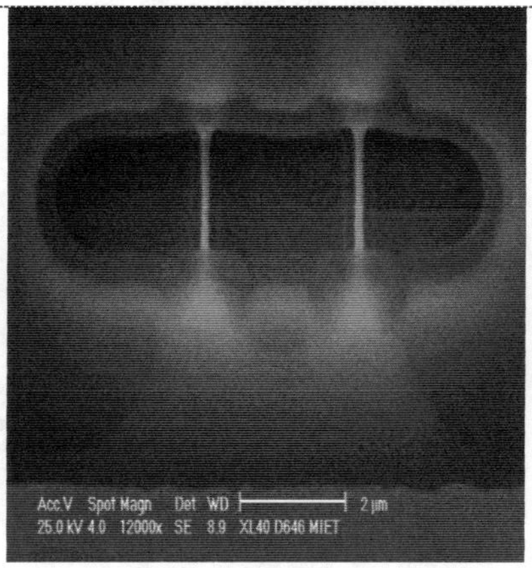

Figure 5. SEM image of nanowire "inverter".

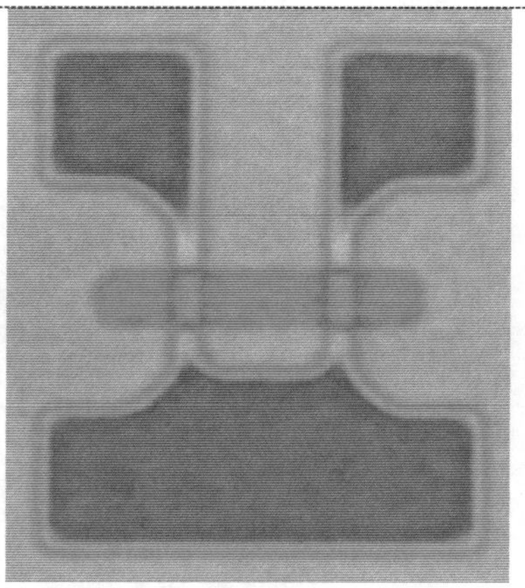

Figure 6. Optical image of nanowire "inverter".

Nanowire SOI and polysilicon "inverter" were investigated as high sensitive sensors to pH changes. Transfer characteristics of nanowire SOI and polysilicon "inverter" for different supply voltage are presented on figures 7, 8. The amplification coefficient of SOI and polysilicon "inverter" (K) varied from 30 to 150 and from 5 to 30 respectively.

The gold reference electrode was used for definition of voltage on the input.

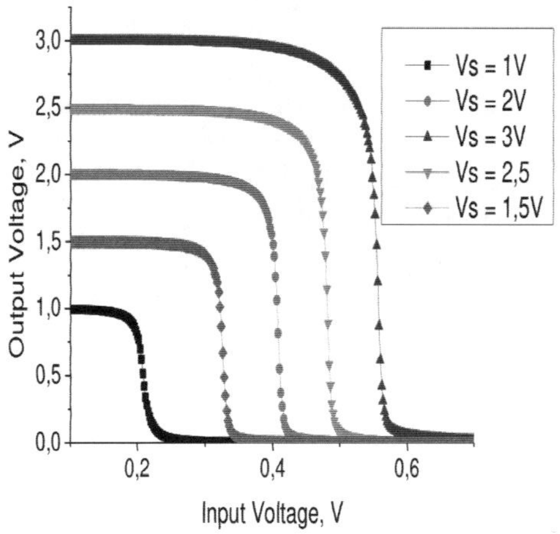

Figure 7. Transfer characteristic of nanowire "inverter" (SOI) for different supply voltage (V_S).

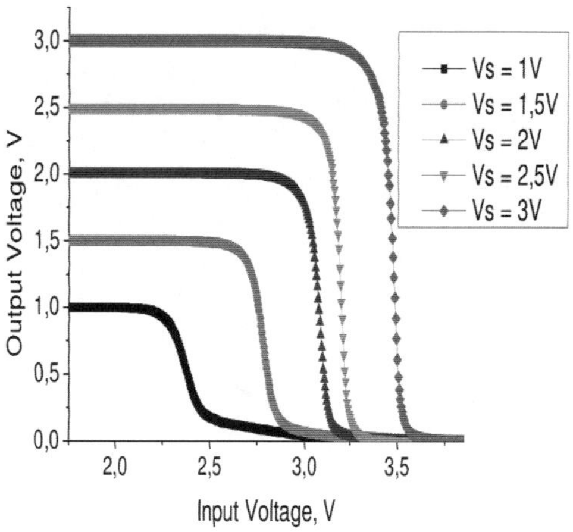

Figure 8. Transfer characteristic of nanowire "inverter" (polySi) for different supply voltage (V_S).

Such elements with high sensitivity 5 V/pH could be used as sensors based on local relative pH changes. Amplification coefficient of "inverter" depended on variation supply voltage.

The amplification coefficient depends on molarity of solution (figure 9). Thus "inverter" could be used for determination of molarity.

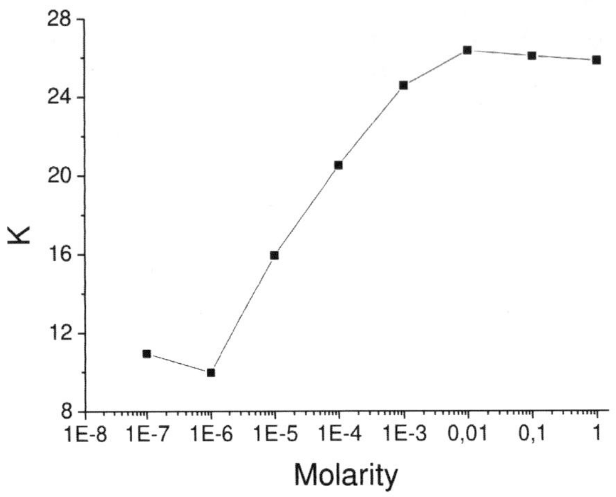

Figure 9. Dependence of K vs molarity of solution.

4. Conclusion

Silicon nanostructures were fabricated with standard semiconductor processing techniques. This demonstrates the feasibility of fabricating ISFET's and associated circuits with a standard CMOS process with minimum modifications.

Recent studies have demonstrated the ability of semiconducting nanowire field-effect transistors to serve as highly sensitive label-free sensors for control of various redox reactions. Research of silicon nanostructures used in case of pH-sensitive elements provided a perspective of its usage in a difficult analytical microsystems «lab on a chip» [1]. Nanowire pH sensors approach appears to have potential for extension to a fully integrated system, with wide use as sensors in molecular and cellular arrays. High sensitivity sensor of local relative change H+ concentration could be used for control biochemical reactions. Miniaturization of sensors opens a broad possibility for studying pH gradient on the active surface at nanoscale or even inside the microobjects like cells [2, 3]. The formation of nanowire systems on polysilicon dramatically reduces the cost of sensors.

5. Reference

[1] Wilson D M, Hoyt S, Janata J, Abando L 2001 *IEEE Sensors* **4** 256–276
[2] Victor V, Ralph K 2005 *Microsystems for Bioelectronics* **33** 157
[3] Bashir R 2009 *Sensors and Actuators* **9** 127

Acknowledgements

The work was supported by ministry of education and science of the Russian Federation, program "Development of electronic component and radio electronics", №11.426.11.0030.

Journal of Physics: Conference Series **345** (2012) 012009 doi:10.1088/1742-6596/345/1/012009

Computer simulation of field ion images of nanoporous structure in the irradiated materials

E.V. Medvedeva[1], **S.S. Alexandrova**[1], **T.A. Belykh**[2]

[1] Institute of Electrophysics, Ural Branch, Russian Academy of Science, Amudsena str., 106, Yekaterinburg, 620016, Russia
[2] Federal State Autonomous Institution of Higher Professional Education "Ural Federal University named after the first President of Russia B.N. Yeltsin", Mira str. 19, Yekaterinburg, 620002, Russia

E-mail: lena_p@bk.ru; alexandrova.swetlana@gmail.com

Abstract. Computer simulation and interpretation of field ion microscopy images of ion irradiated platinum are discussed. Field ion microscopy technique provides direct precise atomic scale investigation of crystal lattice defects of atomically pure surface of material; at the same time it allows to analyze the structural defects in volume by controlled and sequential removal of surface atoms by electric field. Defects identification includes the following steps: at the first stage the type of crystalline structure and spatial orientation of crystallographic directions were determined. Thus, we obtain the data about exact position of all atoms of the given volume, i.e. the model image of an ideal crystal. At the second stage, the ion image was processed used the program to obtain the data about real arrangement of atoms of the investigated sample. At the third stage the program compares these two data sets, with a split-hair accuracy revealing a site of all defects in a material. Results of the quantitative analysis show that shape of nanopores are spherical or cylindrical, diameter on nanopores was varied from 1 to 5 nm, their depth was fond to be from 1 to 9 nm. It was observed that nearly 40% of nanopores are concentrated in the subsurface layer 10 nm thick, the concentration of nanopores decreased linearly with the distance from the irradiated surface.

1. Introduction

It is common knowledge that ion irradiation of metals generates a large number of defects, including vacancies and interstitials, bubbles and dislocation loops [1-3]. The information on the distribution of radiation damages is of great importance for understanding the processes of formation and evolution of defective structure, for an explanation of mechanisms of radiating influence on structure and properties of irradiated material. The aim of this paper is to analyze, using field ion microscopy (FIM), the structure of radiation-induced defects on the atom-free platinum surface and in the platinum subsurface bulk, which were formed by the Ar^{+} ion irradiation. Concerning a possibility of the nanoscale analysis of crystal structure of metals and alloys, the method of field ion microscopy is most informative of existing methods. This method enables one to conduct direct precision study of radiation damages of crystal lattice of a material in the atom scale, to identify large and ultra small pores up to single vacancies and their segregations, to determine their concentration, sizes, shape, and

Journal of Physics: Conference Series **345** (2012) 012009 doi:10.1088/1742-6596/345/1/012009

the distribution over the bulk of irradiated material, to change the specified parameters depending on the radiation fluency.

The obtained data allows us to clarify the mechanism of the effect of radiation on the structure of metal and alloys, which is of great importance for understanding the properties of irradiated materials. As a result, the dynamics and evolution of radiation defects can be related to the macroscopical processes proceeding under the irradiation.

2. Experimental

It is difficult to analyze the modification of the structure and properties of materials subjected to the implantation of ions with average energies of 10–100 keV by using FIM experiments, because the near-surface metal layer with a thickness of tens or hundreds of nanometers (depending on the bombardment regime) is bombarded. Therefore, the samples to be studied were prepared in the form of tip samples with the tip's radius of curvature of 10–30 nm from metal rods by electrochemical polishing. Then, an ion image of Pt in the initial state was obtained. Field emitters certified for ion implantation had atomically smooth tip surfaces close to the spherical one prepared in situ by the field evaporation of surface atoms. If the FIM is used, pure surface of samples is prepared directly in the microscope and is maintained pure during the whole experiment. Irradiation of Pt tip-samples was provided using Ar^+ ions accelerated up to 30 keV with fluencies of $F=10^{18}$ cm^{-2} and ion current density of $j=200$ μA/cm^2 at $T=573$ K. Irradiation was performed in a direction parallel to the axis of the tip-sample. The irradiated tip-samples were inserted again in the field ion microscope and, using photo and video cameras, the field ion micropictures of the surface were recorded during the controlled removal of atomic layers and the experimental results for the subsequent analysis of the defect structure were obtained. The ion images were registered so that the quantity of the evaporated atoms in intervals between pictures was minimum. Such detailed and labor-consuming procedure was necessary for precision revealing of defects of a crystal lattice, which can be overlooked in the case of evaporation of entire atomic plane. The field ion microscope was equipped with a microchannel ion–electron converter increasing the brightness of the surface micropictures by a factor of 10^4, liquid nitrogen (T = 77 K), as a rule, served as a cooling agent, and spectrally pure neon was used as an imaging gas. The information on nanopores and their distribution over the subsurface volume of the irradiated material is obtained by means of a computer program complex for interpretation of field ion images.

3. Results and discussion

The radiation damage in Pt was studied after ion bombardment of preliminarily prepared atomically pure and atomically smooth surfaces of field emitters (Fig. 1).

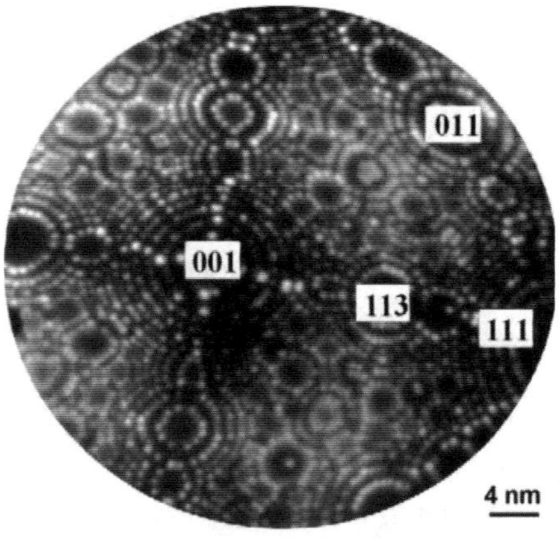

Figure 1. Ion image of the initial atomically pure and atomically smooth platinum surface prior to irradiation.

This is a typical micropicture of the surface of a pure metal with the FCC lattice. The crystal is oriented along the [001], in almost perpendicular direction to the micropicture plane. These contrast points form regular circles (families of planes of the corresponding crystal's region), whose centers are the corresponding poles of the stereographic projection [4]. Figure 2 shows micropicture of surface of irradiated by argon ions Pt. Distortions of the ion contrast and regularities of the pattern of circles were observed in several areas of the surface of the irradiated plate.

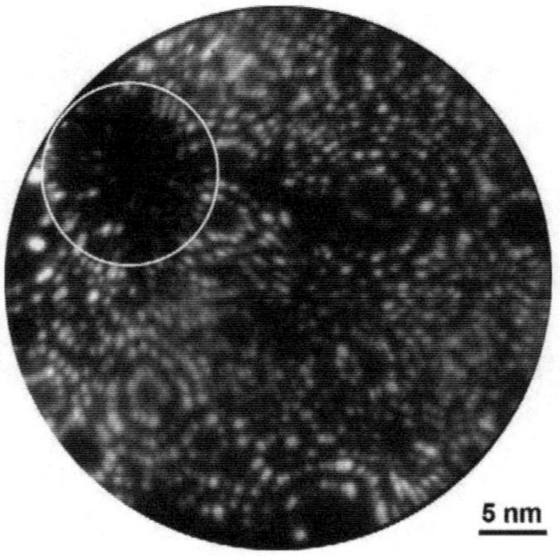

Figure 2. Ion image of pure Pt after irradiation by argon ions (contrast of the nanopore is shown by a circle).

These distortions of the ion contrast in the pattern of circles demonstrate the presence of defects in the crystal structure. Various types of defects induced in materials subjected to some external actions can be identified using contrast available from the literature [5, 6]. In this case, a change in the ion contrast of the irradiated plate compared to the contrast of the initial platinum is due to radiation damage produced by the interaction of ion with the crystal lattice. The structure of defects was analyzed in the material bulk during controlled removal of platinum atoms one by one from the surface by an electric field. The observed radiation damages were the nanopores. Contrast from the nanopores in Pt was registered immediately when the ion image of the surface appeared, and it remained after the surface layers of the irradiated metal were removed (Fig. 2). The ion contrast of nanopores at the moment of the field evaporation of the last atomic layer prior to the formation of a nanopore was registered in the form of craters. Then, as the atomic layers surrounding the defect evaporated, we observed the profile of a defect whose size was a bit less than when the nanopore opened. Finally, the nanopore's exit from the material during the evaporation of atomic layers from the surface ended, as a rule, with a dislocation loop (Fig. 3).

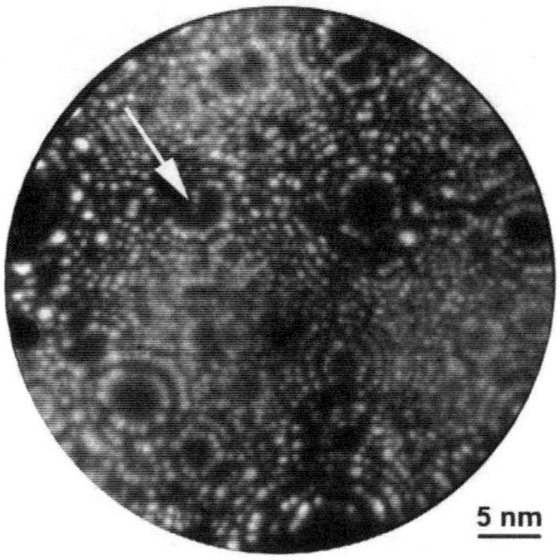

Figure 3. Ion image of Pt after irradiation by ions of Ar^+ (contrast dislocation loop at an exit of a nanopore from the material is shown by an arrow).

It is known that, in the metals, spherical vacancy pores are unstable, they are squeezed in a plane of one of the most dense atom layers of a crystal and form dislocation loops. In the case under consideration, the nanopores represent the bubbles, which are the accumulations of Ar^+ ion gas; they create a pressure upon a cavity wall that compensates a superficial tension and stimulate an increase in a noncompensated flow of vacancies [7].

It should be noted that bulk radiation defects in a layer of 10 nanometers from the irradiated surface have the form of craters. Such surface morphology is a result of erosion. For the given combination of material type of ions, and implantation modes, erosion can proceed by various mechanisms. Firstly, in the course of implantation, the material can be sprayed not as individual atoms, which is typical for beams with the currents of several μA, but in the form of clusters. This is additionally favored by the needle shape of the specimen; a high electric field in the beam plasma is concentrated at the tip of the sample. Secondly, the implantation of inert gases (here, argon) with high fluencies is characterized by the formation of blisters, which will have a shape of craters on the ion image.

For the quantitative analysis of nanoporous structure of the irradiated material, a specially developed computer program complex was used [8]. The principles of defect identification were the following: at the first stage, the type of crystalline structure and spatial orientation of crystallographic directions were determined. Thus, we obtain the data on the exact position of all atoms of the given volume, i.e. the model image of an ideal crystal. At the second stage, the ion image was processed using the program to obtain the data about real arrangement of atoms in sample under investigation. At the third stage, the program compares these two data sets, with a split-hair accuracy revealing the sites of all defects in the material (Fig. 4).

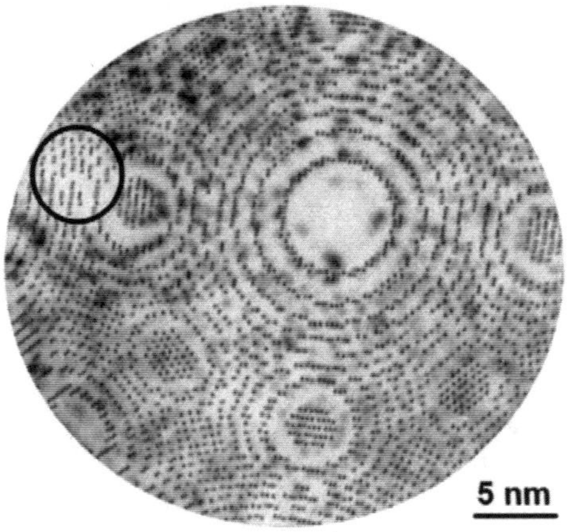

Figure 4. A comparison of the ion image of irradiated Pt with the simulated image of FCC-crystal free of defects (a nanopore is shown by a circle).

The results of the quantitative analysis show that the nanopores were both spherical and cylindrical, the diameter of nanopores varied from 1 to 5 nm, and the depth of nanopores varied from 1 to 9 nm. Successive evaporation by electric field of superficial nuclear layers to depth of 60 nm from the irradiated surface and the application of computer simulation of ion images allowed revealing specific features of formed nanopores. The concentration of nanopores and their distribution over the subsurface volume of the irradiated material were calculated. It was found that nearly 40% of nanopores are concentrated in the subsurface layer 10 nm thick, the concentration of nanopores decreased linearly ($y = -0.084x + 5.154$) with the distance from the irradiated surface (Fig. 5).

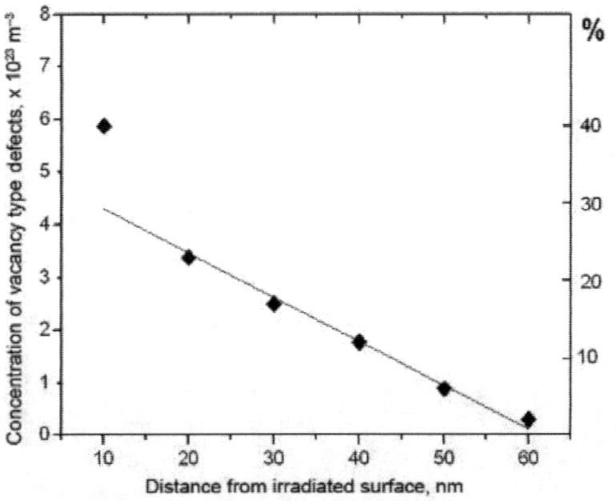

Figure 5. Profile of concentration of defects of vacancy type in Pt irradiated by Ar^+ ions.

These numerical values of coefficients of linear dependence were calculated only for the conditions of above experiment. A deviation of the first experimental point from this dependence can be caused by the fact that, in addition to the mechanism of pore formation, the surface erosion can contribute to the number of volume defects of the vacancy type.

4. Conclusions

Nanoporous structure formed in Pt by ion irradiation was investigated by field ion microscopy. Computer simulation was applied for quantitative analysis of size, shape and concentration of nanopores. It was observed that a decrease in nanopores concentration in the volume with distance from the irradiated surface can be described by linear equation. The analysis of shapes and sizes of nanopores allows us to propose that the nanopores in the samples studied represent gas bubbles formed by argon.

5. Acknowledgments

This work is supported by the Federal Target Program «Scientific and scientific-pedagogical personnel of innovative Russia», State Contract no. P750, May 20, 2010. Also it is supported by the Russian Fund of Basic Researches (the project № 10-02-96003-r_ural_a).

References

[1] Kirsanov V V, Suvorov A L, Trushin Y V 1985 *The processes of radiation defect formation in metals* (Moskow: Energoatomizdat) p 275

[2] Beavan L.A, Scanlam R M, Seidman N 1971 The Defect Structure of Depleted zones in Irradiated Tungsten *Acta Metallurgica* **19** 1339

[3] Gao F, Heinisch H L, Kurtz R J 2009 Migration of vacancies, He interstitials and He-vacancy clusters at grain boundaries in a-Fe *Journal of Nuclear Materials* **386–388** 390–394

[4] *Field-ion Microscopy* 1968 ed Hred J and Ranganathan S (New York: Plenum Press)

[5] Muller E W and Tsong T T 1969 *Field-ion Microscopy (Principles and Applications)* (New York: Elsevier)

[6] Bowkett K M and Smith D A 1970 *Field Ion Microscopy* (Nord-Holland, Amsterdam, London).

[7] Gribkov V A, Grigoryev F I, Kalin B A, Yakushin V L 2001 *Prospective radiation beam technologies of materials working* ed Kalin B A (Moscow) p 528

[8] Moore A J W 1962 The structure of atomically smooth spherical surface *J. Phys. Chem. Solids*, **23** 907

Granulated media for nanoelectronic applications

AV Ognev[1,2], EV Sukovatitsina[1], KS Diga[1], LA Chebotkevich[1,2], AS Samardak[1], SM Janjan[3] and F Nasirpouri[3]

[1]Laboratory of Thin Film Technologies, Far Eastern Federal University, 8 Sukhanova st., Vladivostok, 690095 Russia
[2]Institute of Automation and Control Processes, Far East Branch of Russian Academy of Sciences, 5 Radio st., Vladivostok, 690041 Russia
[3]Department of Materials Engineering and Nanostructured Materials Research Centre, Sahand University of Technology, Tabriz, Iran

E-mail: ognevav@gmail.com

Abstract. We report our experimental results on the structural and magnetic characteristics of nickel films electrodeposited on n-type (111) silicon substrates. Using scanning electron microscopy (SEM), magnetic force microscopy (MFM) and magnetic hysteresis loops, we indeed find a nominal thickness for transition from island to continuous film before which the magnetization of nanoparticles of Ni is represented by vortex states. The evidence of non-uniform ground states in Ni granules has been proofed by micromagnetic simulations. After the formation of a continuous film, stripe magnetic domains are formed with spontaneous in plane magnetization.

1. Introduction

An investigation of magnetic nanostructures is interested from both fundamental and practical points of view. The possible magnetic configurations in such structures are vortex state or single domain state which are stable for a long time in the wide range of applied magnetic fields. Unique physical properties of magnetic nanostructures make them suitable to be used as magnetic recording media [1,2] and logic elements [3,4]. It is well known that magnetic configuration in a nanostructure depends not only on material properties, but on its shape. A lot of studies focus on 2D structures with different geometry, for instance, discs, rings, ellipses and etc. However, one can modify micromagnetic structure and magnetization reversal with manipulating lateral parameters of 3D nanostructures [5,6].

A cheap and effective method for fabrication of magnetic nanostructures based on “bottom-up” approach is electrodeposition of metals in an appropriative electrolyte under an adequate electrical potential between two immersing electrodes. Advantages of the method include fast growth, highly homogeneous structure and good adhesion of material to substrate. Magnetic properties of thin nickel films have been studied on the metal substrates [7-9] and little information is present for electrodeposited films on silicon [10, 11] and GaAs [12, 13].

In this paper we report on an investigation of granular Ni films with different nominal thickness electrodeposited on Si(111) substrate. The deployment of silicon wafer as a substrate allowed excepting any seed layer. We show an evolution of micromagnetic structure and magnetic properties at the crossover from island-like to continuous Ni films.

2. Experimental procedures

Electrodeposition was performed in a conventional three-electrode cell connected to a computer-controlled potentiostat/galvanostat instrument. To fabricate nickel thin films on n-Si(111), the electrochemical deposition was performed in potentiostatic mode using the chronoamperometry technique. Electrodeposition of nickel was carried out in electrolyte with concentration (mole per 1 lit H2O): NiSO4.6H2O = 2.3, NiCl2.6H2O = 0.6, H3BO3 = 0.5 and pH = 2.

Scanning electron microscopy (SEM Supra, Carl Zeiss) and atomic force microscopy (Ntegra Aura, NT-MDT) were used to examine the morphology and structure of Ni electrodeposited on n-Si. Micromagnetic structure was investigated by magnetic force microscopy (MFM, Ntegra Aura, NT-MDT) with variable in-plane field (±1 kOe). MFM images were obtained using CoCr-coated tips (NSG01). Typical scan heights were about 100 nm. The magnetization reversal processes were studied using magneto-optical Kerr effect (NanoMOKE II system). Magnetic field (±3 kOe) was applied in-plane direction of samples. We employed MagPar software [14] for micromagnetic simulation work.

3. Results and discussions

In order to understand the growth behavior, we have examined the sequence of the electrodeposition of nickel from the Watts bath. Figure 1 illustrates SEM micrographs taken during different stages of the electrodeposition characterized by the nominal thickness h_n of the film. It shows that the electrodeposition of nickel starts with the formation of individual Ni granules on silicon at the initial stages of growth, as shown in figure 1(a) for 8 nm thick nickel electrodeposited on Si. As the electrodeposition time increases, the individual Ni granules become bigger in diameter and tend to coalesce in the form of a continuous film (Figure 1b).

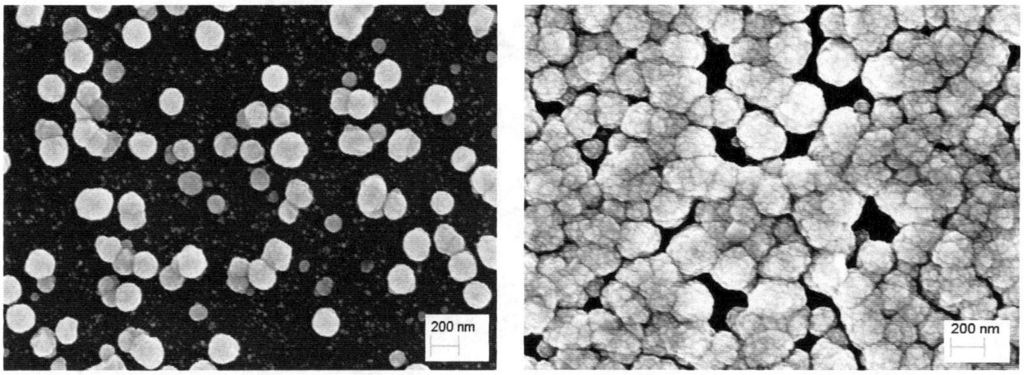

Figure 1. SEM images of granular Ni films with the nominal thickness (h_n), of 8 nm (a) and 1000 nm (b).

The AFM and MFM images are represented in figure 2. As seen for the film with h_n=8 nm, at zero field (figure 2(c)) and in demagnetization state (figure 2(d)) the individual granules or granules separated by a small gap have an uniform magnetic contrast seen with white or black areas. This confirms the out-of-plane magnetization. In case of chains consisting of a few adjoined granules, the single domain states are formed due to shape anisotropy. At magnetic fields near a saturation state the bipolar contrast is observed.

Provided that the density of granular (number per surface area) increases, they become in contact with each other, forming a cluster. For clusters the multidomain magnetic structure is more preferable. This argument is proofed by MFM image in figures 2(h) and 2(i). There is a stripe domain structure. The domain walls spread on a few granules adjoined in one cluster.

In micromagnetic simulation we took into account two cases: (i) when the granules are separated and (ii) after being adjoined to each other as is in a continuous film. We used standard material parameters for Nickel: M_s=484 emu/cm^3, A=9×10^{-7} erg/cm, magnetocrystalline anisotropy constant

K_I=-3.4×10^4 erg/cm3 and the damping parameter α=0.1. Granules were represented by oblate spheroids based on the analysis of the SEM and AFM images. The mesh sizes were chosen to be equal to Δ_x=Δ_y=Δ_z=12 nm.

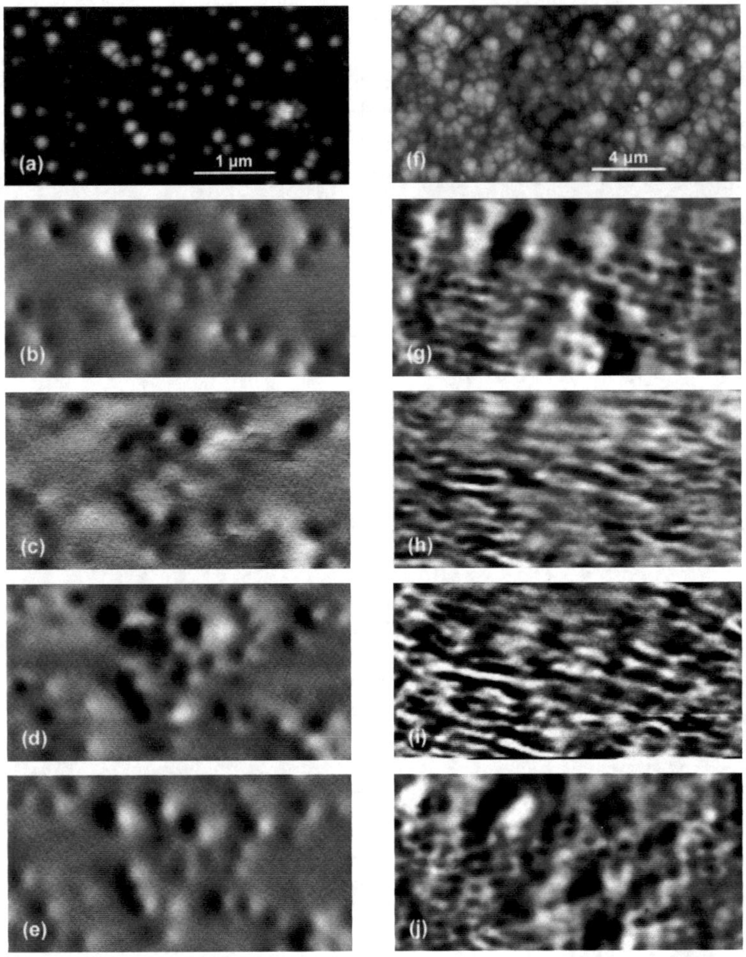

Figure 2. Surface topography (a, f) and magnetic structure in saturated states (b, g, e, j), without field (c, h) and in demagnetizing state (d, f) of Ni electrodeposits with h_n= 8nm (a- e) and 500 nm (f - j).

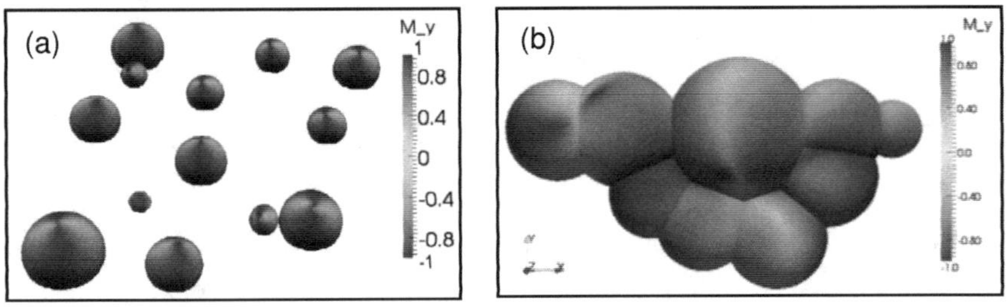

Figure 3. Micromagnetic simulation of ground states for domain structure of isolated granules (a) and cluster (b) from top view. M_y is a vector of the in-plane magnetization direction.

Thus, we can conclude that at the h_n< 50 nm the magnetization reversal is defined by the displacement of the vortex core to granular edges. In samples with h_n> 50 nm the remagnetization process occurs due to domain wall movement. As result, the values of coercive force and remanent magnetization decrease. Further increase in H_c at higher values of h_n is due to the formation of continuous film with structural defects (inter-granular pores) playing role of domain wall pinning centers.

4. Conclusion

The electrodeposition of nickel from Watts bath undergoes two stages including (1) the formation of individual granules and (ii) coalescence of the them into clusters with following growth of a continuous film, below and above a critical nominal thickness 50 nm. The preferable micromagnetic pattern in isolated granules is the vortex state with high value of perpendicular magnetization induced by its spherical shapes. In films with the nominal thickness higher 50 nm we observed stripe domain structure similar to continuous films. Nickel films consisting of granules at vortex ground state are very promising for high density magnetic recording.

Acknowledgments
Support by The grant council of President of Russian Federation (MK-5181.2011.2) and The Federal Program "Scientific and Research-Educational Cadres of Innovative Russia" (contracts 02.740.11.0549 and 16.740.11.0502) is acknowledged.

References
[1] Pigeau B, Loubens G, Klein O, Riegler A, Lochner F, Schmidt G, Molenkamp L, V. Tiberkevich S, Slavin A. N.. 2010 *Appl. Phys. Lett.* **96** 132506
[2] Bohlens S, Krüger B, Drews A, Bolte M, Meier G Pfannkuche D 2008 *Appl. Phys. Lett.* **93** 142508
[3] Nakatani R, Nomura H., Endo Y 2009 *Journal of Physics: Conference Series* **165** 012030
[4] Cowburn R, Welland M 2000 Science **287**, 1466
[5] Diao Z, Abid M, Upadhyaya P, Venkatesan M, Coey J 2010 *J. Magn. Magn. Mater.* **322** 1304
[6] Samardak A, Ognev A, Stebliy M, Chebotkevich L 2011 *Bulletin of the Russian Academy of Sciences. Physics* **75** 193
[7] Kuriki S 1979 *Phys. Stat. Sol. (a)* **51** 227
[8] Fisher J 1970 *Thin Solid Films* **5** R39
[9] Mishra A. C, Thakur A. K, Srinivas V 2009 *J. Mater. Sci.* **44** 3520
[10] Tiberto P, Gupta S, Bianco S, Celegato F, Martino P, Chiolerio A, Tagliaferro A, Allia P 2011 *J. Nanopart. Res.* **13** 245
[11] Tang Y, Zhao D, Shen D., Zhang J, Li B, Lu Y, Fan X 2008 *Thin Solid Films* **516** 2094
[12] Scheck C, Evans P, Zangari G, Schad R 2003 *Appl. Phys. Lett.* **82** 17 2853
[13] Gubbiotti G, Carlotti G, Tacchi S, Liu Y.-K, Scheck C, Schad R, Zangari G 2005 *J. Appl. Phys.* **97** 10J102
[14] Micromagnetic simulations http://www.magpar.net/
[15] Vonsovski S.V 1974 *Magnetism* (New York, Wiley)
[16] Osborn J 1945 *Physical Review.* **67** 351

High-density nickel nanowire arrays for data storage applications

A S Samardak[1], E V Sukovatitsina[1], A V Ognev[1], L A Chebotkevich[1], R Mahmoodi[3], S M Peighambari[2], M G Hosseini[3], F Nasirpouri[2]

[1] Laboratory of thin film technologies, Far Eastern Federal University, Russia
[2] Department of Materials Engineering, Sahand University of Technology, Iran
[3] Department of Physical Chemistry, Tabriz University, Iran

E-mail: asamardak@gmail.com

Abstract. This paper is dedicated to a study of magnetic properties (magnetic anisotropy, coercive force and remanent magnetization) of spatially ordered high-density Ni nanowire arrays. The magnetic nanowires were prepared by electrodeposition of nickel from simple sulfate solutions into anodic aluminum oxide (AAO) nanoporous templates (with diameter of d=20 and 40 nm) fabricated by potentiostatic anodization. We show that Ni nanowires have strong out-of-plane axial magnetic anisotropy (normal to the substrate plane) along with an in-plane anisotropy caused by hexagonal spatial distribution of nanowires in an array. An existence of the six-fold in-plane anisotropy proofs the long-range hexagonal order of nanowires. High quality of nanowire arrays makes possible of its usage as high-density (up to $2Tb/in^2$) magnetic recording media.

1. Introduction

Ferromagnetic nanowires due to unique combination of magnetic, transport and structural properties are in a focus of many scientific groups as potential candidates for applications in microwave electronics as tunable planar devices for very high frequencies [1,2], patterned magnetic recording media [3] with density beyond $2Tb/in^2$ [4] and biomedical sensors [5] for cancer treatments. Shape anisotropy of magnetic nanowires orients the magnetic moments along the nanowires' longitudinal axes forming a high density perpendicular magnetic recording media with small switching fields and high thermal stability. A hexagonal distribution of nanowires induces in-plane magnetic anisotropy with a large degree of magnitude, which is reasonably sufficient for performance required in the information storage media. There are few techniques for wire nanopattering, such as focused ion beam, electron-beam lithography and vapor-liquid-solid technique, but the most promising technique to fabricate highly-ordered magnetic nanowire arrays is porous-alumina template electrodeposition [6]. Electrodeposition into nanoporous templates is cost-effective and suitable for array fabrication over square centimeters. Usually the pores have excellent short-range ordered structure and the pores diameter (normally, ranging between 10 and 200 nm) and length (from a few μm to ~100-150 μm) can be readily controlled [7,8]. Despite good progress in the fabrication and study of ferromagnetic nanowires electrodeposited in nanopores, there are a few unsolved issues connected to understanding the nanoscopic nature of such one- and two-dimensional structures. For instance, an influence of the magnetostatic interaction between nanowires in highly-dense magnetic arrays is still far to be understood. The periodicity and regularity of the arrays, which are crucial properties for the implementation of patterned magnetic media, are still not completely resolved.

In this paper we report on an investigation of high periodic and regular nickel nanowire arrays with different wire diameters fabricated by template electrodeposition. We study on interaction inside the ferromagnetic array structure, spatial distribution and magnetic behavior of nanowires in such arrays.

2. Experimental details

High-purity aluminum foils (99.999%) were anodized after degreasing in acetone for 5 min., chemical surface treatments in KOH 200 g.Lit^{-1}, Na_2CO_3 50 g.Lit^{-1} and HNO_3 50% solutions, and electropolishing under a constant-voltage condition of 16 V for 8 min in a mixture of $HClO_4$: C_2H_5OH (1:4 in vol) at 5°C. Anodization was, then, carried out under a constant cell voltage 40 V in a 0.3 M oxalic acid ($H_2C_2O_4$) and 26 V in a 0.3 M sulfuric acid (H_2SO_4) solutions to obtain nanopore diameter of 40 and 20 nm, respectively. Details of different samples under this study are given in Table 1. The temperature of the electrolyte was maintained at 0-2°C during anodization using a cooling system. The solution was stirred vigorously in order to accelerate the dispersion of the heat evolved from aluminum sheets. The formed alumina was then removed by a mixture of 0.3 M chromic acid and 0.5 M phosphoric acid at 60°C for an appropriate time depending on the anodizing time. Then, the Al foil was reanodized under the same condition as the first step. For nanowire preparation, thickness of the barrier layer of oxide film was produced in sulfuric acid solution decreased by reducing the anodizing voltage at the end of the second stage of anodizing using 2 V/min steps from 26 V to 20 V, 1 V/min steps from 20 V to 10 V, 0.5 V/min steps from 10 V to 8 V and finally 3 min was kept in 8 V. Then, Ni nanowire arrays were using alternating current electrodeposited into the pores of the alumina templates under 14 V_{rms}, 50-100 Hz for an appropriate time to precisely fill up the pores. The electrodeposition bath was containing of $NiSO_4.H_2O$ 0.1 M and H_3BO_3 0.5 M with a pH value of solution was adjusted to 3.5.

Table 1. Characteristics of different nickel nanowire samples under this study prepared by ac electrodeposition in highly ordered nanoporous AAO.

Samples	Anodizing solution	Duration of anodization [min]	Deposition condition
R_1	H_2SO_4	135	14V_{rms}, 100Hz, 60min
R_2	H_2SO_4	60	14V_{rms}, 100Hz, 90min
R_6	$H_2C_2O_4$	300	14V_{rms}, 50Hz
R_7	$H_2C_2O_4$	300	16V_{rms}, 50Hz

The spatial distribution and dimensional parameters (length and diameter) of the nanowires were studied by high resolution scanning electron microscopy (SEM). SEM images were exploited to process spectral Fourier analysis. Magnetic properties of nanowires were investigated by the home-made vibrating sample and magneto-optical Kerr effect magnetometers.

3. Results and discussion

As is evident from the SEM images, nanoporous AAO templates have a spatial ordering of pores (figure 1(a)) with the alignment perpendicular to the substrate surface (figure 1(b)). As seen from figure 1, the template has short-range order hexagonal distribution of pores. The distances l between pores are about 40 nm for sample with d=20 nm and 60 nm for sample with d=40 nm.

To visualize an individual nickel magnetic nanowire, we dissolved the whole AAO template containing nanowires arrays in a mixture of 20g/l CrO3 and 35ml/l H3PO4 [9]. We inspected the free-standing nanowires after dissolving of the porous-alumina template with SEM. The nanowires have continuous structure without visible defects replicating the pore shapes. It means that nickel fills the pores uniformly during ac electrodeposition under the controlled condition of this work.

The hysteresis loops were measured in two geometries: when the applied magnetic field H is rotated in a AAO sample plane (or perpendicular to wire's axis) and out of plane (or parallel to wire's axis). An analysis of hysteresis loops allowed us to plot the angular dependences of coercivity $H_c = f(\varphi)$ and normalized remanent magnetization $M_r/M_s = f(\varphi)$, where φ is an azimuth angle between selected direction in a sample and H (figure 2).

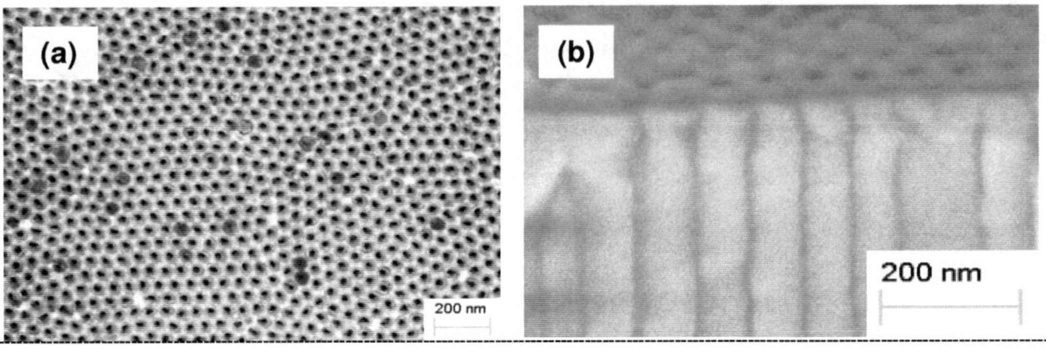

Figure 1. (a) Top view of nanoporous AAO template with d=20 nm. (b) Cross-section of the Ni electrodeposited template.

We studied four different samples marked as R1, R2, R6 and R7 (table 1). The nanoporous AAO templates were prepared using the well known two-step anodic oxidation process [6] at different conditions in order to obtain regular spatial distribution of pores. Figure 2 shows experimental data for nickel nanowires electrodeposited in AAO templates with d=20nm and 40 nm. When H is perpendicular to the sample plane, we found that M_r/M_s = 0.9 and H_c = 600 Oe in all samples irrespective to wire (or pore) diameter, as shown in figures 2(a,b). It means that all samples have strong perpendicular magnetic anisotropy with an easy axis aligned along longitudinal axis of the nanowires caused by shape anisotropy.

More complex behavior of magnetic properties was explored at the in-plane measurement geometry. We revealed an effect of spatial distribution of nanowires on coercive force and magnetization of magnetic arrays and found maximum values for all samples M_r/M_s= 0.2 and H_c = 270 Oe (figures 2(c,d)) which significantly smaller than in case of perpendicular anisotropy. As seen in figure 2(e) sample R2 is characterized by Fourier spectrum indicating near hexagonal spatial distribution of nanowires which induced the six-fold in-plane configurational anisotropy in the array (figure 2(c)). The easy and hard axes have a periodicity of 60° in the substrate plane, in agreement with the hexagonal structure. This is a good proof of the long-range hexagonal order of nanowires in the array. A deviation of spatial distribution from hexagonal configuration can induce four-fold (figure 2(d), sample R7) and two-fold (figure 2(c), sample R1 and figure 2(d), sample R6) in-plane configurational anisotropy.

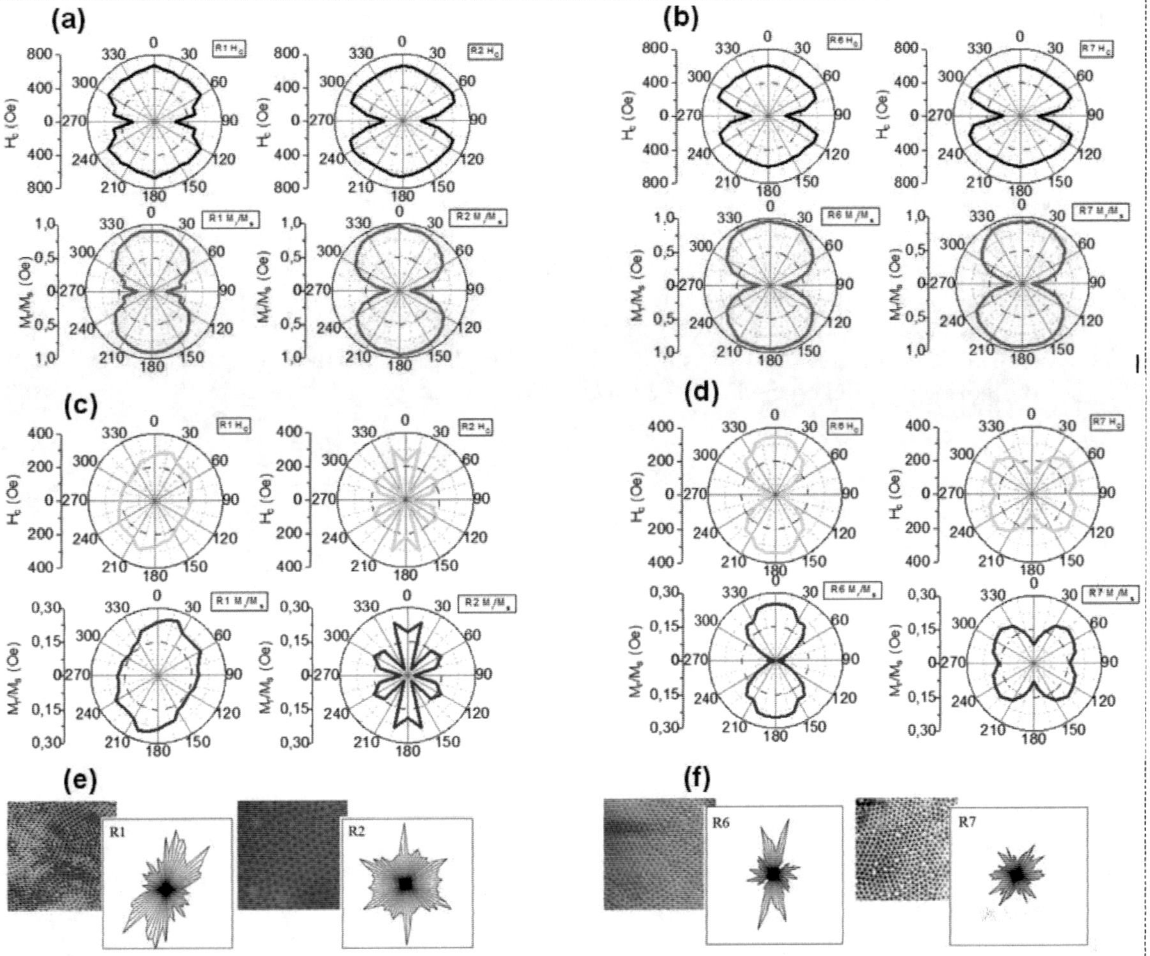

Figure 2. Dependence of magnetic properties on spatial distribution of nanowires in different samples. (a,c,e) Samples R1 and R2 with d=20nm. (b,d,f) Samples R6 and R7 with d=40nm. (a,b) Polar diagrams $H_c=f(\varphi)$ and $M_r/M_s=f(\varphi)$ when H is rotated out-of-plane. (c,d) Polar diagrams $H_c=f(\varphi)$ and $M_r/M_s=f(\varphi)$ when H is rotated in-plane. (e,f) Fourier spectra for samples with different spatial distribution of Ni nanowires.

4. Summary

In conclusion, we have found that all arrays had strong perpendicular magnetic anisotropy induced by shape anisotropy of nanowires. The reason for the multi-fold in-plane anisotropy is a nearly regular hexagonal distribution of nanowires in an alumina substrate. Manipulation by spatial ordering can help to find optimized magnetic properties of nanowires to be used in novel spintronic devises and for magnetic data storage applications with density up to $2Tb/in^2$.

Acknowledgments

Support by The grant council of President of Russian Federation (MK-5181.2011.2) and The Federal Program "Scientific and Research-Educational Cadres of Innovative Russia" (contracts 02.740.11.0549 and 16.740.11.0502) is acknowledged.

References

[1] Spiegel J and Huynen I 2009 *Solid State Phenomena* **152-153** 389
[2] Spiegel J, Eggermont S, Huynen I 2008 *Proc. 2nd Int. Cong. on Adv. Elect. Mat. in Microwaves and Optics* **1** 61
[3] Chou S Y 1997 *Proc. IEEE* **85** 652
[4] Shimizu T, Aoki K, Tanaka Y, Terui T, Shingubara S 2011 *Jap. J. Appl. Phys.* **50** 06GE01
[5] Schmidt H 2001 *Appl. Organomet. Chem.* **15** 331
[6] Nasirpouri F 2007 *Template electrodeposition of magnetic nanowire arrays*, in: Ahmed El Nemr (Ed.), New Dev. in Electrodep. and Pitting Res., Trans. Res. Network, Kerala, pp.1-37
[7] Xiang Y, Lee W, Nielsch K 2008 *Phys. Stat. Sol.* **2** 59
[8] Nielsch K, Choi J, Schwirn K, Wehrspohn R B, Gosele U 2002 *Nano Lett.* **2** 677
[9] Napolskii K S, Eliseev A A, Yesin N V, Lukashin A V, Tretyakov Yu D, Grigorieva D A, Grigoriev S V, Eckerlebe H 2007 *Physica E* **37** 178

Journal of Physics: Conference Series **345** (2012) 012012 doi:10.1088/1742-6596/345/1/012012

An analytical expression of electric potential and field of organic thin film transistors

Sebastian Pankalla and Manfred Glesner

Technische Universität Darmstadt, Microelectronic Systems Research Group, Merckstrasse 25, D-64283 Darmstadt, Germany

E-mail: {pankalla, glesner}@mes.tu-darmstadt.de

Abstract. The two-dimensional electric potential and field of an organic thin-film transistor (OTFT) is derived by conformal mapping using the Schwarz-Christoffel-transformation of the Poisson equation. In this paper we compare this analytical closed-form solution to field simulation results from Silvaco TCAD. Inter alia the potential close to the surface is calculated and we found excellent accordance to the numerical simulations and thus proofed its usability for charge transport calculations. Thus, it is used for calculation of the drain-source-current in the channel.

1. Introduction

The growing development of electronics and its wide use in all kinds of industries has led to the need of decreasing cost per function and area respectively of electronic devices. Until now, more integration of devices, i.e. growing number of transistors in a defined area, integration of functionalities etc., results in a reduction of costs. Another mean to minimize chip costs is organic electronics. The use of low cost fabrication like spin-coating or printing leads to very cheap devices. Still, the development of electronics and also organic electronics would not have been possible without robust CAD and simulation tools. These tools require accurate models to describe a single device and whole circuits made up of them. These models should be both accurate and fast in terms of computing time. Therefore, analytical equations are needed as only these fulfill both claims.

Because OTFTs follow other basic physical principles than inorganic semiconductors, models for circuit simulators must be developed which take the physics of the organic material into account. In order to obtain models that rely on these organic properties information of physical parameters within the channel have to be acquired. We therefore started with a simulation of the electric field and potential of an OTFT and compared the results to an analytical expression obtained by the mean of conformal mapping for the potential within the channel.

2. Conformal Mapping

As it is difficult to solve the Poisson-equation in a two-dimensional area, conformal mapping is used to transfer the two-dimensional channel region of the OTFT into a one-dimensional. In this geometry the solution can be obtained easily. The idea behind conformal mapping is that the problem with a geometry that is rather difficult to solve may be mapped into a problem with simple geometry or with a geometry where the solution is known. The channel region with

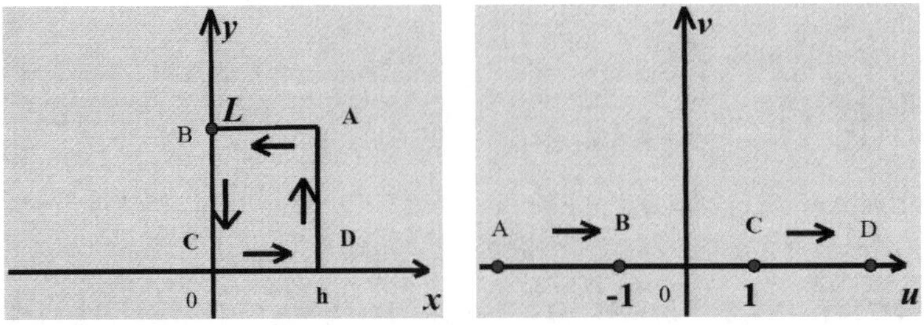

(a) The OTFT structure in z-plane is described by a rectangle with electrode source=BC, drain=AD and AB corresponding to the gate-dielectric-channel capacitor plate below the dielectric

(b) The OTFT structure in w-plane after conformal mapping

Figure 1. OTFT Structure before and after conformal mapping

its polygonal boundaries, i.e. source, drain and gate electrode, can be mapped into an upper half-plane by the use of the Schwartz-Christoffel transformation [1] [2].

We started with an OTFT geometry shown in figure 1(a) with real axis x and y and the channel region with source electrode BC, drain electrode AD and AB corresponding to the gate-dielectric-channel capacitor plate below the dielectric [2]. Applying conformal mapping figure 1(a) is transformed to figure 1(b) with real axis u and imaginary axis v. The channel region within the rectangle, where the Poisson equation will be applied, is mapped into the upper half-plane in figure 1(b). Here the unit length $L = 1$ of the channel is chosen. The Schwarz-Christoffel differential equation for transformation of the w-plane (figure 1(b)) to the z-plane (figure 1(a))

$$\frac{dz}{dw} = S(w-1)^{-\frac{1}{2}}(w+1)^{-\frac{1}{2}} \tag{1}$$

is integrated to obtain the transformation from w- to z-plane:

$$z = S\int_0^t \frac{dt}{\sqrt{(w-1)(w+1)}} + K \tag{2}$$

The integration constant $K = 0$ is obtained from the assumption, that both coordinate systems $w = u + iv$ und $z = x + iy$ have the same origin, so $w = z = 0$. The pre-factor S is derived from the mapping theory and equals L/π, resulting in the final transformation

$$z = f(w) = 2\frac{L}{\pi}\ln(\frac{\sqrt{w-1}\sqrt{w+1}}{\sqrt{2}}) \tag{3}$$

The reverse transformation follows this equation:

$$w = f^{-1}(z) = \frac{e^{\frac{\pi z}{L}} + e^{-\frac{\pi z}{L}}}{2} = \cosh(\frac{\pi z}{L}) \tag{4}$$

The separation of real part u and imaginary part v leads to:

$$u = \frac{e^{\frac{\pi x}{L}}\cos(\frac{\pi y}{L}) + i\sin(\frac{\pi y}{L})}{2} \tag{5}$$

$$v = \frac{e^{\frac{-i\pi x}{L}}\cos(\frac{-\pi y}{L}) + i\sin(\frac{-\pi y}{L})}{2} \tag{6}$$

3. Calculation of electric potential and field

The distribution of the electric potential in the channel of an OTFT can be calculated from the Poisson equation

$$\Delta\Phi(x,y) = \frac{\partial^2\Phi}{\partial x^2} + \frac{\partial^2\Phi}{\partial y^2} = -\frac{\rho}{\epsilon_r\epsilon_0} \tag{7}$$

where ρ denotes the charge densitiy, ϵ_r and ϵ_0 are the relative and vacuum permittivity, respectively. Organic semiconductors usually are unipolar. As they are lightly doped ρ is very small and can be set to zero [3]. Thus, $\rho = 0$ and the Poisson equation becomes the Laplace equation

$$\Delta\Phi(x,y) = \frac{\partial^2\Phi}{\partial x^2} + \frac{\partial^2\Phi}{\partial y^2} = 0 \tag{8}$$

As mentioned in the section before, the Poisson and the Laplace equation respectively are mapped to the upper-half plane for computation with less complexity. Considering an upper-half plane with u as real axis and v as imaginary axis, the boundary condition is $\phi(u,0) = P(u)$. The solution of the Laplace equation in that plane is [1]

$$\Phi(u,v) = \frac{v}{\pi}\int_{-\infty}^{\infty}\frac{P(t)}{(u-t)^2+v^2}dt \tag{9}$$

Assuming the source S to be the boundary for $-\infty < u < -1$, the area below the dielectric for $-1 \leq u \leq 1$, and drain D for $1 < u < \infty$, with boundary conditions ϕ_s, ϕ_{diel}, and ϕ_d, (9) results in

$$\begin{aligned}\Phi(u,v) =& \frac{\phi_d+\phi_s}{2} + \frac{\phi_{diel}-\phi_d}{\pi}\cdot\arctan(\frac{u+1}{v}) \\ &+\frac{\phi_s-\phi_{diel}}{\pi}\cdot\arctan(\frac{u-1}{v})\end{aligned} \tag{10}$$

Taking u and v of (5) and (6) into the solution of the Laplace equation (10), the channel electric potential distribution in the original z-plane (figure 2) is obtained:

$$\begin{aligned}\Phi(x,y) =& \frac{\phi_d+\phi_s}{2} \\ &+\frac{\phi_{diel}-\phi_d}{\pi}\cdot\arctan\left(\frac{cos(\frac{\pi * y}{L})cosh(\frac{\pi x}{L}+1)}{sin(\frac{\pi y}{L})\cdot\sinh(\frac{\pi x}{L})}\right) \\ &+\frac{\phi_s-\phi_{diel}}{\pi}\cdot\arctan\left(\frac{cos(\frac{\pi y}{L})cosh(\frac{\pi x}{L}-1)}{sin(\frac{\pi y}{L})\cdot\sinh(\frac{\pi x}{L})}\right)\end{aligned} \tag{11}$$

The surface potential Φ_{surf} is obtained for $y = 0$. The electric field $\overrightarrow{E}$ is calculated from

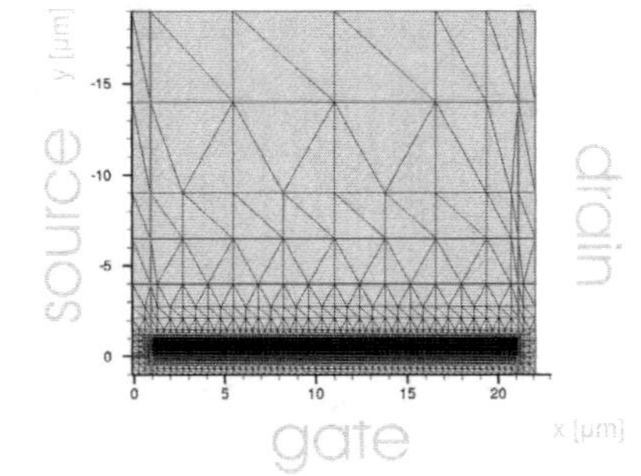

Figure 2. The OTFT structure in z-plane with generated mesh for TCAD simulations

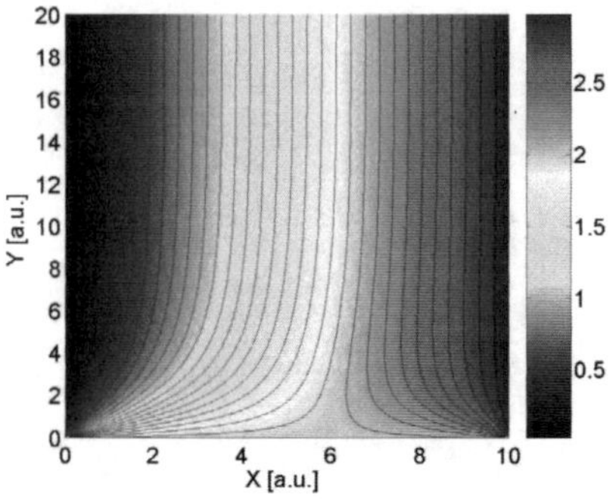

Figure 3. Calculated electric potential distribution for ϕ_s =0, ϕ_{diel} =2, and ϕ_d =3 with equipotential lines

$$\overrightarrow{E}(u, v) = -\nabla \cdot \Phi(u, v) \tag{12}$$

Replacing again u and v of (5) and (6) into (12) results in the electric field distribution in the channel. The total semiconductor surface charge is obtained from Gauss Law

$$Q_{surf} = -\epsilon_{vacuum} E_y. \tag{13}$$

In our model this is the charge responsible for the current conduction. The current I_{DS} from source to drain in the channel is calculated by

$$I_{DS} \propto \int_0^L Q_{surf} E_x \mathrm{d}x \propto \int_0^L E_x E_y \mathrm{d}x \tag{14}$$

4. Comparison to TCAD results

We compared the analytically obtained solution of the Laplace equation with the numerical result of Silvaco TCAD [4]. The device structure has been build according to figure 2, using gold as electrode material for source, drain and below the dielectric in order to emulate the potential at the gate-dielectric-channel capacitor and an organic semiconductor already implemented in the simulation environment of Silvaco TCAD in the channel region. The size of the geometry is chosen in such a way that the simulation results are comparable to the analytical solution for unit length $L = 1$.

TCAD solves coupled Poisson and drift diffusion equations

$$\nabla \cdot [\epsilon \nabla \Phi] = -\, q \left(p - n + N_D^+ - N_A^+ + p_{At} \right) \tag{15}$$
$$\nabla \cdot \mathbf{J}_n = q \left(R - G \right) \tag{16}$$
$$\nabla \mathbf{J}_p = -\, q \left(R - G \right) \tag{17}$$
$$\mathbf{J}_n = -\, q \mu_n \nabla \Phi + q D_n \nabla n \tag{18}$$
$$\mathbf{J}_p = -\, q \mu_p \nabla \Phi - q D_p \nabla p \tag{19}$$

The electric potential for a device as in figure 2 for the boundary condition values $\phi_s = 0$, $\phi_{diel} = 2$, and $\phi_d = 3$ is shown in figure 3 for the calculation of the potential. A cut parallel to the x-axis and perpendicular to the y-axis, i.e. parallel to the dielectric, at $y \rightarrow 0$ plots the potential ϕ_{surf} close to the surface and is shown in figure 4(a). The right y-axis shows the error, i.e. the difference $\Phi_{surf,simulated} - \Phi_{surf,calculated}$ between the simulated and calculated potential close to the surface. It can be clearly seen that our analytical solution fits very precisely to the numerical solution of the Laplace equation. Thus, our analytical expression for the electric potential and field can be used to simulate charge transport in organic semiconductors more accurately and will lead to more precise models for circuit simulation. We have taken the calculated current in the channel of equation (14) to obtain a transfer curve in figure 4(b). In this semilogarithmic plot it can be seen that the current follows the well-known transfer-shape.

5. Conclusion

We have derived an analytical closed-form solution of the Poisson equation to obtain the electric potential and field in the channel of an OTFT. We transformed the rather hard to solve geometry, i.e. the rectangular shaped OTFT channel, to an upper half plane by conformal mapping where the Poisson equation can be solved easily. The obtained analytical expression for the electric potential and field are compared to numerical solutions by Silvaco TCAD and show excellent agreement.

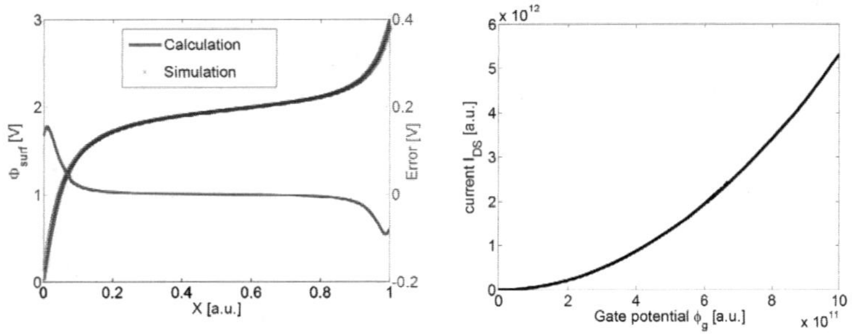

(a) Comparion of simulated (x) and calculated (-) results of surface potential Φ_{surf} with error $\Phi_{surf,simulated} - \Phi_{surf,calculated}$

(b) Calculated transfer curve

Figure 4. surface potential and resultant transfer curve of the current

Acknowledgment

The authors would like to thank the German Federal Ministry of Education and Research (BMBF) for funding this work within the Polytos project in the cluster of excellence "forum organic electronics".

References

[1] Weber, 1950, Electromagnetic fields, *Wiley, 3rd ed.*, **Vol. 1**
[2] Kloes, 1996, A new analytical method of solving 2D Poisson's equation in MOS devices applied to threshold voltage and subthreshold modeling, *Solid-State Electronics*, **Vol. 39, No. 12**, pp. 1761-1775
[3] Schwarz M, Kloes A, and Iniguez B, Analytical 2D Model for the Channel Electric Field in Undoped Schottky Barrier Double-Gate MOSFET, *MIXDES 2010, Wroclaw, Poland*
[4] http://www.silvaco.com/products/device_simulation/atlas.html

Analysis of the mobility of printed organic p-channel transistors depending on the transistor geometry and orientation

D Spiehl[1,*], S Pankalla[2], M Glesner[2], and E Doersam[1]

[1] Technische Universität Darmstadt, Institute of Printing Science and Technology, Magdalenenstrasse 2, 64289 Darmstadt, Germany

[2] Technische Universität Darmstadt, Microelectronic Systems Research Group, Merckstrasse 25, 64283 Darmstadt, Germany

E-Mail: spiehl@idd.tu-darmstadt.de

Abstract. The organic thin film transistor (OTFT) is an elementary part of most organic electronic products. A cost efficient and fast way to produce these circuits is by the use of mass printing techniques like flexography or gravure printing. A huge amount of OTFTs is produced via flexography and via spin-coating as a reference. The morphology of the printed layers and the electrical performance of the produced OTFTs are investigated. Based on these investigations, the dependence of the mobility on the transistor's source/drain geometry is analysed and explained. Analysed geometry parameters are the transistors channel length, channel width and its orientation compared to printing direction.

1. Introduction

Research in the field of printed organic electronics is developing rapidly. Besides organic light-emitting diodes (OLEDs) and organic photovoltaics (OPVs) the third huge topic is about organic thin film transistors (OTFTs). These are part of e-papers, smart packaging, flexible displays and radio frequency identification tags (RFIDs) [1]. The OTFT is an elementary part of most of these applications, especially if several OTFTs are connected to circuits.

A cost efficient and fast way to produce these circuits is by the use of mass printing technologies like flexography or gravure printing. Simulations of circuits consisting of printed OTFTs minimize production costs but must include electrical parameter variations in order to verify the reliability of the circuits. Thus, a huge amount of OTFTs is printed and the performance and variations of the process for OTFTs is investigated by electrical measurements [2] and morphological investigations. Based on these investigations, the dependence of the mobility on the transistor's source/drain geometry is analysed and explained.

2. Materials and production

The OTFTs were built by the use of mass printing techniques for the semiconductor and gate layer; other layers were produced with conventional laboratory techniques. The chosen way to build the OTFTs was a top gate and bottom contact architecture. Figure 1 shows a photograph of a substrate with transistors and the production work flow.

The substrate is a Teonex Q65FA PEN film from DuPont Teijin Films. It has a thickness of 200 µm and a pre-treated side with a surface roughness of 0.6 nm, according to the manufacturer. A source/drain structure consisting of titanium as bonding agent and gold as electrode material was deposited onto the film by a photo-lithographical process. This structure was a layout of 83 transistors with different channel lengths L between 5 µm and 100 µm and with different channel widths W between 500 µm and 20707 µm [3]. There are several transistors with the same ratio of W/L but different W and L. The transistors are ordered in different rows, columns and directions. Every transistor appears several times, so the homogeneity over the whole substrate can be examined.

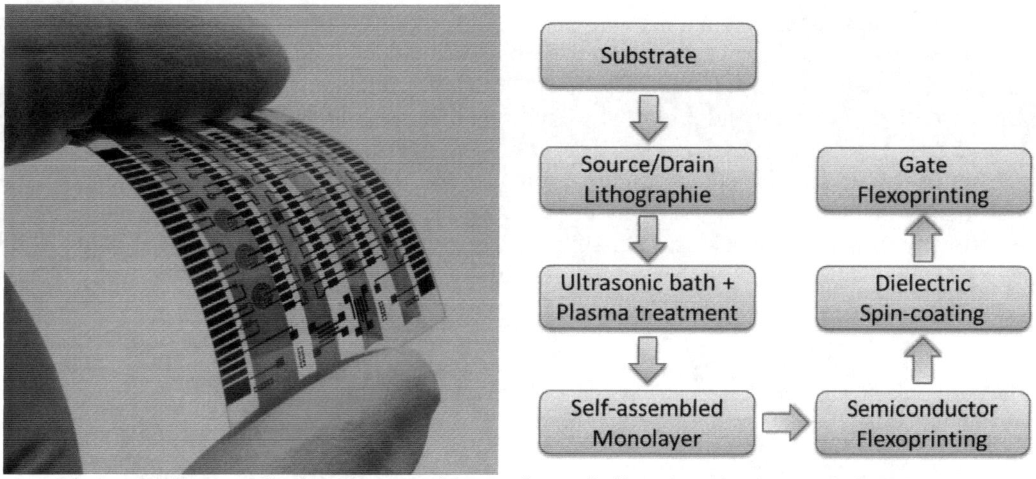

Figure 1. Photograph of a substrate with transistors (left) and production work flow (right)

Before printing, the substrates were cleaned in an ultrasonic bath with isopropyl alcohol and were dried with clean dry air. Then a plasma treatment under argon atmosphere with a power of 300 W followed for two minutes. This leaded to a clean the surface and adjusted its surface tension for the first solution processed layer. Afterwards a self-assembled monolayer (SAM) was used to tune the work function of the golden electrodes to match the semiconductors HOMO-level.

The first printed layer was an organic p-channel polymer semiconductor that we printed by the use of flexography. One of the biggest challenges was the very low shear viscosity of the semiconductor fluid. The literature [4] lists a shear viscosity in the range of 50 mPas to 500 mPas for fluids printed via flexography, whereas here it was 1.7 mPas to 1.9 mPas - depending on the shear rate - for the solution of 1% polymers in the used solvent mesitylene. Viscosity was measured with a Thermo Scientific HAAKE MARS rheometer system. Printing was done with an IGT F1 printability tester at a printing speed of 1.5 m/s to gain a homogeneous layer. The dip volume of the anilox roller was chosen to be 12 ml/m^2 to yield the wanted thickness of the dried semiconductor layer. The chosen printing plate was a photopolymer based printing plate. Printing plates made of elastomers like ethylene-propylene-diene-rubber instead of photopolymer based are advertised as resistive to the aggressive solvents used for organic electronics. But used the solvents, especially toluene for cleaning, even damage these materials and dissolve particles out of the printing plate [5]. So photopolymer based printing plates which are available in the needed format and short delivery times were used in this case. An alternative way to produce the semiconductor layer is via spin-coating. A rotation speed of 1500 rpms leaded to a thickness comparable to the printed layer.

After drying the semiconductor layer for 3 min at 100 °C on a hot plate, a gate-insulator was spin-coated onto the whole substrate and was also dried on a hot plate under equal conditions. The spin parameters were attached to gain the desired layer thickness of 1 µm.

At last a gate consisting of silver nanoparticles was printed onto the insulator. An anilox roller with a dip volume of 16 ml/m^2 and a printing plate made of photopolymers was used. A printing speed of

1.0 m/s led to the best homogeneity of the gate. The silver was dried in a heating oven at 120 °C for 10 min.

No pre-treatments of the underlying layers were done prior to the processing of the gate-insulator and the gate.

3. Measurements

A set of samples with an ascending amount of layers was produced to measure the thickness and topography of each layer. Thickness measurements were done at edges prepared by scratching or wiping off the layers. Measurements were done with a sensofar Plμ Neox profilometer. The used profilometer provides confocal microscopy, white-light interferometry and phase-shift interferometry. The reliability of the measurements was ensured by using all methods provided and identifying the best method for each measuring task. The measured layer thicknesses of the semiconductor were 40 nm in case of spin-coating and 35 nm to 60 nm when printed. In both cases the surface roughness was about 5 nm. The measured thickness variations in the printed layer seem to be caused by swelling of the printing plate and dewetting of the semiconductor layer before drying. Figure 2 shows a confocal image of the thickness variations in the flexo-printed semiconductor layer with part of the source/drain structure beneath. The spin-coating of the gate-insulator was adjusted to result in a layer thickness of 1 μm. The printed gate had a thickness of 2 μm.

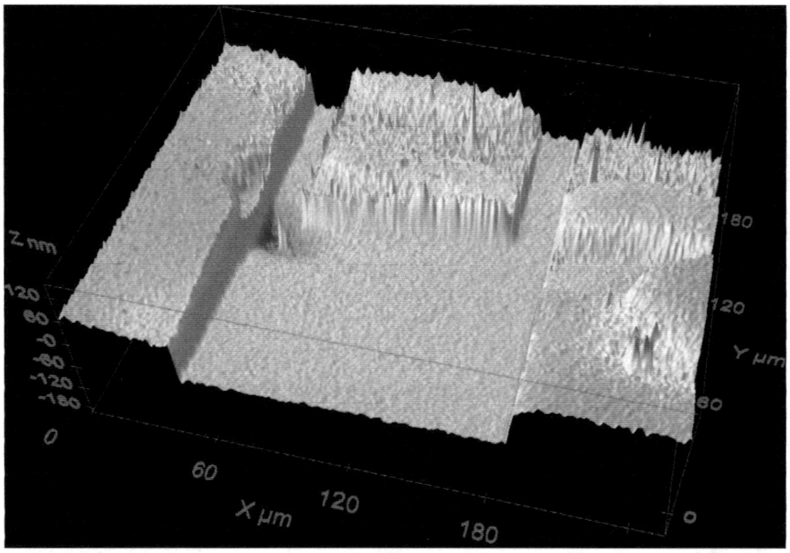

Figure 2. Confocal image of a flexo-printed semiconductor layer with S/D beneath

Output and transfer characteristics of 244 transistors with spin-coated semiconductor and 183 transistors with flexo-printed semiconductor were automatically measured according to a previously defined protocol [2]. This was done on a x-y-table using spring contact probes and a Keithley 2636A source-meter unit. Automatic evaluation of that immense data was realised by a VBA script within MSEXCEL [7], that determines the threshold voltage Vt and the mobility derived from the well-known simple MOS-Modell [6]. The calculated parameters are plotted arranged in rows and columns according to the substrate's layout.

4. Results

Altogether, the mobility of transistors with a spin-coated semiconductor is comparable to the mobility of transistors with flexo-printed semiconductor (Figure 3). For the investigations, we only took transistors with a mobility μ of 1E-3 cm^2/Vs or higher into account. Here it has to be announced that the reachable mobility in the chosen top gate and bottom contact architecture is, according to the manufacturer of the semiconductor, 5E-2 cm^2/Vs when semiconductor and insulator are produced via spin-coating and the conducting layers are evaporated. The thickness variations of the printed semiconductor layer shown in Figure 2 have no negative influence on the mobility.

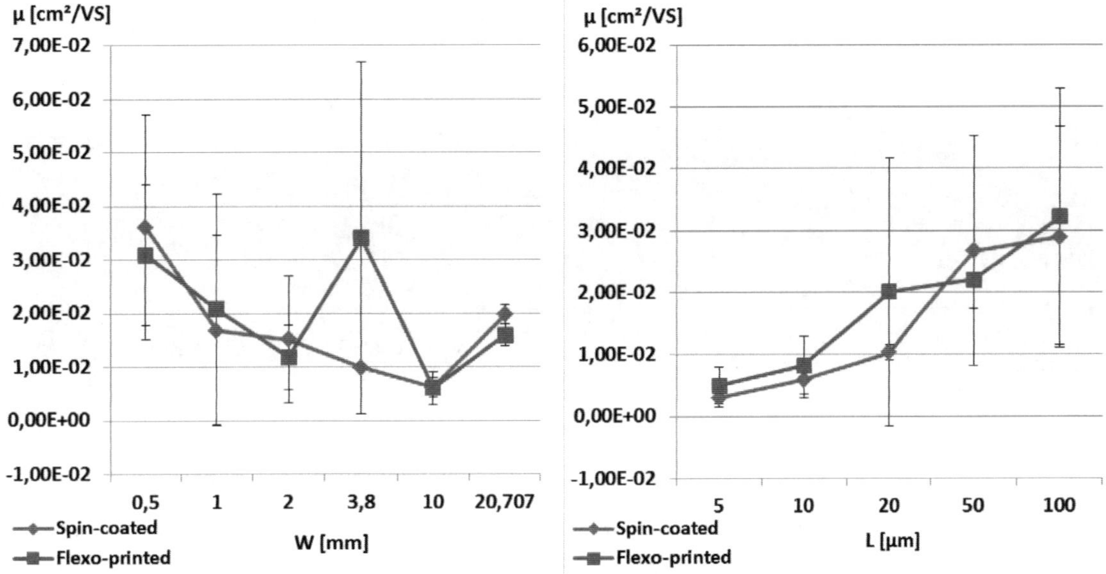

Figure 3. Dependence of the mobility μ on the transistor channel width W (left) and on the transistor channel length L (right)

Overall we have seen same behaviour for transistors with both, spin-coated and flexo-printed semiconductor. Regarding to Figure 3 (left) the mobility decreases with greater W. The reason is an increasing number of defects and traps over the larger area within the channel. Transistors with a channel width W of 3.8 mm and 20.707 mm drift a little of. This could be due to their round geometry [3], which leads to additional influences on the mobility. Figure 3 (right) shows the dependence of the mobility μ on the transistor channel length L. We previously showed that it is more probable that a transistor with the shown geometries works if its L is greater than approx. 10 μm [7]. According to this, the mobility μ also increased with greater L. This can be addressed to the contact resistance that has major influence on the current for small channel lengths [8]. Thus, those two effects superimpose and transistors with small ratio of W/L show higher mobility.

There is no clear dependence of the mobility on the orientation of the transistor channel relative to the printing direction (Figure 4). The mobility of transistors with flexo-printed semiconductor reaches its climax for uniaxial transistors. Uniaxial transistors were realised with circular, concentric source/drain structures [3]. But the value is within the error of measurement, so no prediction is possible. In addition to that, round transistors gained a higher mobility, as described on the previous page.

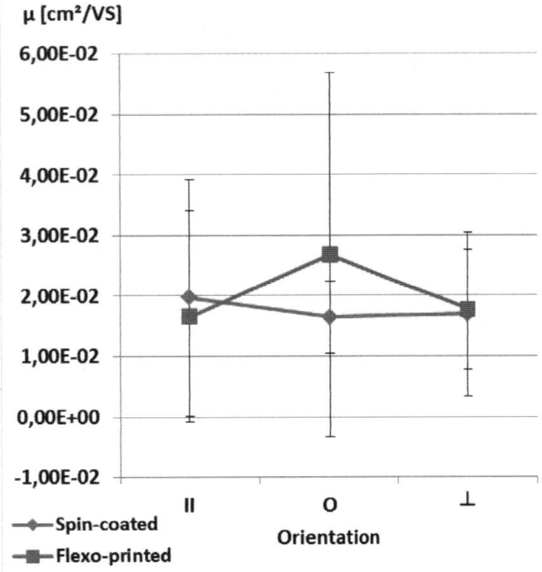

Figure 4. Dependence of the mobility μ on the transistor orientation compared to printing direction

5. Conclusions

We showed the production of partly printed OTFTs with flexo-printed or spin-coated semiconductor layers, to be able to compare between these two, usually chosen production techniques. With a measurement routine we were able to compare performance parameters of more than four hundred transistors. This we have done for the transistors mobility depending on the production technique of the semiconductor layer, the transistor channel width W, the transistor channel length L and the transistor orientation compared to printing direction. We showed, that for our transistor design the mobility μ increases with decreasing channel width and the mobility increases with the channel length. The transistors orientation relative to the printing direction has no influence on the mobility.

Acknowledgments

This work was funded by the Bundesministerium für Bildung und Forschung (BMBF), Germany, within the Polytos project in the cluster of excellence "Forum Organic Electronics".

References

[1] Klauk H (Ed.) 2006 *Organic Electronics: Materials, Manufacturing and Applications* WILEY-VCH Verlag
[2] Hengen S, Pankalla S, Spiehl D, Dörsam D, Glesner M and Giehl J 2010 *Proc. of the 1st Winter School of Organic Electronics* **1** TB-1
[3] Pankalla S, Hollstein T and Glesner M 2010 *Proc. of the 2th Lope-C* **2** 233-236
[4] Kipphan H (Ed.) 2001 Handbook of print media: technologies and production methods Springer Verlag
[5] Theopold A and Dörsam E 2011 *Proc. of the 3rd Lope-C* **3** 330-334
[6] Shichman H and Hodges D 1968 *IEEE J. of Solid-State Circuits* **3** 285-289
[7] Pankalla S, Hengen S, Spiehl D, Dörsam E, Giehl J and Glesner M 2011 *Proc. of the 4th Int. Symp. on Flexible Organic Electronics* **4** 97
[8] Zschieschang U, Ante F, Kälblein D, Yamamoto T, Takimiya K, Kuwabara H, Ikeda M, Sekitani T, Someya T, Blochwitz-Nimoth J and Klauk H 2011 *Organic Electronics* **12** 1370-1375

The influence of δ-<Mn>-layer's magnetization on polarization of photoluminescence of quantum well in singular and vicinal InGaAs/GaAs/δ-<Mn> heterostructures

A I Dmitriev[1], A D Talantsev[1], S V Zaitsev[2], O V Koplak[3], R B Morgunov[1]
[1] Institute of problems of chemical physics RAS, 1 ac. Semenova street, 142432, Chenogolovka, Russia
[2] Institute of solid state physics RAS, 2 Ac. Ossipyan str.,142432, Chenogolovka, Russia
[3] Kiev National University, 64, Volodymyrs'ka St., 01601 Kyiv, Ukraine

E-mail: artgtx32@mail.ru

Abstract. Magneto-optical properties and spin dynamics in a p-type GaAs-based heterostructures containing an InGaAs quantum well and 2 nm thick GaAs layer doped with 5 at. % Mn (GaAs:Mn) on flat and vicinal substrates were studied between 2 and 300 K. A significant enhancement of photoluminescence circular polarization P_C = 15 % was found at temperatures T < 30 K. In the non-magnetic quantum well P_C = 0 %. The polarization of the InGaAs quantum well photoluminescence follows the temperature dependence of the GaAs:Mn layer magnetization. Two types of ferromagnetism were found. In the heterostructures grown on the flat substrate parallel to the (001) GaAs plane the magnetization obeys the Bloch $T^{3/2}$ temperature dependence while for the structures grown on the vicinal surface (disoriented by 3°) the magnetization follows percolation dependence. The later is attributed to the formation of strongly disordered islands of GaAs:Mn.

1. Introduction

Heterostructures containing a quantum well and a thin layer of a diluted magnetic semiconductor (DMS) like GaAs:Mn have attracted great attention due to their possible application for injecting spin-polarized carriers in light emitting diodes [1]. The implantation of Mn into the quantum well leads to low mobility of charge carriers and suppresses radiative recombination [2]. The spatial separation of charge carriers and the GaAs:Mn layer results in a considerable increase in the mobility of charge carriers and the luminescence quantum yield [3]. This combination of optically and magnetically active structures appears promising for controlling optical properties by spin-dependent magnetic field control over the luminescence.

The presence of charge carriers in the GaAs:Mn layer facilitates the indirect Ruderman-Kittel-Kasuya-Yosida (RKKY) exchange interaction between Mn^{2+} ions in hybrid structures [4-8]. Measurements of the luminescence polarization [8], the electric resistance and the Hall voltage [4-7] indirectly confirmed the existence of ferromagnetism in GaAs:Mn layers. The interpretation of the magnetization data, however, is difficult and usually requires additional assumptions for the separation of the contributions of the diamagnetic crystal lattice and paramagnetic structural defects, substrate, and interlayer boundaries. This problem can be solved by the electron spin resonance (ESR) technique [9] allowing the separation of contributions of above mentioned subsystems to the magnetic susceptibility of the samples.

It will be shown here that the circular polarization of the photoluminescence of a InGaAs quantum well is sensitive to the ferromagnetic ordering in the GaAs:Mn layer. The combinations of optical measurements with superconducting quantum interference device (SQUID) magnetometry and electron spin resonance (ESR) spectroscopy allowed us to identify fine details of electronic transitions, their dependence on spin polarization and their correlation to the observed increase of the circular polarization in the ferromagnetic ordered state of GaAs:Mn.

Here, we present results on:

a) the ferromagnetic GaAs:Mn layer contribution to the magnetic moment of the samples,

b) the effect of the GaAs:Mn layer magnetization on the polarization of the photoluminescence of the InGaAs quantum well,

c) the effect of different GaAs substrate orientations on the type of magnetic ordering,

d) exchange parameters, magneto-optical properties and spin dynamics in the GaAs-based heterostructures with a InGaAs quantum well and a GaAs:Mn layer.

2. Experimental setup

P-type GaAs-based heterostructures containing a InGaAs quantum well and GaAs:Mn layer (Mn delta-doped GaAs thin layer) (figure 1) were prepared by stepwise epitaxial growth [10].

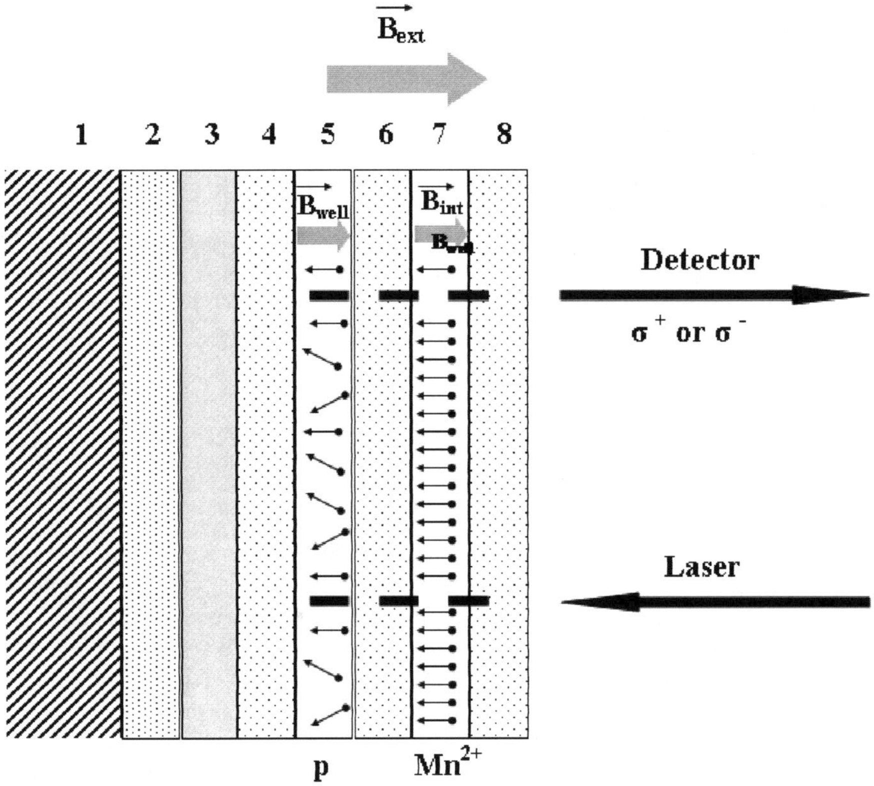

Figure 1. Scheme of the heterostructure grown on a 0.5 mm GaAs substrate (1). Heterostructure consists of 0.5 μm thick GaAs buffer layer (2), 1 nm thick GaAs:C layer (3), 15 nm thick GaAs separating layers (4), 10 nm thick InGaAs quantum well (5), 3 nm thick GaAs separating layers (6), 2 nm thick GaAs:Mn layer (7) and 30 nm thick GaAs cover layer (8). External magnetic field B_{ex}, magnetic field B_{int} in the GaAs:Mn layer with spin-polarized Mn^{2+} ions, and magnetic field B_{well} in the InGaAs quantum well with spin-polarized holes p are shown.

InGaAs quantum well was grown by metal-organic hydride epitaxy on a GaAs substrate, as described in detail earlier [10]. The GaAs:Mn layer (concentration of Mn at. 5 %) was grown in the same reactor by laser ablation of Mn and GaAs targets [10]. Moreover, in order to compensate the effects of depletion of charge carriers in the quantum well, the acceptor GaAs:C layer was deposited on the side of the buffer layer in all studied heterostructures.

We studied three types of heterostructures:
(1) Three samples of singular heterostructures containing a InGaAs quantum well and a 2 nm thick ordered GaAs:Mn layer on a precisely oriented GaAs (001) substrate. The angle between (001) crystallographic plane and the substrate plain is $\varphi < 0.2^{o}$.
(2) Three samples of vicinal heterostructures containing InGaAs quantum well and disordered GaAs:Mn layer on a vicinal GaAs (001) substrate, with an angle $\varphi = 3 \pm 0.2\,^{o}$ between the (001) crystallographic plane and the substrate plane.
(3) Two samples of reference heterostructure containing a InGaAs quantum well and a GaAs:C layer instead of the GaAs:Mn layer. GaAs:C layer was used to determine the contribution of boundaries between layers and paramagnetic defects in addition to the ferromagnetic contribution in the GaAs:Mn layer. Thus, reference heterostructures contain two GaAs:C layers.

The temperature dependence of the magnetization was measured using a SQUID magnetometer (MPMS 5XL, Quantum Design) at temperatures T = 2 - 300 K and in the magnetic field B = 0.1 T. ESR measurements were carried out at 9.5 GHz with a modulation frequency of 100 kHz in the temperature range T = 4 - 300 K using a Bruker ESR-500 spectrometer and Oxford Instrument cryostat. The magnetic field was applied along the [001] direction perpendicular to the plane of the heterostructure. Technical details of the experiment have been described elsewhere [11]. The circular polarization, P_C, of photoluminescence was measured in a magnetic field of B = 3 T in an Oxford Instruments cryostat.

Polarization of photoluminescence was determined by formula $P_C = (I_+ - I_-)/(I_+ + I_-)$, where $I_+(I_-)$ is the intensity of components with the right (left) polarization obtained by integrating the part of the spectrum corresponding to the optical transition in a quantum well. Photoluminescence was excited with a 632.8 nm He-Ne laser and the magnetic field oriented along the [001] direction perpendicular to the substrate and recorded with a spectral resolution better than 0.05 meV.

3. Experimental results

3.1. Static magnetization of singular heterostructures

We have studied three singular samples to be sure in reproducibility of the experimental data. All three samples possessed same magnetic properties. Typical temperature dependence of the magnetization M(T) for one of the samples is shown on the figure 2.
Decrease of the magnetization of δ-<Mn>-layer caused by heating obeys Bloch formula below Curie temperature as it usually observed in the bulk ferromagnetic materials. Curie temperature T_C can be estimated in the frame of Weiss mean field theory. In ferromagnets the magnetization, M, is expected to decrease with temperature, T, obeying the Bloch $T^{3/2}$ law in the low-temperature range [12]:

$$M(T) = M_0(1 - \beta T^{3/2}) \tag{1}$$

where M_0 is the magnetic moment at $T \to 0$ K, and β the spin-wave parameter. The Bloch $T^{3/2}$ law has experimentally been found to be applicable for thin ferromagnetic GaMn:As films [13]. We approximated the experimental M(T) dependence for the singular heterostructures using Eq. (1) in the low-temperature range 2 – 10 K. In our case $M_0 \approx M(T = 2\text{ K})$ is the experimental value while β is a fitting parameter. We have fond $\beta = (2.1 \pm 0.1) \cdot 10^{-2}\ K^{-3/2}$. In our case β value exceeds one obtained in [13] by order of value because Mn^{2+} concentration is higher and J value in our heterostructure is larger at the approximately same Curie temperature T_C.
In conventional spin-wave theory [12], the spin-wave parameter, β, and spin-wave stiffness, D, are conjugated by the equation [13]:

$$\beta = 2.612\frac{g\mu_B}{M_0}\left(\frac{k_B}{4\pi D}\right)^{3/2} \tag{2}$$

where $g \approx 2$ is the g-factor of a Mn^{2+} ions, μ_B is the Bohr magneton, k_B is the Boltzmann constant. The value of spin-wave stiffness determined from the Eq. (2) is $D = 1.5 \pm 0.3$ T · nm^2. The value of exchange integral, $J = 0.18 \pm 0.01$ meV, was calculated using the following expression [14]:

$$J = \frac{Dg\mu_B}{2Sr_S^2} \tag{3}$$

where $r_S \approx 7$ Å – average distance between Mn^{2+} ions, $S = 5/2$ is the spin of Mn^{2+} ions. The obtained values of β, D, J are in good agreement with the results reported earlier for similar systems by other authors [15, 16].

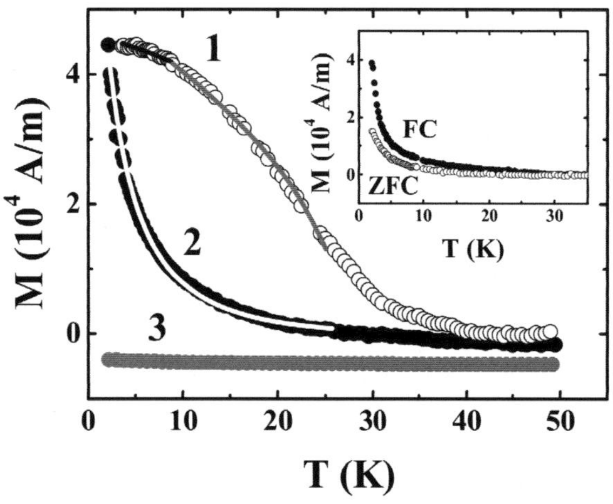

Figure 2. Temperature dependencies of the magnetization of the singular (1) and vicinal (2) heterostructure with the GaAs:Mn layer and the reference heterostructure without GaAs:Mn layer (3) at B = 0.1 T. The solid lines are approximations (see the text). The inset shows temperature dependencies of the magnetization of a vicinal heterostructure containing GaAs:Mn layer cooled in zero magnetic field (ZFC) and B = 1 T magnetic field (FC). The magnetic field was applied in the (001) heterostructures plane. The diamagnetic contribution from the GaAs substrate was subtracted. The error bar ($\pm 0.2 \cdot 10^4$ A/m) is equal to the symbol size.

It means that quantum well does not effect on the magnetization of the GaAs:Mn layer. Experimental value of Curie temperature, $T_C = 35 \pm 5$ K, was defined by minimum of the dM/dT derivative. We used mean-field approximation $M(T) \sim M_0(1\text{-}T/T_C)^{1/2}$ of the experimental M(T) dependence in the middle-temperature range 10 – 25 K. In the frame of the mean-field theory, the Curie temperature, T_C, is defined as [16]:

$$T_C = \frac{zS(S+1)}{k_B}J. \tag{4}$$

This equation allows us to estimate coordination number z. Assuming $T_C = 35$ K , one can obtain $z = 4.38 \pm 0.6$. This value is very close to expected value $z = 4$ typical for tetrahedral symmetry. Coordination number does not mean the number of nearest Mn^{2+} ions. The z value corresponds to number of exchange channels including indirect exchange interaction through Ga or As atoms. Because in GaAs indirect exchange is provided by charge carriers, $z = 4$ is reasonable value indicating tetrahedral Mn^{2+} environment.

3.2. Static magnetization of vicinal heterostructures

We have studied three vicinal samples to be sure in reproducibility of the experimental data. All three samples possessed same magnetic properties. Typical temperature dependence of the magnetization M(T) for one of the samples is shown on the figure 2. In vicinal heterostructures the magnetization, M, decreases gradually with temperature, T, in the low-temperature range 2 – 35 K (in contrast to a Bloch and mean- field type M vs. T curve in the singular heterostructure). As it was shown previously this type of the temperature dependence is typical for the percolation ferromagnetic ordering [17-19]. The percolation model is based on the assumption that indirect exchange between magnetic centres (Mn^{2+} ions) is carried out by charge carriers (holes in the case of GaAs:Mn). The exchange interaction between holes and magnetic ions leads to the formation of crystal regions with local ferromagnetic ordering. At high temperature, there is no correlation between such individual ferromagnetic regions, their macro spins are oriented arbitrarily and the net magnetic moment obeys the Curie law for superparamagnets. However, as temperature decreases, the radius of ferromagnetic regions increases. The increasing interaction between different ferromagnetic regions leads to macroscopic correlation between magnetic moments of the individual magnetic ions and appearance of the net magnetization. When such correlated clusters of ferromagnetic regions have extended and coupled through the whole sample, the ferromagnetic ordering is complete. From the temperature dependence of the magnetization of the vicinal heterostructures cooled in zero magnetic field (ZFC) and in 1 T magnetic field (FC) (Figure 2) we determine the Curie temperature, $T_C = 30 \pm 5$ K as a point which below the ZFC and FC curves are differ. We approximated the experimental M(T) dependence for the vicinal heterostructures using the expression obtained by the percolation model [17]:

$$M(T) = M_0\{1 - \exp[-\left(\frac{R}{r_S}\right)^3 \ln^3 \frac{JS(S+1)}{T}]\} \tag{5}$$

where R – radius of indirect exchange interaction. A close correlation between the experimental dependence M(T) and theoretical curve (figure 2), indicates a percolation nature of the ferromagnetism in the vicinal heterostructures.

In our case $M_0 \approx M$ (T = 2 K), $r_s = 7$ Å and S = 5/2 are fixed values while R and J are variable fitting parameters. The value of exchange integral determined from approximation is J = 0.72 ± 0.04 meV. The parameter R was also extracted from the approximation is $R = (0.21 \pm 0.03) \cdot r_{S,} = 1.47$ Å. This value is practically equal to Bohr radius $a_B = 1.5$ Å of the Mn^{2+} ion in GaAs.

This remarkable fact seems to be not casual. Coincidence of the radius of indirect exchange interaction R with localization length indicates contribution of the localized charge carriers to the indirect exchange interaction. Most probably holes being captured by shallow traps near top of the valence band has admixture of d-orbital of Mn^{2+} wave functions. This localization of the holes provides additional channels of exchange interaction between Mn^{2+} ions.

3.3. High-frequency dynamical magnetic properties of singular and vicinal heterostructures

The ESR spectrum of the singular heterostructure shows: an non-angular dependent (isotropic) line near the zero magnetic field (line **a**), isotropic lines **b**, **d**, **e** and angular dependent anisotropic line **c**. Isotropic lines **b**, **d**, **e** are also observed in heterostructures without the GaAs:Mn layer (figure 3) and in a GaAs substrate. Consequently lines **b**, **d**, **e** corresponds to the paramagnetic centres of an uncontrollable paramagnetic impurity in all samples.

The intensity of line **c** drastically decreases with increasing temperature, disappearing at temperatures above 40 K. Resonance line **c** was observed only in samples with the GaAs:Mn layer (figure 3). The numbers of magnetic moments was determined from the calculated area under the absorption curve (second integral) of ESR spectrum of the line **c**. The spin number normalized to the known number of spins of a reference $CuSO_4 \cdot 5H_2O$ sample was 100 times higher than the number of spins corresponding to noninteracting Mn^{2+} ions at 4 K. Therefore, line **c** corresponds to the enhanced susceptibility due to the long-range ferromagnetic order, that is ferromagnetic resonance in the GaAs:Mn layer. This fact is supported by the temperature dependence of line **c**, which disappears above $T_C = 35 \pm 5$ K. This Curie temperature was determined by ESR technique.

Coincidence of the ESR determined T_C value with one determined by SQUID magnetometer confirms ferromagnetic long ranged nature of this magnetic transition, because in the ensemble of superparamagnetic clusters these temperatures are different and frequency dependent [20].

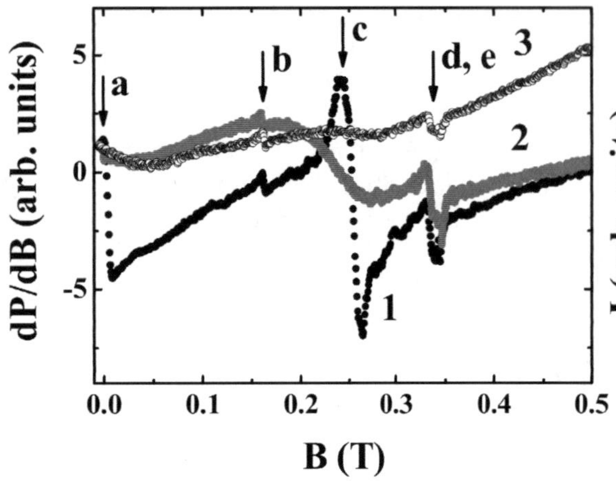

Figure 3. ESR spectra in the singular (1) and vicinal (2) heterostructure with the GaAs:Mn layer and the reference heterostructure without GaAs:Mn layer (3) at T = 4 K. The dc magnetic field was applied in the (001) heterostructures plane. The error bar is equal to the symbol size.

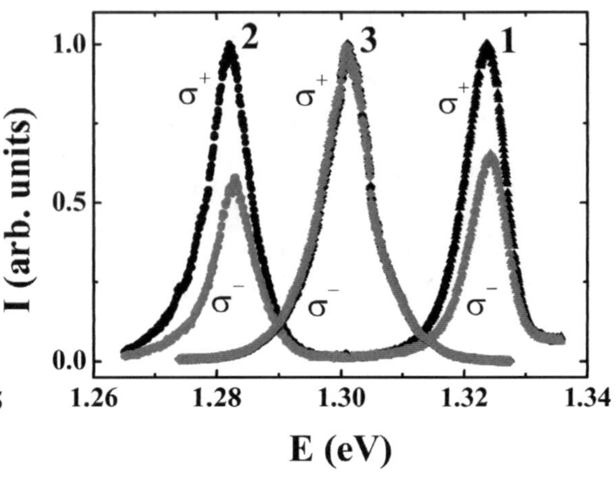

Figure 4. Photoluminescence spectra at T = 2 K in the: singular (1) and vicinal (2) heterostructures with the GaAs:Mn layer and reference (3) heterostructure without GaAs:Mn layer. The magnetic field B = 3 T was applied perpendicular to the (001) heterostructures plane. The error bar is equal to the symbol size.

This conclusion is also confirmed by the axial anisotropy of line **c** with the easy magnetization axis lying in the (001) plane, which is consistent with the results of magneto-optical studies [8, 10]. The intensity of line **a** is drastically decreases with increasing temperature, and they are not observed at temperatures above 40 K. Line **a** locates near zero magnetic field. This line can not correspond to ferromagnetic resonance because 0.06 T anisotropy field determined from the angular dependence of the anisotropic line **c** should control resonant field of the all ESR lines. Isotropy observed in our experiments indicates another origin of the line **a**. Similar lines were observed in [21] where it was shown that this line corresponds to magnetoresistance. In our opinion, line **a** is caused by the microwave magnetoresistance [11].

ESR spectra of vicinal heterostructures differ from the ones in singular heterostructures. First, nonresonant line **a** is not observed in vicinal heterostructures (Figure 3). In vicinal heterostructures quasi-one-dimensional conducting channels are formed [22], while in singular heterostructures the electron gas is quasi-two-dimensional. Therefore, dimension limitations suppress the magnetotransport properties of the GaAs:Mn. Similar results were reported earlier for Ge:Mn films [23].
Second, ESR spectra of the line **c** in the vicinal sample are broader then spectra in singular samples. Reason of broadening of the line **c** is the distribution of resonant fields corresponding to separated ferromagnetic GaAs:Mn islands of different sizes and orientations in local crystal fields. The observed ferromagnetic resonance line is the sum of the magnetically inhomogeneous local components. Thus line **c** confirms the assumption of magnetic disorder in the GaAs:Mn layer in vicinal heterostructures. Because isotropic lines **b**, **d**, **e** were the same in the singular and vicinal heterostructures we used these lines as reference signals.

Very often DMS crystals consist from two magnetic subsystems: dispersed transition metals ions and ferromagnetic secondary phase in the nanoclaster form [19]. The explanation of

Journal of Physics: Conference Series **345** (2012) 012014 doi:10.1088/1742-6596/345/1/012014

ferromagnetism of the GaAs:Mn layer and corresponding **a** and **c** lines in ESR spectra by the presence of the known ferromagnetic secondary phase (MnAs or MnGa clusters) can be excluded, because:

- These materials exhibit different ferromagnetic resonance spectra (another lineshape, resonance field, linewidth) [24].

- The Curie temperatures of these alloys (T_C = 320 K for MnAs [25] and T_C = 600 K for MnGa [26]) are noticeably higher than the Curie temperature $T_C = 35 \pm 5$ K observed in our experiments. It was found that, as the size of crystals decreases Curie temperature, T_C, of ferromagnetic nanoclusters or nanofilms usually decreases no more than 3 – 5 times [27]. Estimation of the size effect on the Curie temperature can be performed by formula: $T_C(d) = T_C(\infty)[1\text{-}3\Delta L/2d]$ [27], $T_C(d)$ is the Curie temperature of nanoclusters, d is diameter, $T_C(\infty)$ is the Curie temperature of bulk sample, ΔL is the thickness of surface layer of nanoclusters, which characterizes the influence of the surface layer on the Curie temperature. If we assume reasonable sizes d = 1 nm and $\Delta L \sim 0.5$ nm, Curie temperature becomes T_C = 80 K in MnAs alloy clusters. This value is noticeably higher than Curie temperature 35 K observed in our heterostructures.

From the other hand, photoluminescence spectra indicate magnetic long range ordering and Curie temperature coinciding with SQUID and ESR data. Since optical experiments can not correspond magnetic ordering in clusters (because wavelength is higher than particles sizes) one can assume these experiments as an evidence of long range magnetic order.

3.4. Magneto-optical properties of singular and vicinal heterostructures

Figure 4 shows the photoluminescence spectra of the InGaAs quantum well for singular and vicinal heterostructures with the GaAs:Mn layer and heterostructures with the GaAs:C layer instead of the GaAs:Mn layer. The emission band of the quantum well in heterostructures with the GaAs:Mn layer was split into two circularly polarized components σ^+ and σ^- due to the Zeeman interaction of the electron and hole states (figure 4). Small difference in concentration and correspondent quantum well depth results in energy shift of photoluminescence line (figure 4). This shift doesn't influence the results of the magnetic experiments, described in this paper.

In all heterostructures circular polarization, P_C, was 0 % in zero magnetic field. Position of the sample was not changed during magnetic field application. For that reason the sample in the absence of the magnetic field was considered as reference one. Circular polarization, P_C, was 15 % in the 3 T magnetic field at 2 K (figure 5).

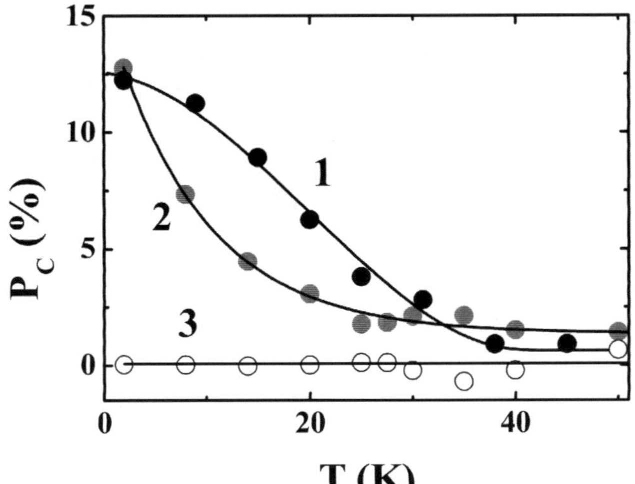

Figure 5. Temperature dependencies of the circular polarization, P_C, of a InGaAs quantum well photoluminescence in a singular (1) and vicinal (2) heterostructure with the Mn-doped GaAs layer and with the C-doped GaAs layer instead of the Mn-doped GaAs layer (3). The solid lines are guides for eyes. The error bar (± 1 %) is equal to the symbol size.

On the contrary, the polarization in heterostructures without GaAs:Mn layer was $P_C \approx 0$ % within the experimental error (figure 4). The circular polarization of the InGaAs quantum well effected by the GaAs:Mn layer magnetization because ferromagnetism results in spin polarization of the electrons and holes [5]. Same results were obtained in similar heterostructures [28]. Magnetic field dependencies of photoluminescence polarization were discussed in paper [28]. In our work we will discuss temperature dependencies of photoluminescence polarization. Temperature dependencies of photoluminescence polarization, P_C, (figure 5) qualitatively reproduce the temperature dependencies of GaAs:Mn layer magnetization (figure 2). In the frame of standard two-level Boltzmann model we suppose polarization of charge carriers in the quantum well due to magnetic field generated by GaAs:Mn layer. Temperature dependencies of photoluminescence polarization, P_C, (figure 5) are in a good agreement with temperature dependencies of GaAs:Mn layer magnetization (figure 2). Observed differences in magneto-optical data can be explained by the different types of magnetic ordering in GaAs:Mn layer.

4. Summary

Magnetic field of GaAs:Mn layer induces spin polarization of charge carriers in InGaAs quantum well. The polarization of quantum well photoluminescence follows the variations of GaAs:Mn layer magnetization induced by change of temperature or GaAs substrates orientations. Effects of GaAs substrates orientations on magnetization, spin dynamics and photoluminescence in the heterostructures were found. In singular heterostructures temperature dependence of magnetization follows the mean field theory allowing us to extract reasonable parameters of exchange interaction. In vicinal heterostructures temperature dependence of magnetization follows the percolation theory for disordered ferromagnets. Disorder in vicinal heterostructures leads to broadening of ferromagnetic resonance line and varies of magneto-photoluminescence characteristics.

5. References

[1] Ohno Y, Young D K, Beschoten B, Matsukura F, Ohno H and Awschalom D D 1999 *Nature* **402** 790–2.
[2] Poggio M, Myers R C, Stern N P, Gossard A C and Awschalom D D 2005 *Phys. Rev.* B **72** 235313–28.
[3] Pankov M A, Aronzon B A, Rylkov V V, Davydov A B, Meilikhov E Z, Farzetdinova R M, Pashaev E M, Chuev M A, Subbotin I A, Likhachev I A, Zvonkov B N, Lashkul A V and Laiho R 2009 *Journal of Experimental and Theoretical Physics* **109** 293–301.
[4] Nazmul A M, Sugahara S, Tanaka M 2002 *Appl. Phys. Lett.* **80** 3120–3.
[5] Nazmul A M, Sugahara S, Tanaka M 2003 *Phys. Rev.* B **67** 241308–12.
[6] Nazmul A M, Amemiya T, Shuto Y, Sugahara S and Tanaka M 2005 *Phys. Rev. Lett.* **95** 017201–5.
[7] Nazmul A M, Lin H T, Tran S N, Ohya S, Tanaka M 2008 *Phys. Rev.* B **77** 155203–11.
[8] Zaitsev S V, Dorokhin M V, Brichkin A S, Vikhrova O V and Danilov Yu A, Zvonkov B N, and Kulakovskii V D 2009 *Lett. to J. of Experimental and Theoretical Physics* **90** 658–62.
[9] Poole C P 1967 *Experimental Techniques of Electron Spin Resonance* (Wiley, New York)
[10] Dorokhin M V, Danilov Yu A, Demina P B, Kulakovskii V D, Vikhrova O V, Zaitsev S V and Zvonkov B N 2008 *J. Phys.* D **41** 245110–18.
[11] Dmitriev A I, Morgunov R B and Zaitsev S V 2011 *Journal of Experimental and Theoretical Physics* **112** 317–26.
[12] Chikazumi S 1998 *Physics of Ferromagnetism* (Berlin, Clarendon Press).
[13] Sperl M, Singh A, Wurstbauer U, Das S K, Sharma A, Hirmer M, Noltimg W, Back C H, Wegscheider W and Bayreuther G 2008 *Phys. Rev.* B. **77** 125212–19.
[14] Kittel C 1971 *Introduction to Solid State Physics*, *4th ed.* (Wiley, New York).
[15] Callen H B 1963 *Phys. Rev.* **130** 890–8.

[16] Goennenwein S T B, Graf T, Wassner T, Brandt M S, Stutzman M, Philipp J B, Gross R, Krieger M, Zurn K, Ziemann P, Koeder A, Frank S, Schoch W and Waag A 2003 *Appl. Phys. Lett.* **82** 730–3.
[17] Korenblit I Ya and Shender E F 1978 *Sov. Phys. Usp* **21** 832–51.
[18] Kaminski A and Das Sarma S 2002 *Phys. Rev. Lett.* **88** 247202–6.
[19] Morgunov R B, Dmitriev A I and Kazakova O L 2009 *Phys. Rev. B* **80** 085205–10
[20] Antoniak C, Farle M 2007 *Mod. Phys. Lett.* B **21** 1111–31.
[21] Golosovsky M, Monod P, Muduli P K, Budhani R C, Mechin L and Perna P 2007 *Phys. Rev.* B **76** 184414–23.
[22] Visser A D, Kadushkin V I, Kul'bachinskii V A, Kytin V G, Senichkin A P and Shangina E L 1994 *Letters to Journal of Experimental and Theoretical Physics* **59** 363–8.
[23] Kazakova O, Morgunov R, Kulkarni J, Holmes J and Ottaviano L 2008 *Phys. Rev.* B. **77** 235317–23
[24] Cubukcu M, von Bardeleben H J, Khazen Kh 2009 *J. Appl. Phys.* **105** 07C506–9.
[25] Akeura K, Tanaka M, Ueki M, and Nishinaga T 1995 *Appl. Phys. Lett.* **67** 3349–52.
[26] Tanaka M, Harbison J P, DeBoeck J, Sand T, Philips B, Cheeks T L and Keramidas V G 1993 *Appl. Phys. Lett.* **62** 1565–8.
[27] Lang X Y, Zheng W T and Jiang Q 2006 *Phys. Rev.* B **73** 224444–52.
[28] Zaitsev S V and Zvonkov B N 2011 *Phys. Status Solidi* B **248** 1526–31.

6. Acknowledgments

This work was partly funded under the Grant of the President of Russia MK-1764.2011.3 and RFBR project 10-03-00314 a. Authors are grateful to Danilov Yu. A., Dorokhin M. V., Kudrin A. V. and Zvonkov B. N. for samples providing.

Journal of Physics: Conference Series **345** (2012) 012015 doi:10.1088/1742-6596/345/1/012015

The built-in electric field in P-HEMT heterostructures with near-surface quantum wells $Al_xGa_{1-x}As/In_yGa_{1-y}As/GaAs$

R A Khabibullin[1], I S Vasil'evskii[1], D S Ponomarev[1], G B Galiev[2], E A Klimov[2], L P Avakyanz[3], P Yu Bokov[3], A V Chervyakov[3]

[1]National Research Nuclear University «MEPHI», Moscow, Russia
[2]Institute of Ultra High Frequency Semiconductor Electronics of Russian Academy of Sciences, Moscow, Russia
[3]Moscow State University, Moscow, Russia

E-mail: khabibullin_r@mail.ru

Abstract This study is concerned with the built-in electric field of heterostructures with the quantum wells grown at different depths L_b with respect to the surface. All samples had the same concentration of the two-dimensional electron in the quantum wells AlGaAs/InGaAs/GaAs. The built-in electric field was determined from the photoreflectance spectra of the samples. In case of L_b decreasing, the intensity ratio of two peaks in the photoluminescence spectra increases. The calculation shows that occurred effect originates from the redistribution of doping atom and variation of the surface potential due to the fact that the quantum well is set closer to the surface.

1. Introduction

Field-effect transistors with a two-dimensional (2D) high-mobility and high-density electron gas are extensively used for production of present-day devices in microwave electronics. At present, AlGaAs/InGaAs/GaAs pseudomorphous high-electron-mobility transistor (P-HEMT) nanoheterostructures, in which the mobility μ_e and the concentration n_s of electrons in the 2D electron gas are higher than the corresponding parameters in the AlGaAs/GaAs heterostructures, are widely used for development of advanced microwave electronic devices [1,2].

In the $Al_xGa_{1-x}As$ alloy, the Fermi level is pinned at surface states with a large capacitance near the midgap. Because of this, a rather high built-in electric field is produced in the depletion region between the surface and the quantum well (QW) filled with electrons. As the QW with the 2D electron gas is approved closer and closer to the surface, the effect of the built-in electric field in the heterostructure becomes more and more pronounced, which influences the electrical and optical properties of such structures. Therefore, consideration of these factors is of great importance in the development of advanced transistor heterostructures [3].

In this study, we investigate the built-in electric field of heterostructures with the quantum wells grown at different depths L_b with respect to the surface using a specially developed approximation. In this approximation, the total concentration of electrons in the subbands of the QW remains unchanged when the QW is set closer to the surface.

2. Experimental

The samples were grown at Institute of Ultra High Frequency Semiconductor Electronics by molecular-beam epitaxy on semiinsulating GaAs (100) substrates 2″ in diameter. The structure of the samples corresponds to the standard of P-HEMT structure with one side δ-doping [4]. The width of the InGaAs QW was 11 nm in all of the samples. The barrier layer thickness L_b was 18, 15, 13, 11 and 9 nm in samples #1, 2, 3, 4 and 5, respectively. Such barrier layer thicknesses are the typical values used in the production of low-noise microwave transistors. The molar fractions of components in the ternary $Al_xGa_{1-x}As$ and $In_yGa_{1-y}As$ compounds corresponded to $x = 0.23$ and $y = 0.20$ for all of the grown samples. The concentration of 2D electrons were determined from Hall measurements. The measurements were accomplished for mesa structures shaped as Hall bridges. The photoluminescence (PL) spectra were recorded by the standard method [4]. The photoreflectance (PR) spectra were registered by 0.3 m double grating monochromator equipped with Si photodiode in the spectral range of 550 – 1100 nm. The pitch angle of the probe beam was equal to 8°. The modulation of reflectance was made by irradiation of LED (maximum wavelength 540 nm) flashed at 370 Hz. All experiments were carried out at room temperature. The detailed description of the experimental technique is scribed in [5].

3. Results and discussion

By self-consistently solving the set of the Schrodinger and Poisson equation, we calculated and analyzed the spatial profile of the conduction band edge $U(z)$, the electron energy levels E_i, the envelope wave functions of electrons in subbands $\psi_i(z)$, the distribution of the electron density $n(z)$ at different barrier layer thickness, and the concentration of donors N_d. For a certain level of doping, the calculations showed that, as barrier layer thickness L_b is increased, the size-quantization levels E_i shift to lower energies, thus leading to increase of the difference $E_F - E_i$. We calculated the compensating increase in the degree of doping of the Si-doped δ layer at smaller L_b and, thus, determined the increase, necessary for keeping the constant concentration of electrons in the QW. The results of the calculations are shown in figure 1. With regard to these results, we grew five MBE samples with close concentrations of the 2D electron gas in the QW. The samples differed in barrier layer thickness (L_b = 18, 15, 13, 11 and 9 nm) and degree of doping $N_d = (4.3, 5, 5.7, 6.6 \text{ and } 8.25) \times 10^{12}$ cm^{-2}.

The experimental data determined from Hall measurements showed that in all of the five samples concentrations of 2D electrons are very close to each other: $n_H = (1.54–1.77) \times 10^{12}$ cm^{-2}. Some of the parameters of the samples are listed in the table 1. The PR spectra of investigated samples are shown at figure 2. There are four spectral peculiarities in the spectra: with energies less than 1.4 eV is connected with the interband transitions in the InGaAs quantum well, in the spectral range of 1.4-1.5 eV is connected with the fundamental transitions in the GaAs layer, in the spectral range of 1.6-1.8 eV is connected with the interband transitions in the AlGaAs/GaAs buffer superlattice and

Table 1. Sample parameters. L_b is barrier thickness, n_H - Hall concentration at 300 K, E - built-in electric field determined from theory and PR spectra

Samp. #	L_b, nm	n_H, $\times 10^{12}$ cm^{-2}	E, kV/cm	
			theory	Experiment
1	18	1.54	350	340
2	15	1.69	523	510
3	13	1.70	613	594
4	11	1.71	752	652
5	9	1.77	890	692

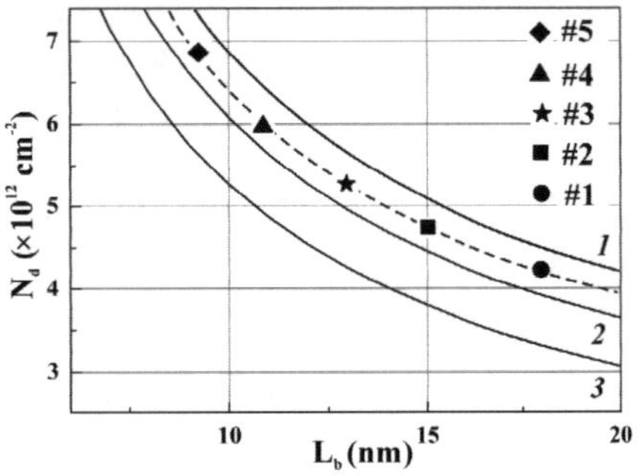

Figure 1. Isolines corresponding to constant concentrations of electrons in the QW. Isolines 1, 2 and 3 correspond to the concentrations (1.7, 1.4 and 1.1)×10^{12} cm^{-2}. The dashed line indicates the concentration chosen for the experimental series of samples – 1.6×10^{12} cm^{-2}.

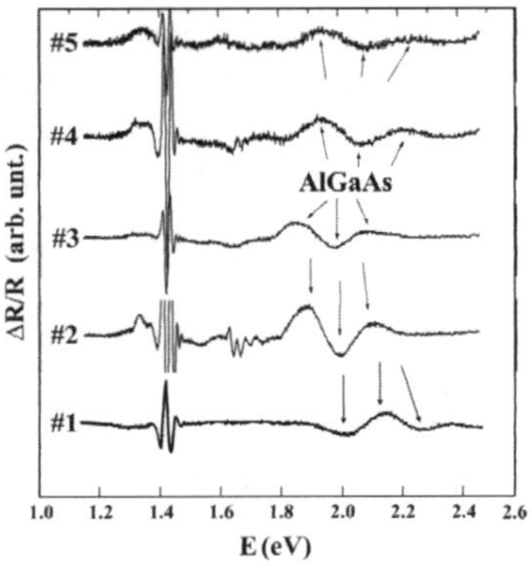

Figure 2. PR spectra of samples #1-#5 at room temperature.

in the spectral range of 1.8-2.3 eV is connected with the fundamental transition in the δ-doped $Al_xGa_{1-x}As$ barrier layer. The electric field strength E in the delta-doped $Al_xGa_{1-x}As$ barrier layer was calculated from the results of linear regression in accordance with the Aspnes model [6] and have been shown at the table 1.

The PL spectra of the samples exhibited two bands, at the photon energies $\hbar\omega_1$ = 1.28–1.30 and $\hbar\omega_2$ = 1.35–1.38 eV. These bands are identified as defined by transitions from the first and second electron subbands to the first subband of heavy holes, *e1–h1* and *e2–h1*, respectively. The PL spectra shown in figure 3 are typical for the samples where the Fermi level is located between the first (E_0) and the second (E_1) size-quantization subbands (figure 4). In addition, as L_b decreases, the intensity of the peak corresponding to the transitions *e2–h1* increases. Changes in the rate of optical transitions are commonly related to two factors. The one is the change in the matrix elements of transitions and the breakage of the selection rules under changes in the built-in electric field. This factor is associated with the symmetry of the QW, since the overlapping of the wave functions of electrons and holes depend on the built-in electric field. The other factor is the change in the relation between the concentrations of electrons in the subbands of the QW. In other studies, a change in the spectrum was observed if the Fermi energy and the concentration of electrons in the QW were

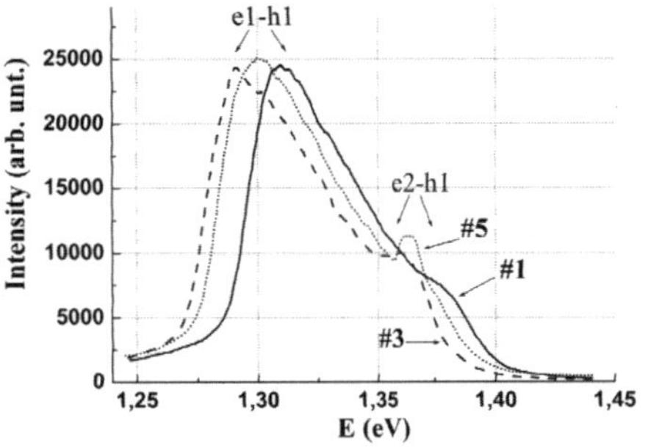

Figure 3. PL spectra of samples #1, #3, #5 at 77 K.

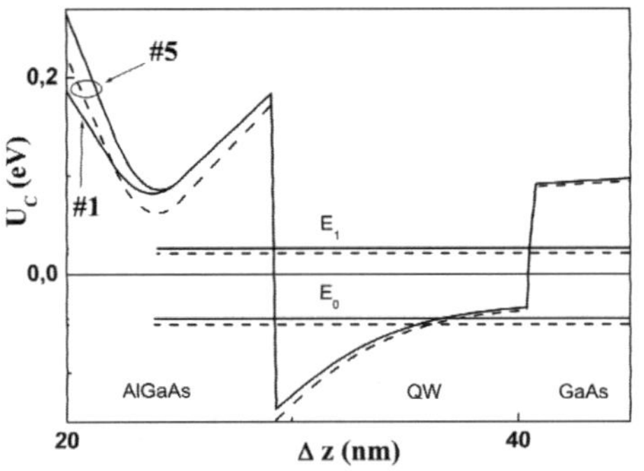

Figure 4. The potential profiles of the conduction band U_C for samples #1 and #5. The profiles are reduced to one point Δz corresponding to the boundry of the QW.

increased and the upper size-quantization band E_1 was filled with electrons. The two above factors always "worked" in combination.

In this study, the total concentration of electrons and the position of the Fermi level in the QW were chosen to be constant and the built-in electric field and the symmetry of the QW were varied only because of variations in the potential profile in the AlGaAs barrier layer and the δ-layer with donors. As the calculations of the band structure showed, the rate of the transitions *e2–h1* increases with decreasing L_b, since the built-in electric field increases as the QW approaches the surface. A decrease in L_b yields an increase in the energy of the conduction band in the QW region, resulting in a decrease in the concentration of the 2D electron gas in the QW. To keep the constant concentration of electrons in the QW, we increased the strength of the built-in electric field in the AlGaAs barrier region by increasing the concentration of donors in the Si δ-layer. For convenient comparison of the changes in the band structure in figure 4, the potential profiles of the conduction band in the samples were shifted with respect to each other to have one common point, the upper boundary of the QW. In this case, the coordinates of the surface Δz of different samples do not coincide, but the surface potential $\varphi_s = 0.7$ eV is the same for all of the samples.

The electrostatic dipole interaction between carriers localized in surface states and their parent donors causes the donor to segregate along with the growing surface. Thus, when the QW is set closer to the surface, spatial profile of the doping increases. In addition, when L_b decreases, the surface potential φ_s decreases as well. The above effects for #5 are showed on the figure 4 by dash line. In this case, the curvature of the bottom of the conduction band in the QW is enhanced. As the curvature of the bottom of the QW increases, the overlapping of the wave functions of electrons *e2* and holes *h1* becomes more efficient, resulting in an increase in the intensity of the PL peak corresponding to the transitions *e2–h1*. Thus, the experimentally observed changes in the positions of the PL peaks and the change in the rate of the transitions *e2–h1* should be attributed to the change in the doping profile and variation of the surface potential.

4. Conclusion

In this study, the built-in electric field of heterostructures with the quantum wells grown at different depths L_b with respect to the surface is analyzed. The effect of variations of the AlGaAs barrier layer thickness L_b and of Si-doping level on the energy band spectrum is simulated. To keep the concentration of electrons in the QW constant, we increase the strength of the built-in electric field in the AlGaAs barrier region by increasing the concentration of donors in the Si δ-layer. As the parameter L_b is decreased, variations in the potential profiles of the conduction band occur in the regions of the *V*-shaped potential of the δ-layer and QW. Comparison of determination the built-in electric field from PR spectra and the band structure calculations shows that altering in the intensity ratio of two peaks in the PL spectra originates from the redistribution of doping atoms and variation of the surface potential.

Acknowledgements

The work was supported by Ministry of education and science of the Russian Federation, Project 2010-218-02-136

References

[1] Gaquiere C, Grünenütt J, Jambon D, Delos E, Ducatteau D, Werquin M, Treron D, Fellon P 2005 *IEEE Electr. Dev. Lett.* **26** 8 533
[2] Lee D Y, Leem J Y, Kang S K, Kim J S, Son J S, Bae I H 2003 *Phys. E.* **19** 349
[3] Ambrazevifius G, Marcinkeviflus S, Lideikis T, Naudzius K 1992 *Semicond. Sci. Technol.* **7** 818
[4] Khabibullin R A, Vasil'evskii I S, Galiev G B, Klimov E A, Ponomarev D S, Gladkov V P, Kulbachinskii V A, Klochkov A N, Uzeeva N A 2011 *Semicond.* **45** 5 657
[5] Avakyants L P, Bokov P Yu, Chervyakov A V 2005 *Technical Physics* **50** 1316
[6] Aspnes D E 1974 *Phys Rev B.***10** 4228

Diffraction of polarized light on periodic structures

V. Bukanina[1], D. Divakov[2], A. Tyutyunnik[3], A. Hohlov[4]

[1] Student, Peoples Friendship University of Russia, Moscow, Russia

[2] Student, Peoples Friendship University of Russia, Moscow, Russia

[3] Student, Peoples Friendship University of Russia, Moscow, Russia

[4] PhD in physics, Peoples Friendship University of Russia, Moscow, Russia

E-mail: aaxoxlov@sci.pfu.edu.ru

Abstract. Periodic structures as photonic crystals are widely used in modern laser devices, communication technologies and for creating various beam splitters and filters. Diffraction gratings are applied for creating 3D television sets, DVD and Blu-ray drives and reflective structures (Berkley mirror). It is important to simulate diffraction on such structures to design optical systems with predetermined properties based on photonic crystals and diffraction gratings. Methods of simulating diffraction on periodic structures uses theory of Floquet-Bloch and rigorous coupled-wave analysis (RCWA). Current work is dedicated to analysis of photonic band gaps and simulating diffraction on one-dimensional binary diffraction grating using RCWA. The Maxwell's equations for isotropic media and constitutive relations based on the cgs system were used as a model.

1. One-dimensional photonic crystal

The photonic crystal is a dielectric structure with periodic refractive index along one or more directions. A distinctive feature of these structures is the presence of so-called photonic band gaps, preventing the propagation of waves of a certain frequency. Due to this property with the help of photonic crystals it's possible to create devices that can reflect or transmit light with fixed wavelength.

We consider the formation of photonic band gaps by example of monochromatic waves propagating along z-axis, which is perpendicular to the direction of periodicity of one-dimensional photonic crystal. Its dielectric permittivity is a periodic function of z:

$$\varepsilon(z)=\begin{cases}\varepsilon_1, & (n-1)d<z<nd-a\\ \varepsilon_2, & nd-a<z<nd\end{cases}, \tag{1}$$

where n is a number of the cell, a is the width of layer with dielectric permittivity ε_2 and $\varepsilon(z)=\varepsilon(z+d)$, where d is a period.

Since the mediums are isotropic, the TE and TM polarizations propagate independently and they can be considered separately. Consider the case of waves of TE polarization. We can reduce Maxwell's equations to the following wave equation:

$$\frac{\partial^2 E_y}{\partial z^2}+k_0^2\beta E_y=0,\ (\beta=\varepsilon-\frac{k_x^2}{k_0^2}), \tag{2}$$

From this equation, we find a general solution for the field components in a uniform layer, we obtain considering

$$H_x = -\frac{i}{k_0}\frac{\partial E_y}{\partial z}$$

$$\begin{cases} E_y \ z \ = Ae^{ik_0\sqrt{\beta}(z-nd)} + Be^{-ik_0\sqrt{\beta}(z-nd)} \\ H_x \ z \ = A\sqrt{\beta}e^{ik_0\sqrt{\beta}(z-nd)} - B\sqrt{\beta}e^{-ik_0\sqrt{\beta}(z-nd)} \end{cases}. \quad (3)$$

Then we obtain the system considering the condition that the tangential field components at the interface between two media with $z = \ n-1 \ d$ and $z = nd - a$

$$\begin{cases} A_{n-1}^{\ 2} + B_{n-1}^{\ 2} = A_n^{\ 1} e^{-ik_0\sqrt{\beta_1}d} + B_n^{\ 1} e^{ik_0\sqrt{\beta_1}d} \\ A_{n-1}^{\ 2}\sqrt{\beta_2} - B_{n-1}^{\ 2}\sqrt{\beta_2} = A_n^{\ 1}\sqrt{\beta_1}e^{-ik_0\sqrt{\beta_1}d} - B_n^{\ 1}\sqrt{\beta_1}e^{ik_0\sqrt{\beta_1}d} \\ A_n^{\ 1} e^{-ik_0\sqrt{\beta_1}a} + B_n^{\ 1} e^{ik_0\sqrt{\beta_1}a} = A_n^{\ 2} e^{-ik_0\sqrt{\beta_2}a} + B_n^{\ 2} e^{ik_0\sqrt{\beta_2}a} \\ A_n^{\ 1}\sqrt{\beta_1}e^{-ik_0\sqrt{\beta_1}a} - B_n^{\ 1}\sqrt{\beta_1}e^{ik_0\sqrt{\beta_1}a} = A_n^{\ 2}\sqrt{\beta_2}e^{-ik_0\sqrt{\beta_2}a} - B_n^{\ 2}\sqrt{\beta_2}e^{ik_0\sqrt{\beta_2}a} \end{cases}. \quad (4)$$

$A^{\ 1}$, $B^{\ 1}$ and $A^{\ 2}$, $B^{\ 2}$ are indefinite coefficients in the layers with dielectric permittivity ε_1 and ε_2.

From (4) we find equations relating the indefinite coefficients from (n-1) –th cell with n-th cell in a layer with a dielectric permittivity ε_2:

$$\begin{pmatrix} A_{n-1}^{\ 2} \\ B_{n-1}^{\ 2} \end{pmatrix} = \begin{pmatrix} m_{11} & m_{12} \\ m_{21} & m_{22} \end{pmatrix}\begin{pmatrix} A_n^{\ 2} \\ B_n^{\ 2} \end{pmatrix} \quad (5)$$

where

$$\begin{aligned} m_{11} &= e^{-ik_0\sqrt{\beta_2}a}\left(\cos k_0\sqrt{\beta_1}b - \frac{i}{2}\left(\frac{\beta_1+\beta_2}{\sqrt{\beta_1\beta_2}}\right)\sin k_0\sqrt{\beta_1}b\right) \\ m_{12} &= e^{ik_0\sqrt{\beta_2}a}\left(\frac{i}{2}\left(\frac{\beta_2-\beta_1}{\sqrt{\beta_1\beta_2}}\right)\sin k_0\sqrt{\beta_1}b\right) \\ m_{21} &= e^{-ik_0\sqrt{\beta_2}a}\left(-\frac{i}{2}\left(\frac{\beta_2-\beta_1}{\sqrt{\beta_1\beta_2}}\right)\sin k_0\sqrt{\beta_1}b\right) \\ m_{22} &= e^{ik_0\sqrt{\beta_2}a}\left(\cos k_0\sqrt{\beta_1}b + \frac{i}{2}\left(\frac{\beta_1+\beta_2}{\sqrt{\beta_1\beta_2}}\right)\sin k_0\sqrt{\beta_1}b\right) \end{aligned} \quad (6)$$

For the case of TM polarization done a similar action.

Next we use the Floquet-Bloch theorem, according to which the periodic solution in a layered periodic medium has the form $E \ z \ = \tilde{E} \ z \ e^{iKz}$, where $\tilde{E} \ z$ is a periodic function with the period of d. Constant K is called the Bloch wave number.

Considering $E_{n-1}(z-d)e^{iK(z-d)} = E_n(z)e^{iKz}$ we write the condition of periodicity for the Bloch wave in the form:

$$\begin{pmatrix} A_n^2 \\ B_n^2 \end{pmatrix} = e^{-iKd} \begin{pmatrix} A_{n-1}^2 \\ B_{n-1}^2 \end{pmatrix}. \tag{7}$$

From (3) and (7) we obtain the following system:

$$\begin{pmatrix} m_{11} & m_{12} \\ m_{21} & m_{22} \end{pmatrix} \begin{pmatrix} A_n^2 \\ B_n^2 \end{pmatrix} = e^{iKd} \begin{pmatrix} A_n^2 \\ B_n^2 \end{pmatrix} \tag{8}$$

In this system the factor of e^{iKd} is an eigenvalue of the coefficient matrix. Then solving the characteristic equation of this system with respect to factor e^{iKd}, given that the coefficient matrix is unimodular, we obtain:

$$2\cos Kd = m_{11} + m_{22} \tag{9}$$

Substituting the values of the coefficient matrix to the (9), we obtain the dispersion equation in implicit form, which establishes the relationship between the Bloch wave number K, frequency ω, and x-component k_x of wave vector. The dispersion equation for TE waves:

$$\cos Kd = \cos\, k_0\sqrt{\beta_1}b\, \cos\, k_0\sqrt{\beta_2}a - \frac{1}{2}\left(\frac{\beta_1+\beta_2}{\sqrt{\beta_1\beta_2}}\right)\sin\, k_0\sqrt{\beta_1}b\, \sin\, k_0\sqrt{\beta_2}a \tag{10}$$

The dispersion equation for TM waves has the following form:

$$\cos Kd = \cos\, k_0\sqrt{\beta_1}b\, \cos\, k_0\sqrt{\beta_2}a - \frac{1}{2}\left(\frac{\varepsilon_2}{\varepsilon_1}\sqrt{\frac{\beta_1}{\beta_2}} + \frac{\varepsilon_1}{\varepsilon_2}\sqrt{\frac{\beta_2}{\beta_1}}\right)\sin\, k_0\sqrt{\beta_1}b\, \sin\, k_0\sqrt{\beta_2}a\ . \tag{11}$$

If $|(m_{11}+m_{22})/2| > 1$, K takes complex values, the Bloch wave is evanescent, and the propagation of electromagnetic waves is impossible in this region, it is called band gap. If $|(m_{11}+m_{22})/2| < 1$, K is real, this is the case of permitted zones, in this area Bloch waves will propagate.

So the reflection coefficient tends to 1 in the band gap. The graph below represents the reflection coefficient for a one-dimensional photonic crystal of 125 layers (fig. 1).

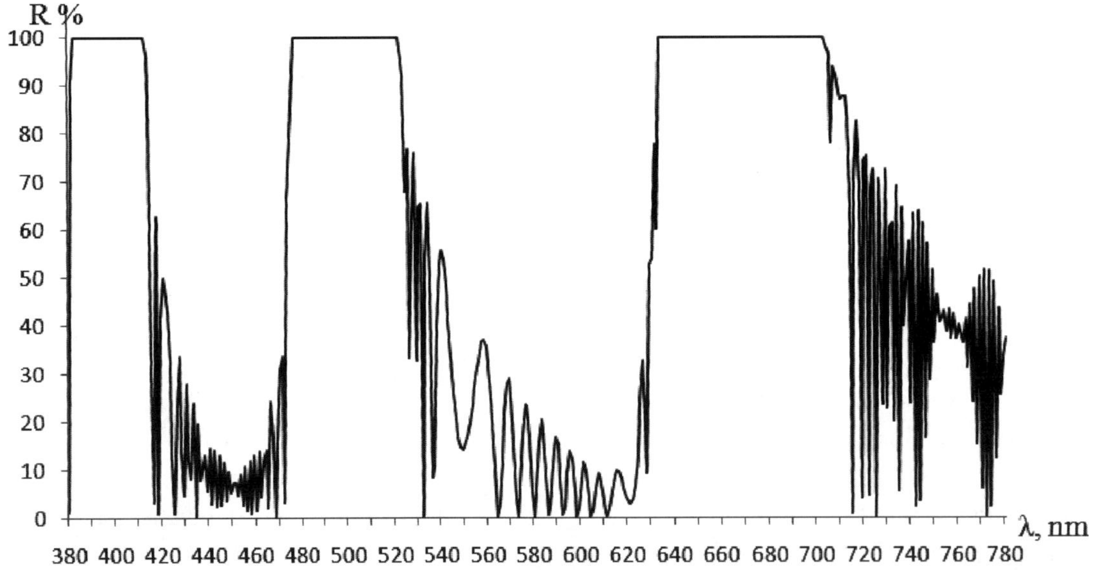

Fig. 1 The reflection coefficient for a one-dimensional photonic crystal of 125 layers

The wavelengths for which the reflection coefficient reaches 100% correspond to the band gaps. The reflection coefficient is the closer to unit the more is number of layers.

2. One-dimensional binary diffraction grating

Let us assume that a linearly polarized electromagnetic field with wavelength λ incidents at Λ-periodic binary grating at an angle θ (fig. 2).

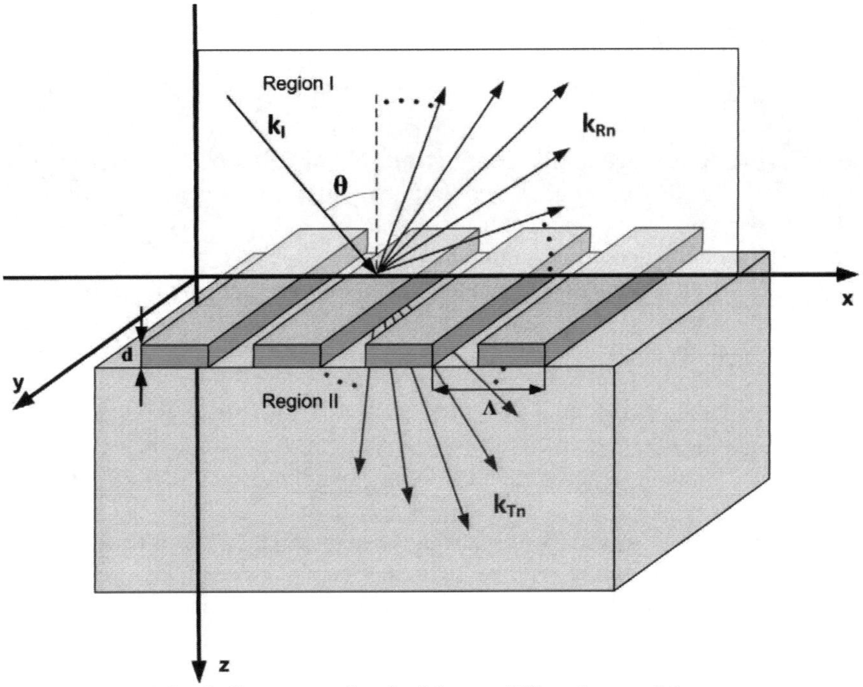

Fig. 2 Geometry for the binary diffraction problem

Wave-vector of the incident field can be obtained from the geometry as $\mathbf{k_I} = k(\sin\theta, 0, \cos\theta) = n_I k_0 (\sin\theta, 0, \cos\theta)$, where $k_0 = 2\pi / \lambda_0$, n_I - refractive index of region I (fig.2). Wave-vectors of the reflected and transmitted fields are defined from Floquet-condition and geometry:

$$k_{xj} = k_0 (n_I \sin\theta - j\lambda_0 / \Lambda), \tag{12}$$

$$k_{L,zj} = \begin{cases} \sqrt{(k_0 n_L)^2 - k_{xj}^2}, & \operatorname{Re}((k_0 n_L)^2 - k_{xj}^2) \geq 0 \\ -i\sqrt{k_{xj}^2 - (k_0 n_L)^2}, & \operatorname{Re}((k_0 n_L)^2 - k_{xj}^2) < 0 \end{cases}, L = I, II. \tag{13}$$

Grating region is bounded by two media with constant relative permittivity. Region I is the input media with $\varepsilon_I = n_I^2 = 1$. Region II corresponds to the output media, where $\varepsilon_{II} = n_{II}^2 > 1$. Relative permittivity in the grating region is a Λ-periodic function $\varepsilon(x) = \varepsilon(x + m\Lambda)$, $m = 0, \pm 1, \pm 2, \ldots$.

Electromagnetic fields in regions I and II are satisfied to Rayleigh expansion:

$$u = \exp(-ik_0 n_I (\sin\theta x + \cos\theta z)) + \sum_{j=-\infty}^{\infty} R_j \exp(-i(k_{xj}x - k_{I,zj}z)), \tag{14}$$

$$u = \sum_{j=-\infty}^{\infty} T_j \exp\ -i\ k_{xj}x - k_{II,zj}(z-d)\ \ . \tag{15}$$

where u corresponds to E_y in TE-case and H_y in TM-case.

Electromagnetic field in the grating region satisfies to Maxwell's equations, which can be reduced to the system:

$$-ik_0 H_x = \frac{\partial E_y}{\partial z},\quad -ik_0 H_y = \frac{\partial E_z}{\partial x} - \frac{\partial E_x}{\partial z},\quad -ik_0 H_z = \frac{\partial E_y}{\partial x}, \tag{16}$$

$$ik_0\varepsilon(x)E_x = \frac{\partial H_y}{\partial z},\ ik_0\varepsilon(x)E_y = \frac{\partial H_z}{\partial x} - \frac{\partial H_x}{\partial z},\ ik_0\varepsilon(x)E_z = \frac{\partial H_y}{\partial x}. \tag{17}$$

The system **Error! Reference source not found.-Error! Reference source not found.** defines electromagnetic field in the grating region. Binary diffraction problem consists of definition electromagnetic fields in regions I, II and grating region.

We will solve the problem according to the plan:

Expansion of the tangential electric and magnetic fields in Fourier series in the grating region;

Substitution of the expanded tangential fields into the system (16), (17) to become two systems differential equations of the first order with normalized amplitudes of the tangential fields in the grating region as a variable;

Reduction of two systems differential equations of the first order to one system differential equations of the second order;

General solution of system differential equations of the second order through eigenvalues and eigenvectors;

Using boundary conditions at the input and output boundaries to obtain quantities for the general solution of system differential equations of the second order;

Substitution quantities into general solution to determine electromagnetic field in the grating region;

2.1. TE-case

Fourier series for tangential electric and magnetic fields:

$$E_y = \sum_j S_{yj}\ \ z\ \ \exp\ -ik_{xj}x\ \ , \tag{18}$$

$$H_x = -i\sum_j U_{xj}\ \ z\ \ \exp\ -ik_{xj}x\ \ . \tag{19}$$

Substitution of the expanded tangential fields into the system (18), (19) to become two systems differential equations of the first order:

$$\frac{\partial S_{yj}}{\partial z} = k_0 U_{xj}, \tag{20}$$

$$\frac{\partial U_{xj}}{\partial z} = \left(\frac{k_{xj}^2}{k_0}\right) S_{yj} - k_0 \sum_p \varepsilon_{j-p} S_{yp}. \tag{21}$$

Equations (20), (22) can be written in matrix form using $z' = k_0 z$:

$$\begin{bmatrix} \partial \mathbf{S}_y / \partial z' \\ \partial \mathbf{U}_x / \partial z' \end{bmatrix} = \begin{bmatrix} \mathbf{0} & \mathbf{I} \\ \mathbf{A} & \mathbf{0} \end{bmatrix} \begin{bmatrix} \mathbf{S}_y \\ \mathbf{U}_x \end{bmatrix}, \tag{22}$$

where $\mathbf{A} = \mathbf{K}_x^2 - \mathbf{F}$;

$\mathbf{I}$ - identity matrix;

$$\mathbf{K}_x = \begin{bmatrix} k_{x-N}/k_0 & 0 & \cdots & & 0 \\ 0 & \ddots & & & \vdots \\ \vdots & & k_{x0}/k_0 & & \\ & & & \ddots & 0 \\ 0 & \cdots & & 0 & k_{xN}/k_0 \end{bmatrix};$$

$\mathbf{F}$ - matrix for the coefficients of Fourier series for $\varepsilon\ x$.

Previous systems can be reduced to:

$$\begin{bmatrix} \partial \mathbf{S}_y / \partial z' \\ \partial \mathbf{U}_x / \partial z' \end{bmatrix} = \begin{bmatrix} \mathbf{0} & \mathbf{I} \\ \mathbf{A} & \mathbf{0} \end{bmatrix} \begin{bmatrix} \mathbf{S}_y \\ \mathbf{U}_x \end{bmatrix}, \tag{23}$$

General solution of system differential equations of the second order:

$$S_{yj}\ z\ = \sum_{m=1}^{n} w_{jm}\ c_m^+ \exp\ -k_0 q_m z\ + c_m^- \exp\left[k_0 q_m\ \ z-d\ \right], \tag{24}$$

$$U_{xj}\ z\ = \sum_{m=1}^{n} v_{jm}\ -c_m^+ \exp\ -k_0 q_m z\ + c_m^- \exp\left[k_0 q_m\ \ z-d\ \right], \tag{25}$$

where w_{jm} and q_m are the elements of the eigenvector matrix $\mathbf{W}$ and the positive square root of the eigenvalues of the matrix $\mathbf{A}$, v_{jm} - elements of the matrix $\mathbf{V} = \mathbf{WQ}$, $\mathbf{Q}$ - diagonal matrix with elements q_m.

Using conditions on the boundary $z=0$:

$$\delta_{j0} + R_j = \sum_{m=1}^{n} w_{jm} \left[c_m^+ + c_m^- \exp\ -k_0 q_m d\ \right], \tag{26}$$

$$i\left[n_I \cos\theta \delta_{j0} -\ k_{I,zj}/k_0\ \ R_j\right] = \sum_{m=1}^{n} v_{jm} \left[c_m^+ - c_m^- \exp\ -k_0 q_m d\ \right], \tag{27}$$

where $\delta_{j0} = \begin{cases} 1, j=0 \\ 0, j \neq 0 \end{cases}$.

Using conditions on the boundary $z=d$:

$$\sum_{m=1}^{n} w_{jm} \left[c_m^+ \exp\ -k_0 q_m d\ + c_m^-\right] = T_j, \tag{28}$$

$$\sum_{m=1}^{n} v_{jm} \left[c_m^+ \exp\ -k_0 q_m d\ - c_m^-\right] = i\ \ k_{II,zj}/k_0\ \ T_j. \tag{29}$$

Eliminating R_j and T_j we become system of linear equations to obtain c_m^+, c_m^-:

$$\begin{aligned} i\delta_{j0}\ \ k_{I,zj}/k_0 + n_I \cos\theta\ &= \sum_{m=1}^{n} c_m^+ w_{jm} \left[i\ \ k_{I,zj}/k_0\ + q_m\right] + \\ &+ \sum_{m=1}^{n} c_m^- w_{jm} \exp\ -k_0 q_m d\ \left[i\ \ k_{I,zj}/k_0\ - q_m\right], \end{aligned} \tag{30}$$

$$\sum_{m=1}^{n} c_m^{+} w_{jm} \exp\left(-k_0 q_m d\right)\left[-i\left(k_{II,zj}/k_0\right)+q_m\right]+ \\ +\sum_{m=1}^{n} c_m^{-} w_{jm}\left[-i\left(k_{I,zj}/k_0\right)-q_m\right]=0. \tag{31}$$

Substituting c_m^{+}, c_m^{-} into (26), (28) we determine R_j and T_j. The efficiencies of the diffracted orders are given by:

$$\mathbf{R}_j = \left|R_j\right|^2 \operatorname{Re}\left(\frac{k_{I,zj}}{k_0 n_I \cos\theta}\right), \tag{32}$$

$$\mathbf{T}_j = \left|T_j\right|^2 \operatorname{Re}\left(\frac{k_{II,zj}}{k_0 n_I \cos\theta}\right). \tag{33}$$

2.2. TM-case

Fourier series for tangential electric and magnetic fields:

$$H_y = \sum_j U_{yj}(z) \exp\left(-ik_{xj}x\right), \tag{34}$$

$$E_x = i\sum_j S_{xj}(z) \exp\left(-ik_{xj}x\right). \tag{35}$$

Substitution of the expanded tangential fields into the system (16), (17) to become two systems differential equations of the first order:

$$\sum_p \bar{\varepsilon}_{j-p} \frac{\partial U_{yp}}{\partial z} = k_0 S_{xj}, \tag{36}$$

$$\frac{1}{k_0}\frac{\partial S_{xj}}{\partial z} = \frac{k_{xj}}{k_0}\left(\sum_p \bar{\varepsilon}_{j-p} \frac{k_{xp}}{k_0} U_{yp}\right) - U_{yj}. \tag{37}$$

Equations (26), (28) can be written in matrix form using $z' = k_0 z$:

$$\begin{bmatrix} \partial \mathbf{U}_y / \partial z' \\ \partial \mathbf{S}_x / \partial z' \end{bmatrix} = \begin{bmatrix} \mathbf{0} & \mathbf{F} \\ \mathbf{B} & \mathbf{0} \end{bmatrix} \begin{bmatrix} \mathbf{U}_y \\ \mathbf{S}_x \end{bmatrix}, \tag{38}$$

where $\mathbf{B} = \mathbf{K}_x \mathbf{F}^{-1} \mathbf{K}_x - \mathbf{I}$;

$\mathbf{I}$ - identity matrix;

$$\mathbf{K}_x = \begin{bmatrix} k_{x-N}/k_0 & 0 & \cdots & & 0 \\ 0 & \ddots & & & \vdots \\ \vdots & & k_{x0}/k_0 & & \\ & & & \ddots & 0 \\ 0 & \cdots & & 0 & k_{xN}/k_0 \end{bmatrix};$$

$\mathbf{F}^{-1}$ - matrix for the coefficients of Fourier series for $1/\varepsilon(x)$;

$\mathbf{F}$ - inverse to $\mathbf{F}^{-1}$.

Previous systems can be reduced to:

$$\left[\partial^2 \mathbf{U}_y / \partial \ z'^{\ 2}\right] = \left[\mathbf{FB}\ \right]\left[\mathbf{U}_y\right] \tag{39}$$

General solution of system differential equations of the second order:

$$U_{yj}\ z\ = \sum_{m=1}^{n} w_{jm}\ \ c_m^+ \exp\ -k_0 q_m z\ + c_m^- \exp\left[k_0 q_m\ \ z - d\ \right], \tag{40}$$

$$S_{xj}\ z\ = \sum_{m=1}^{n} v_{jm}\ \ -c_m^+ \exp\ -k_0 q_m z\ + c_m^- \exp\left[k_0 q_m\ \ z - d\ \right], \tag{41}$$

where w_{jm} and q_m are the elements of the eigenvector matrix $\mathbf{W}$ and the positive square root of the eigenvalues of the matrix $\mathbf{A}$, v_{jm} - elements of the matrix $\mathbf{V} = \mathbf{F}^{-1}\mathbf{WQ}$, $\mathbf{Q}$ - diagonal matrix with elements q_m.

Using conditions on the boundary $z = 0$:

$$\delta_{j0} + R_j = \sum_{m=1}^{n} w_{jm}\left[c_m^+ + c_m^- \exp\ -k_0 q_m d\ \right], \tag{42}$$

$$i\left[\frac{\cos\theta}{n_I}\delta_{j0} - \frac{k_{I,zj}}{k_0 n_I^2}R_j\right] = \sum_{m=1}^{n} v_{jm}\left[c_m^+ - c_m^- \exp\ -k_0 q_m d\ \right], \tag{43}$$

where $\delta_{j0} = \begin{cases} 1, j = 0 \\ 0, j \neq 0 \end{cases}$.

Using conditions on the boundary $z = d$:

$$\sum_{m=1}^{n} w_{jm}\left[c_m^+ \exp\ -k_0 q_m d\ + c_m^-\right] = T_j, \tag{44}$$

$$\sum_{m=1}^{n} v_{jm}\left[c_m^+ \exp\ -k_0 q_m d\ - c_m^-\right] = i\left(\frac{k_{II,zj}}{k_0 n_{II}^2}\right)T_j. \tag{45}$$

Eliminating R_j and T_j we become system of linear equations to obtain c_m^+, c_m^-:

$$\begin{aligned} i\delta_{j0}\left(\frac{k_{I,zj}}{k_0 n_I^2} + \frac{\cos\theta}{n_I}\right) &= \sum_{m=1}^{n} c_m^+\left[i\frac{k_{I,zj}}{k_0 n_I^2}w_{jm} + v_{jm}\right] + \\ &+ \sum_{m=1}^{n} c_m^- \exp\ -k_0 q_m d\ \left[i\frac{k_{I,zj}}{k_0 n_I^2}w_{jm} - v_{jm}\right], \end{aligned} \tag{46}$$

$$\begin{aligned} \sum_{m=1}^{n} c_m^+ \exp\ -k_0 q_m d\ &\left[-i\frac{k_{II,zj}}{k_0 n_{II}^2}w_{jm} + v_{jm}\right] + \\ &+ \sum_{m=1}^{n} c_m^- w_{im}\left[-i\frac{k_{II,zj}}{k_0 n_{II}^2}w_{jm} - v_{jm}\right] = 0. \end{aligned} \tag{47}$$

Substituting c_m^+, c_m^- into (42), (44) we determine R_j and T_j. The efficiencies of the diffracted orders are given by:

Journal of Physics: Conference Series **345** (2012) 012016 doi:10.1088/1742-6596/345/1/012016

$$\mathbf{R}_j = |R_j|^2 \operatorname{Re}\left(\frac{k_{I,zj}}{k_0 n_I \cos\theta} \right), \tag{48}$$

$$\mathbf{T}_j = |T_j|^2 \operatorname{Re}\left(\frac{k_{II,zj}}{n_{II}^2} \right) / \left(\frac{k_0 \cos\theta}{n_I} \right). \tag{49}$$

2.3. Simulation

The results of comparison between simulation and spectrophotometric data for copper binary grating with period 200000nm, spike width 110000nm and spike height 187nm are presented below.

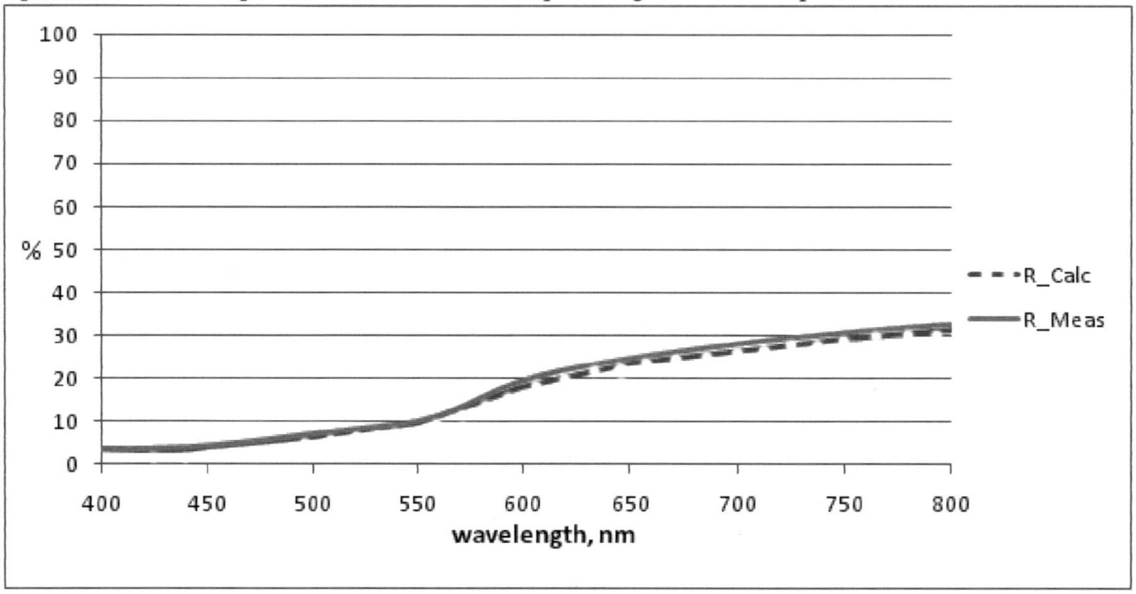

Fig. 3 TE-case, angle of incidence 8°

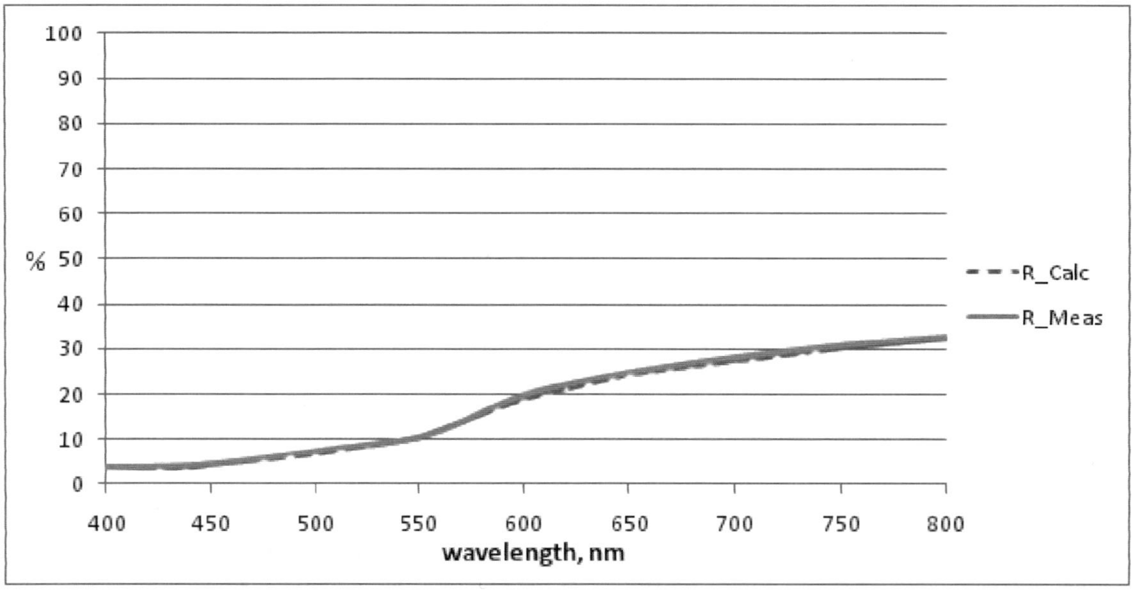

Fig. 4 TM-case, angle of incidence 8°

Journal of Physics: Conference Series **345** (2012) 012016 doi:10.1088/1742-6596/345/1/012016

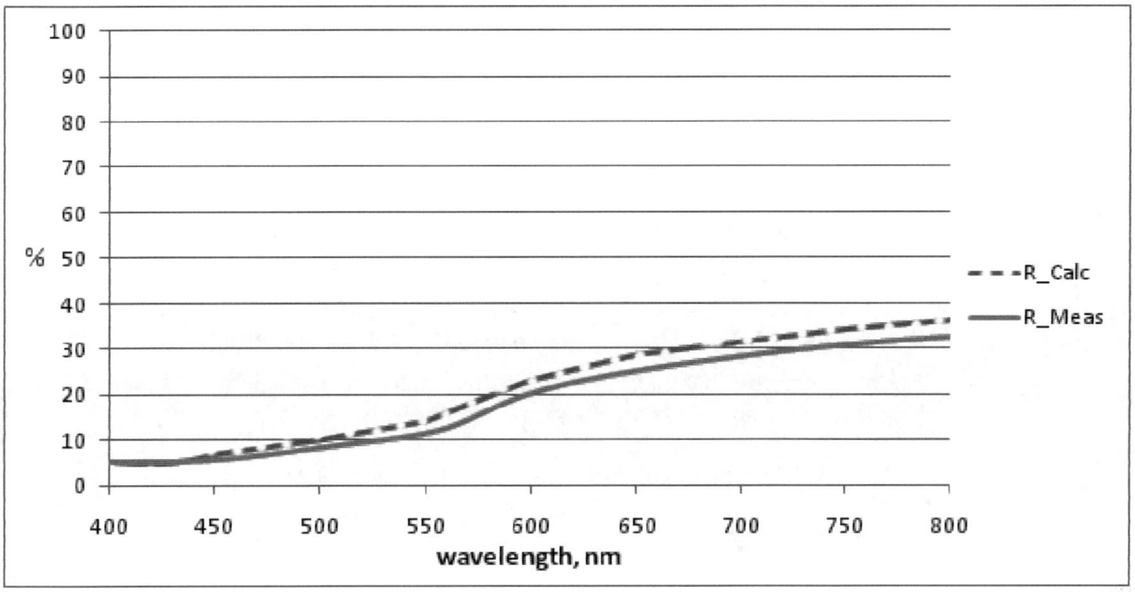

Fig. 5 TE-case, angle of incidence 30°

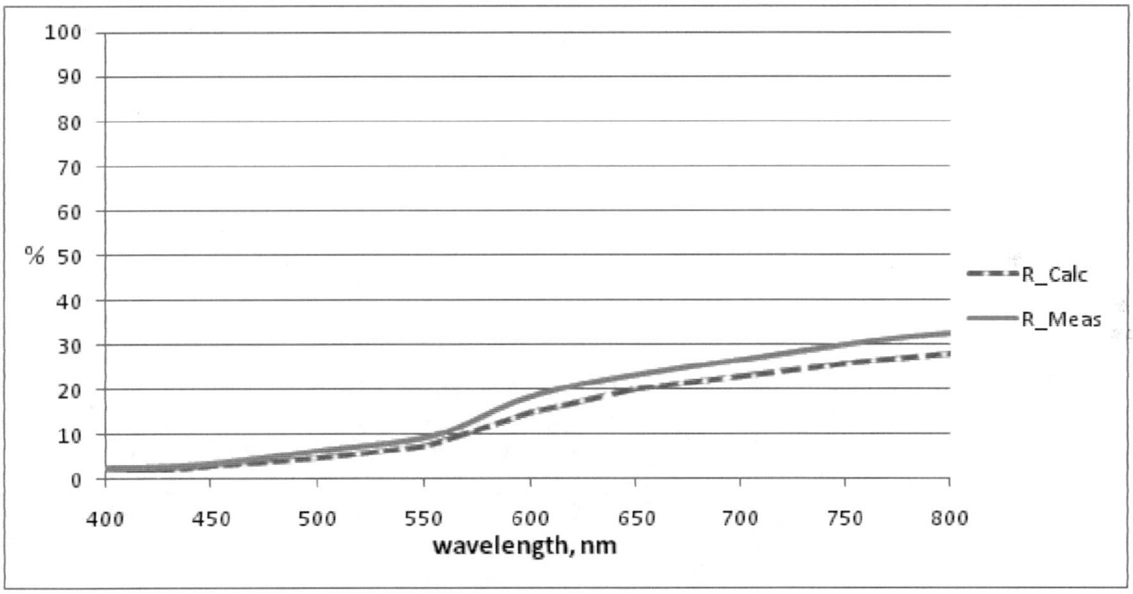

Fig. 6 TM-case, angle of incidence 30°

According to the results (fig. 3 – 6) we can conclude that simulation results and experimental data differs for less than 4.5% which proves the accuracy of the method.

Journal of Physics: Conference Series **345** (2012) 012017 doi:10.1088/1742-6596/345/1/012017

Fluoride laser nanoceramics

P. P. Fedorov*, V.V. Osiko*, S.V. Kuznetsov*, E.A. Garibin**

*A.M. Prokhorov Institute of General Physics, Russian Academy of Sciences, 38 ul. Vavilova, Moscow, 119991, Russian Federation.

**INCROM Ltd, 36 (1) ul. Babushkina, S.-Peterburg, 192171, Russian Federation

E-mail: ppf@lst.gpi.ru

Abstract. Because of the recent technology progress in the synthesis of fluoride ceramics by so-called hot-pressing method, several cooperating research groups have produced and meticulously described various polycrystalline ceramic samples with optical characteristics similar to the properties of corresponding single crystals. Their achievements made fluoride optical materials less expensive, more diverse and easier available for the use in solid-state lasers. The present review covers studies of fluoride ceramic materials and their properties in the last fifty years. It describes the most significant results obtained by leading scientific groups and indicates the most promising directions where research of fluoride laser ceramics should proceed in near future.

1. Introduction

Originally, word "ceramics" was applied to thermally treated mixtures of clay, water and some additional supplements. In the present time, this term is used for polycrystalline inorganic materials obtained by the thermal treatment of precursor batches in order to assure mechanical integrity and property continuity of the manufactured item. Ceramics can contain X-ray amorphous or glassy component, however, one would use more specific name such as "glass ceramics" only in the case of domination of non-crystalline components over crystalline ones in the described sample.

Optical ceramic materials, and in particular fluoride optical ceramics, have been known for several decades [1]. Since the time of their discovery in the middle of the 20-th century, many researchers from various laboratories have worked on the application of these polycrystalline transparent optical materials in various areas. Both American and Soviet techniques for preparing of ceramic magnesium, calcium and barium fluorides (US trademark IRTRAN and USSR trademark CO) for use in IR-optics were developed in 1960-s to 1970-s [2-5]. Bulk ceramic samples were produced by a hot pressing technique via sintering powders of the corresponding substances under pressure. The first remarkable preparations of optical ceramics for use as laser materials were carried out specifically with metal fluorides by Eastman Kodak [2, 6, 7]. This included the first hot-pressed polycrystalline $CaF_2:Dy^{2+}$ laser body. It is worth noting that these laser materials should not have optical losses at the generation wavelength of more than 10^{-3} cm^{-1}, while gradients of the refractive index should not exceed 10^{-5}-10^{-6} cm^{-1}. In order to achieve such quality product, the earliest technology for manufacturing laser materials included the preparation of starting components from their melts, followed by their pulverization and pressing. Complex procedures of powder surface treatment and ceramics translucence were also utilized. These fluoride samples exhibited sustaining lasing in them, but power of the generated beams (7 mW) was much lower than produced with single crystal laser bodies. Similar results were obtained with the use of oxide laser ceramic materials [8, 9].

A new phase of laser ceramics development begins at the end of the 20-th to the beginning of the 21-st centuries, when the high quality neodymium-doped yttrium-aluminum garnet ceramic materials were prepared [10-13]. In contrast with preceding experiments of the 1970-s, the lasing properties of these products were comparable or superior to single crystals. These results renewed interest in the study of fluoride laser ceramics, including its starting materials and preparation. Various ideas concerning the manufacture of fluoride laser ceramics, as well as raw materials for its preparation, were expressed in [14-17]. However, preparation of laser fluoride ceramics has been successfully carried out by the different research groups from A. M. Prokhorov General Physics Institute (Moscow, Russia), State Optical Institute (St. Petersburg, Russia) and INKROM Company (St. Petersburg, Russia), cooperating under the auspices of Russian Federation state contracts [18-25].

Compared to single crystals and glasses, ceramic materials possess many advantages. First of all, ceramic optical products have an essential degree of freedom in choosing the shape and the size of manufactured specimens. In particular, use of ceramics allows manufacturing of the large size samples. Secondly, one can produce ceramic materials with high and uniform (or otherwise specifically designed) distribution of dopants, whereas the similar feature is not so easily achievable (or even possible) in the case of the single crystals. Thirdly, ceramic preparations have substantial tenacity and durability, whereas crystal and glass optical materials are notoriously fragile. Single crystals become more prone to cracking at the cleavage planes with an increase in their chemical purity and perfection of their structure, while ceramics does not have this tendency.

For example, the growth of single crystals of fluorite-type $M_{1-x}R_xF_{2+x}$ (M – alkali earth metal, and R – rare earth element) solid solutions from the melt is hampered by instability of the crystallization front. As a result, the obtained crystals have honeycombed substructure and are optically inhomogeneous.

There are some areas of chemical composition, where one cannot grow single crystals of laser quality from the melt by directed crystallization, or such growth becomes a very difficult procedure despite the existence of the corresponding thermodynamically stable phase [26, 27]. Thus fluoride ceramics technology becomes a very appealing alternative option.

2. Problems of fluoride powders' chemistry and relevant technology processes

Achieving 100% optical density is a necessary (but insufficient) requirement for laser optic materials. It is a serious problem for oxide ceramic materials. However, physical and chemical properties of fluorides (binary and complex as well) are different from those of the corresponding oxides, and, therefore, the above target can be achieved much easier by the use of the fluoride systems. Unfortunately, there is an additional obstacle in the case of fluoride ceramics: in order to obtain highly optically transparent fluoride ceramics, one has first to overcome some very complex chemical problems.

One such difficulty stems from the fact that fluorides are prone to pyrohydrolysis, *i.e.*, reaction with water and/or water vapor at elevated temperatures [26, 28], leading to the formation of the corresponding oxides and oxofluorides. These products are not isomorphous to the initial fluorides and, being accumulated in the initial material, usually form the second phase as finely dispersed inclusions. This results in the irreversible destruction of optical homogeneity in pyrohydrolyzed materials. Therefore, one must take all necessary precautions to prevent hydrolysis (pyrohydrolysis) at all stages of the manufacturing process. The well-developed surfaces of fluoride particles create additional obstacles in overcoming pyrohydrolysis because these surfaces adsorb various molecules, including water. [29] estimate that fluorides precipitated from aqueous solutions retain up to 15 adsorbed monolayers of water at their surface. These water molecules, which are strongly bound to the particle surface, are not easily removed. This situation becomes even more intractable in the case of nanoparticles with their vast surface area making their pyrohydrolysis much easier than that of macro-particles [30].

One may minimize effect of hydrolysis by using various active fluorinating atmospheres to convert the formed oxides and/or oxofluorides back to the desired fluorides [26]. Gaseous carbon tetrafluoride, CF_4, is the reagent most frequently used for this purpose. This or similar fluorinating technologies have been developed to improve the optical quality of fluoride ceramics by conversion of oxides impurities to fluorides (reference), but combining batch fluorination and ceramics manufacturing is a very complicated technical problem. It requires hermetically-sealed vacuum furnace and complete absence of air in the system.

In order to avoid the above complications, researchers developed different protocols to synthesize fluoride ceramics. One the much simpler process is hot pressing. This method turns polycrystalline powders into a monolithic body via particle agglomeration (Figure 1a). Usually, this process performed at $T = (0.5\text{-}0.8)*T_{melting}$ and up to 300 MPa load under $10^{-2}\text{-}10^{-3}$ torr vacuum. This technique also has its shortcomings, including reducing conditions which may lead to the appearance of color centers or transition of ions-activators to their lower valence states. Such samples usually exhibit yellow, grey or even black coloration, and they may have inclusions, too. Their treatment in fluorinating atmosphere can result in discoloration and, therefore, remedy this problem.

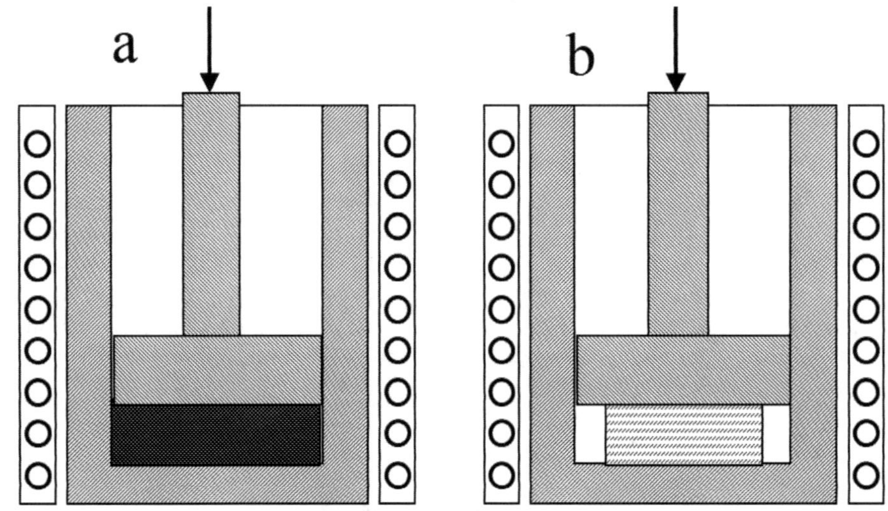

Figure 1. Two methods of ceramics preparation: hot pressing with agglomeration of powder particles (a) and hot forming via crystal deformation (b).
1 – cylindrical chamber, 2 – heating jacket, 3 – piston, 4 – sample

The quality of the starting materials is one of the most important factors affecting the synthesis of the optical ceramics. Transparency data for different fluorite ceramics samples prepared from different quality batches are presented in Figure 2a: a proper choice of the initial ingredients allowed the creation of products of the same quality as in [2].

Laser materials usually need doping with activator admixtures (unless they work exclusively utilizing color centers). Therefore, authors [31-34] attempted synthesis of two-component activated batches that provided activated ceramics of the appropriate quality.

Preliminary experiments [34] have shown that mixing powders of calcium and erbium fluorides does not lead to products which have an acceptable level of composition homogeneity, for diffusion in the samples is incomplete. Product samples are optically inhomogeneous and do not undergo lasing. This highlights the obvious obstacle with samples of complex chemical composition and batches with multiple precursor components Kuznetsov [32, 33], and Bensalah [31] simultaneously attempted similar approaches to address the latter problem by implementing a co-precipitation from aqueous solutions via slow dropwise addition of starting nitrate solutions to aqueous hydrofluoric acid:

$$0.95\ Ca(NO_3)_2 + 0.05Yb(NO_3)_3 + 2.05\ HF \rightarrow Ca_{0.95}Yb_{0.05}F_{2.05}\downarrow + 2.05\ HNO_3.$$

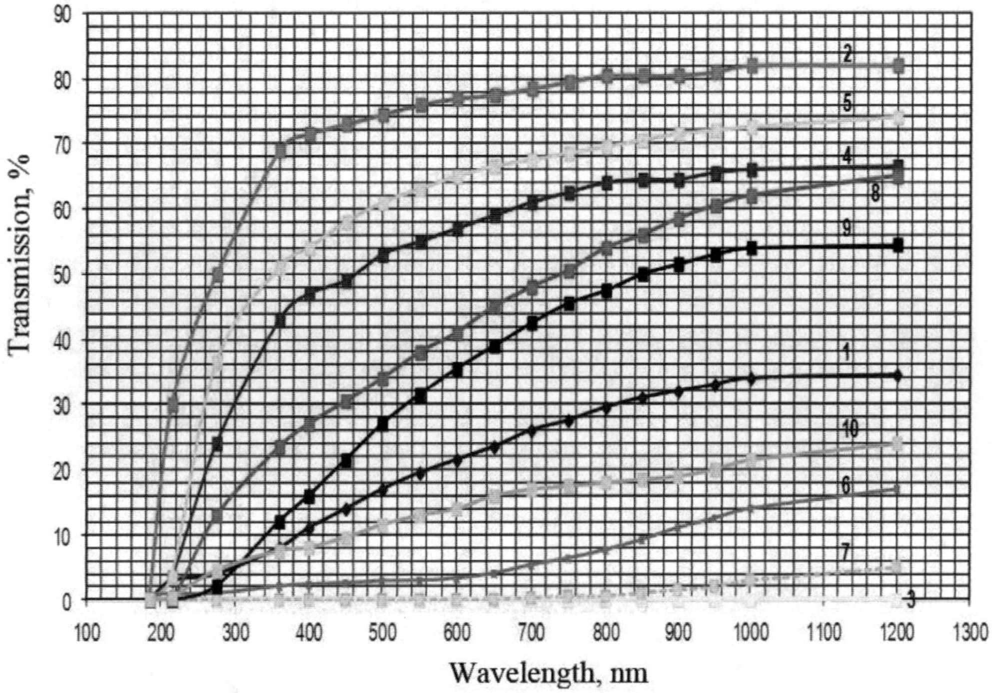

Figure 2. Optical transparency of CaF_2 ceramics prepared by hot-pressing from different starting materials [34] (a); appearance of high optical quality ceramics (external outlook) [2] (b).

Whereas target single-phase $M_{1-x}R_xF_{2+x}$ solid solutions have been easily precipitated by addition of HF to the aqueous systems containing calcium and/or strontium ions along with lanthanide ions, similarly treated barium-lanthanide fluoride samples appeared to be a mixture of BaF_2 and $Ba_4R_3F_{17}$ phases instead [35]. The latter circumstances prohibited the synthesis of optically homogeneous ceramics due to the fluctuation of the chemical composition within the sample. In contrast, $Ca_{1-x}R_xF_{2+x}$ batches – after necessary treatment – produce transparent ceramics (Figure 3a and 3b). Some of these $Ca_{1-x}R_xF_{2+x}$ specimen, e.g., CaF_2:Yb ceramics (Figure 3a and 3b), in accordance with their UV spectra, accommodate significant Yb^{2+} concentrations.

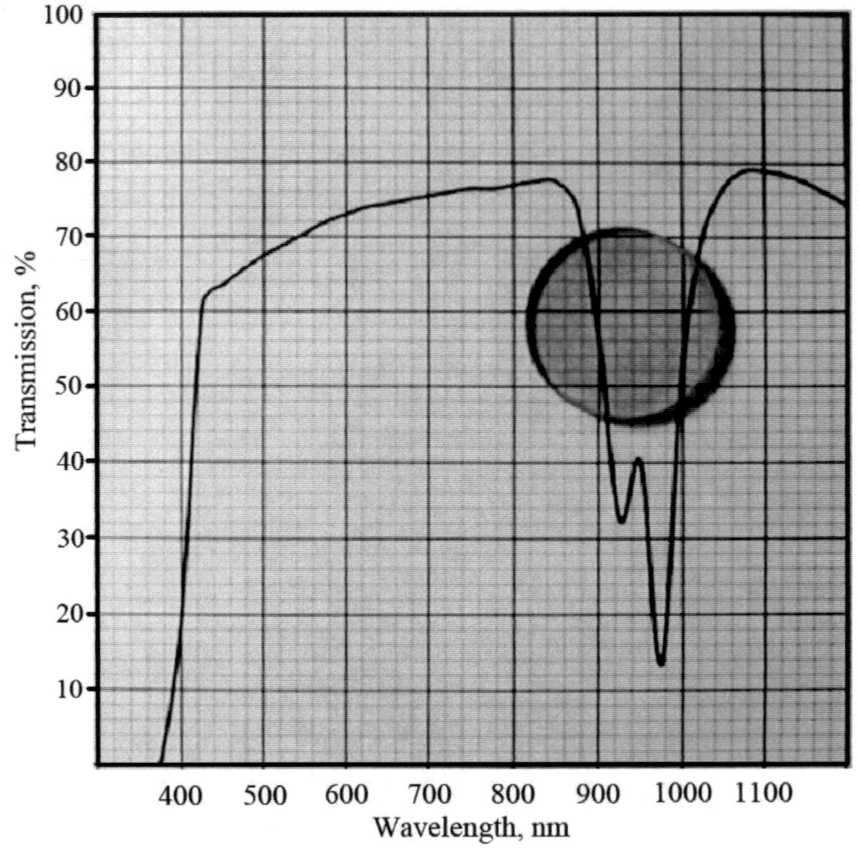

a	b

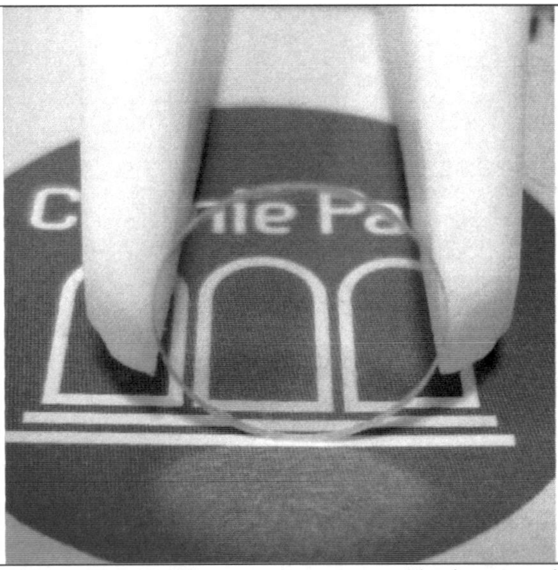

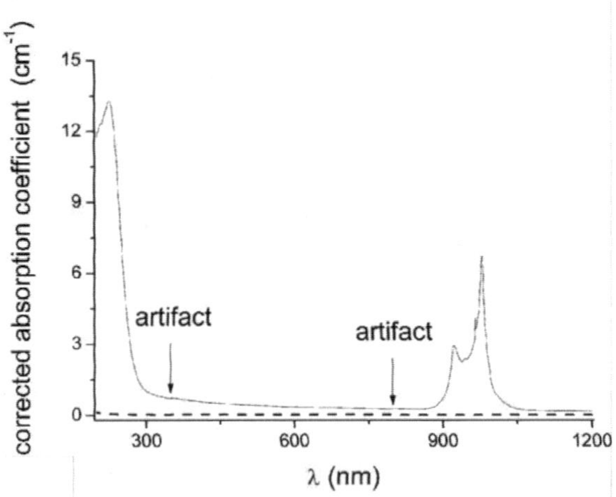

Figure 3. CaF_2:Yb ceramics prepared by hot-pressing according to [34, 36] (a) and [37] (b)

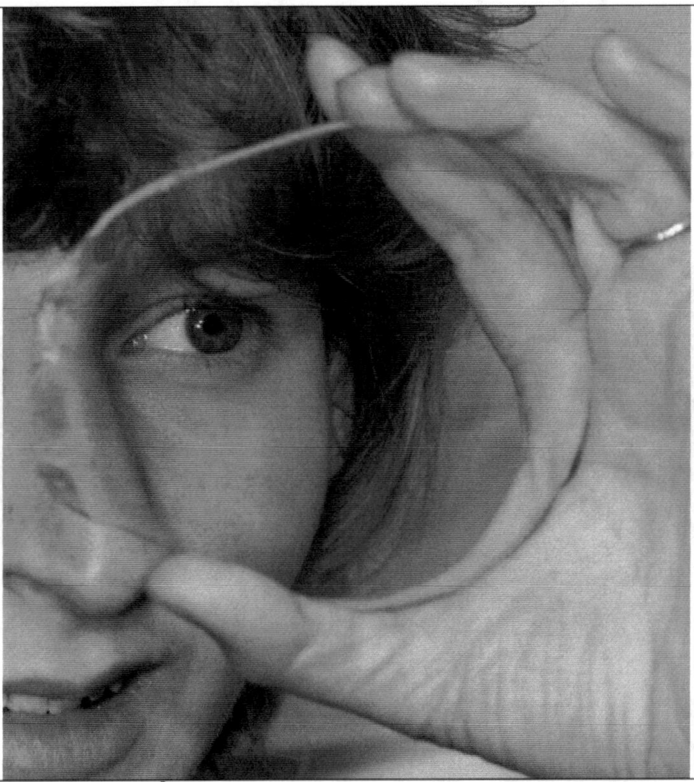

Figure 3c. Respectively; $CaF_2:Yb^{3+}$ laser ceramics prepared by hot-forming.

However, ceramics, prepared from batches initially co-precipitated from aqueous solutions, do not generate lasing because of the strongly adsorbed water layer on the surface of the batch particles [34, 37]. This layer can be hardly removed under a fluorinating atmosphere or under translucence treatment. Lanthanide fluorides undergo hydrolysis much more easily than alkali earth metal fluorides [26, 30], generating oxygen-containing phases between particles and thus causing the light beam scattering.

The deformation of heated crystal under pressure, or hot-forming process for the synthesis of ceramics mentioned above, led to successful avoidance of product contamination with oxygen via pyrohydrolysis (Figure 1b). Whereas Hatch, Carnall et al. [2, 6, 7] hot-pressed preliminary pulverized melts, Basiev, Fedorov et al. [19-25] used the single crystals as starting materials for our hot-pressing protocols. Indeed, parameters of the above hot-forming and hot-pressing processes are different due to the different physical mechanisms of single crystal deformation and consolidation of powder particles in the course of the formation of polycrystalline optic materials.

3. Microstructure

Microstructure of ceramics samples depends on their method of synthesis (Figures. 4 and 5).

Hot-pressing fluoride powders leads to homogeneous microstructure with relatively large (80-150 nm) domains. This has been observed for pure CaF_2 as well as for doped ceramics, such as CaF_2:Yb. The layered fine nanostructure within these domains has been detected by electron microscopy and atomic force microscopy (AFM): fine layer width for CaF_2 is ca. 25 nm [38]. The most probable guess is that this nanostructure stems from crystal twinning within the grains of sample.

Deformed at higher temperatures, single crystal samples have different microstructure. For low degrees of deformation, one can observe a fractal crack-type structure of the borders of crystalline domains (Figure 5a). This structure is most likely not stable and should relax with sample aging. When the degree of deformation is increased, sample microstructure becomes inhomogeneous and contains relatively large areas (ca. 1 mm) with single crystal-type structure along with areas of grain-type structure, typical for classic ceramics. Only at higher degrees of deformation does the structure of the deformed single crystal turns into a completely homogeneous typical ceramic structure. One can observe layered nanostructure of the ceramics domains of the deformed single crystals very similar to the nanostructure of the hot-pressed powders [38] (Figure 5c).

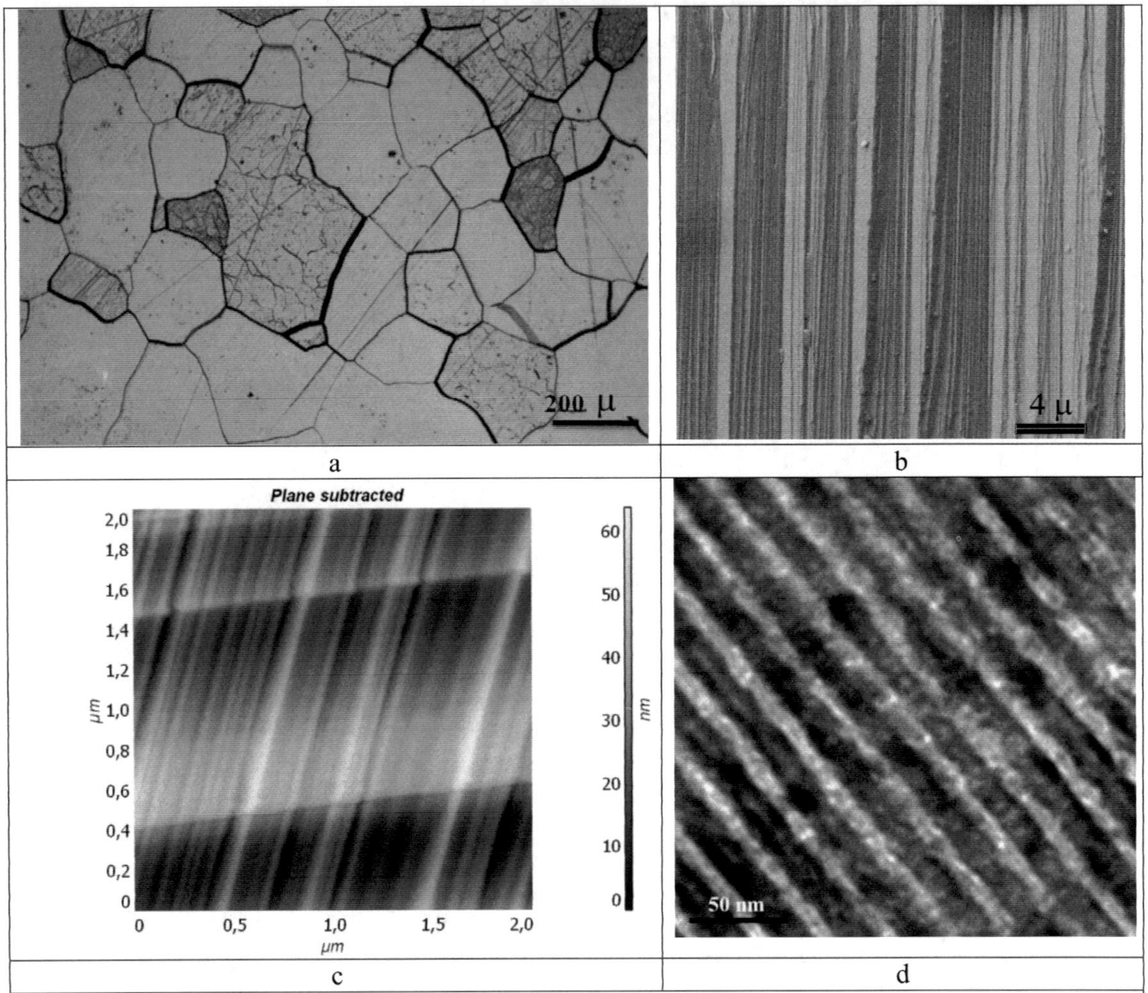

Figure 4. Microstructure of hot-pressed CaF_2 ceramics by optical microscopy (a), electron microscopy (shearing cut) (b), atomic force microscopy (shearing cut) (c), high resolution electron microscopy (d) [38].

The above observation can be illustrated with a typical correlation between the average size of BaF_2 crystalline domains and degree of deformation of the single crystal (Figure 6); only highly-deformed samples correspond to the "completely" ceramic state of the studied material.

Figure 5. Microstructure of hot-formed $Ca_{0.65}Sr_{0.30}Yb_{0.05}F_{2.05}$ (a, [19]) and CaF_2:Tm (b,c, [44]) samples.

4. Fluoride ceramics as optical medium

Samples of CaF_2 optical ceramics prepared by hot-pressing were studied by [24, 34, 39-43] as the model object. Their CaF_2 samples appeared to be practically the same as in [2] (Figure 2b). For comparison, [39-43] have also used standard CaF_2 single crystals from the State Optical Institute, St. Petersburg, Russia, and CaF_2 natural optical ceramics specimen (Suranskoe deposit, Urals Mountains; average crystalline domain size about 37 nm). The latter ceramic sample was unique natural object formed approximately one billion years ago.

UV/visible and IR absorption spectra of the synthesized and natural CaF_2 ceramics (curves *1* and *2*) as well as CaF_2 single crystal (curves *3*) are presented in Figure 7(*a*, *b*). The transmission of the ceramics samples is somewhat inferior to that of the single crystal in the near-UV part of the spectrum, however, at visible and IR wavelengths, the corresponding transmission values coincide or almost coincide.

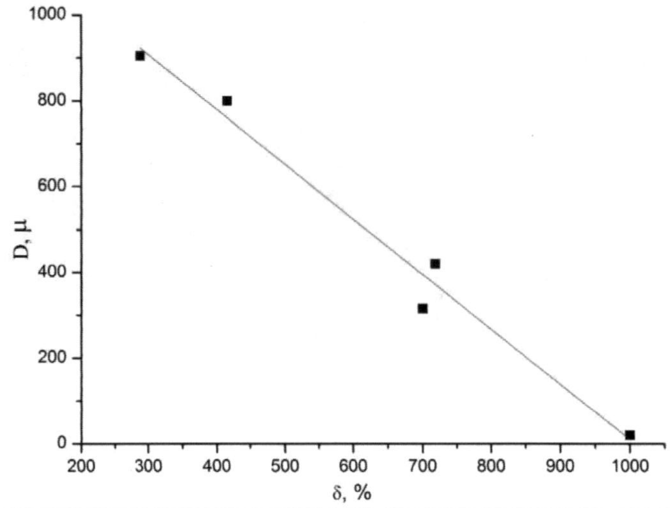

Figure 6. Size of crystalline domains vs. deformation degree of BaF_2 single crystal (δ) at 800°C; $\delta = (h_o-h)/h\cdot 100$, %, where h_o – height of the initial sample and h – height of the formed plate.

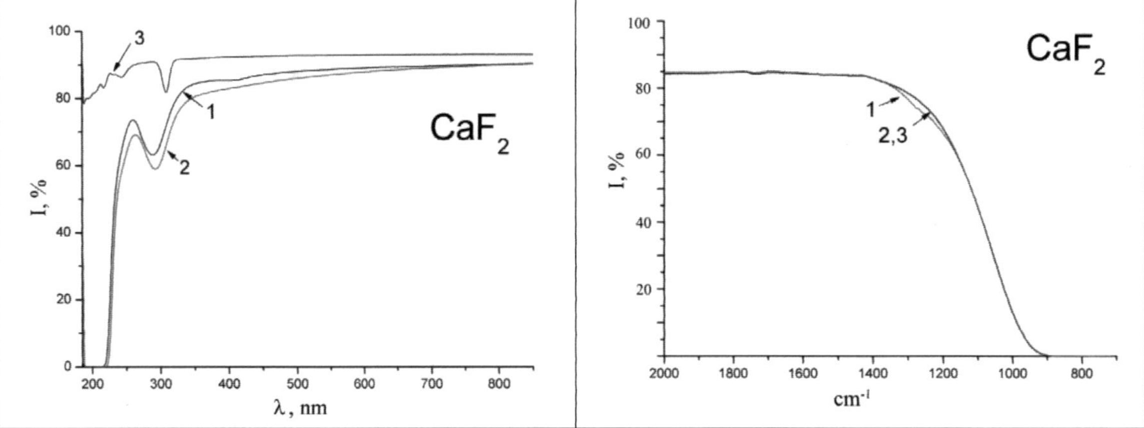

Figure 7. Optical transparency of synthesized CaF_2 ceramics (thickness = 8.8 mm) (1, 2) and CaF_2 single crystal (thickness = 8.3 mm) (3) [34].

Two different methods were used to measure small optical losses in synthesized CaF_2 ceramics under monochromatic laser radiation. One method was specifically designed to estimate the losses in optical fibers and was based on measuring the evolving heat from the weakly absorbing medium when a laser beam passed through it [24]. The second method, used by Palashov et al. [41, 42], measured the thermal depolarization of laser radiation in an absorbing medium. Both types of measurements were carried out at 1.06 micron wavelength. The first method gave a loss value of $9 \cdot 10^{-3}$ cm^{-1}, whereas the second method produced a value of $(1.1\text{-}1.5) \cdot 10^{-3}$ cm^{-1}. Both numbers have indicated a high transparency of the prepared ceramic samples suitable for laser materials. Low values of the optical losses in synthesized CaF_2 ceramics was determined by the specific structure of the boundaries between crystalline grains.

The study of transport of sub-THz heat acoustic phonons in CaF_2 single crystals and ceramics showed the length of their free path at 3.82 K to be equal to $1.2 \cdot 10^{-1}$; $6.6 \cdot 10^{-2}$; $1.4 \cdot 10^{-2}$ cm respectively for single crystal, synthesized and natural ceramic samples [43]. The actual thickness of the above boundary layers between crystalline domains in natural ceramics has been estimated to be about 1.5 ± 0.3 nm. At the same time, actual samples of the synthesized CaF_2 ceramics exhibited properties with zero or almost no boundary layer.

A comparative study of the thermal conductivity of CaF_2 single crystal, synthesized and natural ceramic samples has shown that all have practically the same properties over the 50-300 K temperature interval [39] (Figure 8).

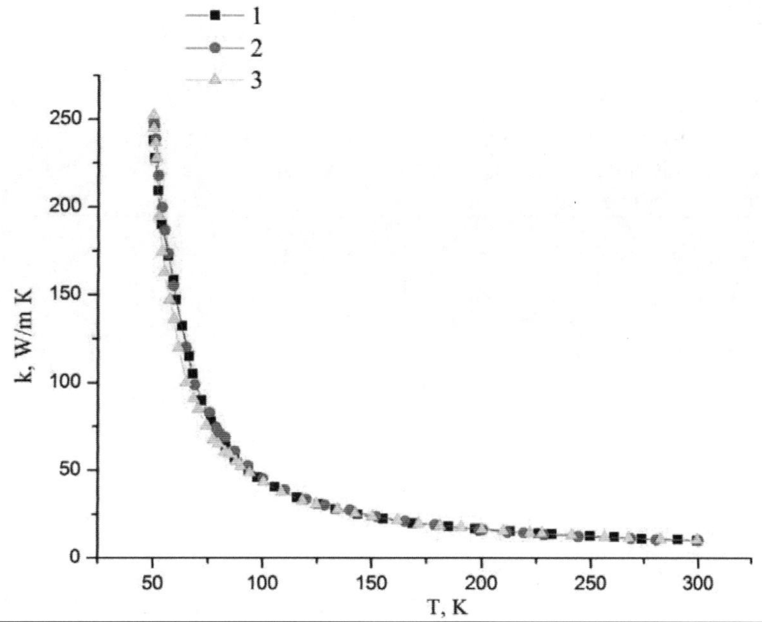

Figure 8. Thermal conductivity of CaF_2 samples: natural sample from Suranskoe deposit, Urals Mountains (1), single crystal (2) and synthesized optical ceramics (3) [40].

Carnall [2] has found that the refraction and optical dispersion coefficients of CaF_2 single crystal and synthesized ceramic specimens were practically the same. This was confirmed for the pure CaF_2 as well as doped CaF_2 single crystals and ceramics samples [19].

In order to characterize mechanical properties of the prepared CaF_2 ceramic samples, their microhardness and fracture toughness were studied by Akchurin et al. [23, 39] (Table 1). These samples were subjected to the mechanical load, followed by measuring the linear size of the formed radial cracks (*C*), which appeared near the site of load application, as a function of the load on the indenter. Calculations of the coefficient of stress intensity (fracture toughness) K_{1C} exhibited the following correlation:

$$K_{1C} = 0.016(E/H)^{1/2}P/C^{3/2},$$

where H is the microhardness, P is the load, and E is Young's modulus (Table 1). K_{1C} values for the synthesized and natural CaF_2 ceramic specimens were more than 3 and 4 times higher than K_{1C} values for single crystals.

In general, fluoride ceramic samples appear to have sufficiently good optical properties for use in laser manufacturing, while their mechanical properties absolutely exceed those of single crystals.

5. Spectral luminescence and generation properties

Lithium fluoride crystals with aggregate color centers are among the most promising laser materials for smoothly tunable generation of pico- and femto-second pulses in the near-IR region. The broad enhancement band in the 1-1.3 micron range, high cross-section of luminescence transition (ca. 10^{-17} cm^2), broad absorption band for pumping by lasers irradiating near 1 micron and high thermal conductivity make these crystals virtually ideal for application in tunable lasers and lasers with mode synchronization. Since laser diodes with radiation wavelength of 960-980 nm have become accessible recently, efficient and compact solid state tunable lasers can now be created with the use of lithium fluoride crystals. However, the latter crystals have some serious drawbacks, including, first of all, their low mechanical strength that restricts the limiting power of generated radiation. The creation of lithium fluoride optical ceramics with color centers can eliminate this obstacle, and LiF laser materials with the F_2^- color centers were produced by the irradiation of pre-synthesized LiF optical ceramics.

Comparison of lasing of LiF-F_2^- ceramics and single crystals of the similar optical density was carried out [20, 21].

Absorption, luminescence and generation spectra of LiF-F_2^- ceramics are shown in Figure 9. Figure 10 shows the dependence of the average lasing output power of ceramic and single crystal LiF F_2^- samples [21]. Ceramic samples were characterized by lower lasing thresholds than single crystals. The maximum differential lasing efficiency of 26% has been observed for a 4 mm thick ceramic sample (vs. 18% for the single crystal sample).

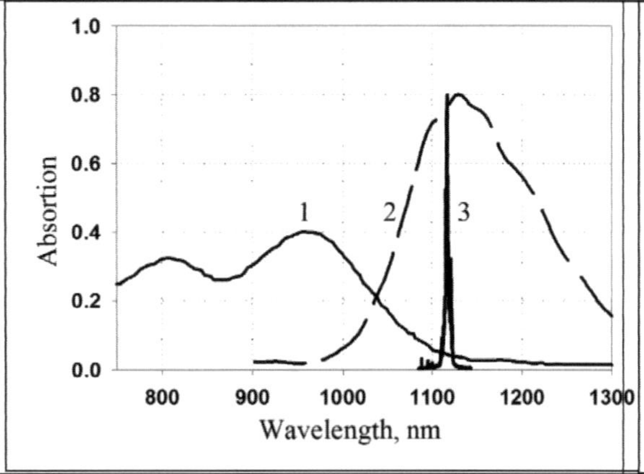

Figure 9. Absorption (1), luminescence (2) and induced (3) spectra of the LiF ceramics with color centers F_2^- [20, 21].

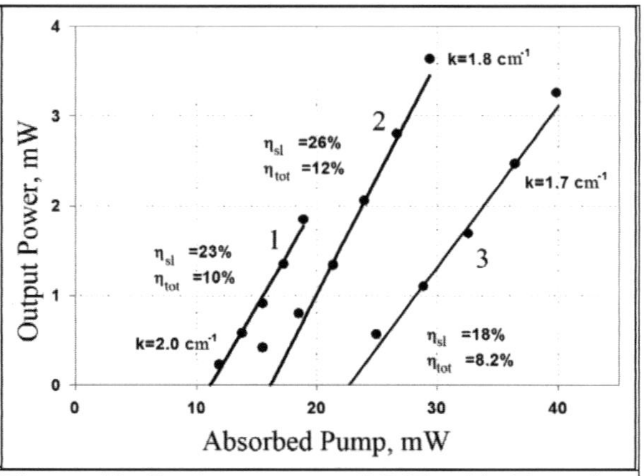

Figure 10. Lasing output of LiF with F_2^- color centers (laser diode pumping, λ=967 nm, τ=1ms, f=5Hz): 1 – 1mm ceramics, 2 – 4 mm ceramics, 3 – 7 mm single crystal [20, 21].

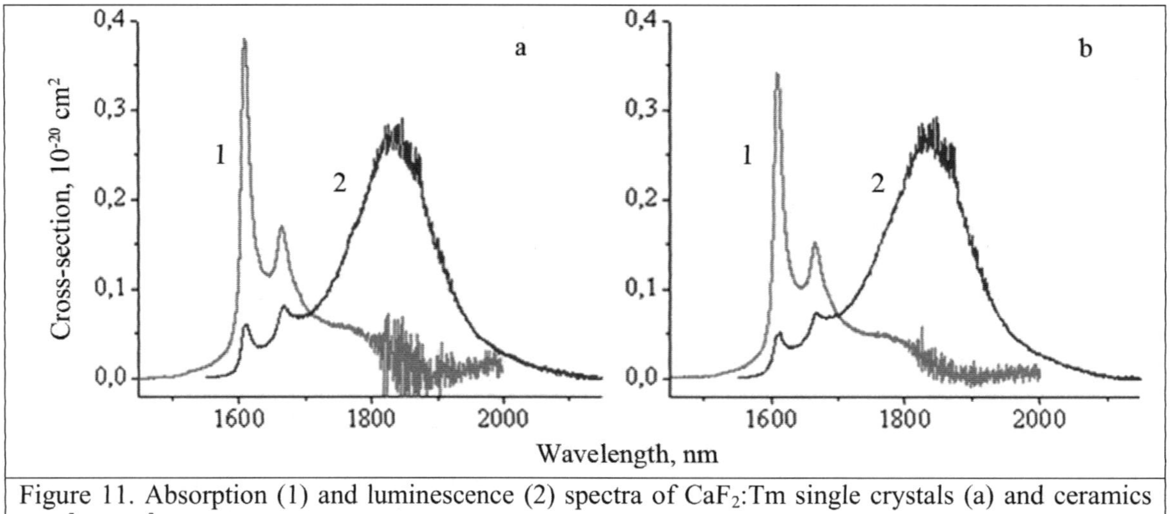

Figure 11. Absorption (1) and luminescence (2) spectra of CaF_2:Tm single crystals (a) and ceramics (b) ($^3H_6 \leftrightarrow {}^3F_4$ transition, T=300K) [44].

Absorption and luminescence spectra of fluoride ceramics and single crystals doped with rare-earth ions (Er, Yb, Tm) are essentially identical [34, 44] (Figure 11). Their optical centers' spectral compositions, lifetimes of the excited states, absorption coefficients, wavelengths of absorption bands and luminescence lines are very close for the samples of the same chemical composition. Excellent quality of rare earth-doped fluoride ceramic specimens, prepared by hot-forming of crystals, allows achieving lasing parameters almost equivalent to the properties of the single crystals of the same chemical composition (Figure 12).

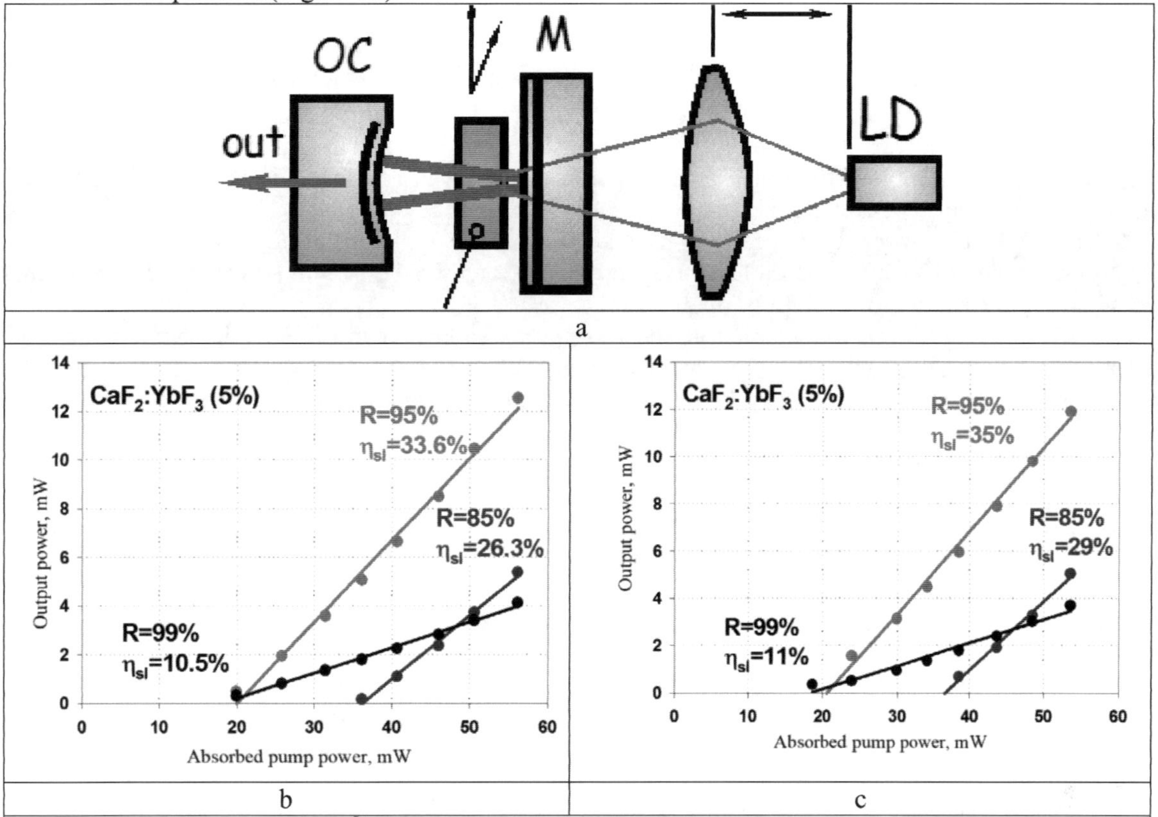

Figure 12. Lasing of $CaF_2:Yb^{3+}$ (5%) samples [19]. Optical scheme for studying of the generation properties (a). Output power vs. input (utilized) power for single crystal (b) and ceramics (c) for various reflection coefficients of the output mirror.

Generation properties of $CaF_2:Yb^{3+}$ single crystals and ceramics were studied by Doroschenko [19] with the use of IPG laser diode pumping (100 micrometer fiber radiation output; impulse mode; 2 msec duration of impulse at 10 GHz repetition frequency). Radiation wavelength (967 nm at room temperature), generated by laser diode, was thermally adjusted to coincide with the absorption maximum of Yb^{3+} ions in CaF_2 matrix (972 nm). Output beam was focused by the single lens (focus distance = 5 mm). Semispherical resonator for measuring of generation properties was formed by flat dichroic mirror (maximum transparency at the generation wavelength of 972 nm and maximum reflection at the generation wavelength of 1020-1080 nm) and spherical output mirror (50 mm radius) (Figure 12a). $Ca_{0.95}Yb_{0.05}F_{2.05}$ experimental sample, used for studying lasing generation, was doped with 5 mol. % Yb^{3+} ions [19].

Output parameters of $Ca_{0.95}Yb_{0.05}F_{2.05}$ single crystal (Figure 12b) and ceramic (Figre 12c) specimens for the different reflection coefficients (as a function of the absorbed pump power) are very close. The highest differential efficiency coefficient of 35% has been obtained for $Ca_{0.95}Yb_{0.05}F_{2.05}$ ceramics at 95% reflection coefficient of the output mirror.

Usually, the grains of fluoride ceramics are too large to produce some kind of size effects and affect spectral kinetic properties related to path lengths and trajectories of electrons, excitons and phonons. However, borders between domains as well as twinning borders within crystalline grains can cause the difference in the properties of ceramic and crystal samples. Such differences have been observed for LiF lasing [21], CaF_2 color center formation [45], and scintillation features of pure BaF_2 and Ce-doped BaF_2:Ce: ceramic materials produced higher photon yield in the case of X-ray luminescence [46, 47].

6. $CaF_2:Yb^{3+}$ System

By now, lasing has been observed for $CaF_2:R^{3+}$, where R = Yb, Nd, Tm [19, 25, 44, 48], $SrF_2:R^{3+}$, where R = Nd ([49], Pr [50] and $Ca_{0.65}Sr_{0.30}Yb_{0.05}F_{2.05}$ [23] ceramic (polycrystalline) samples. However, most efforts are concentrated on $CaF_2:Yb^{3+}$ ceramics. Ytterbium(III) ions, compared to the widely utilized neodymium(III) ions, have numerous advantages as ions-activators, when they are used in diode-pumped laser materials [51]:

- Yb^{3+} ions have lower difference between pumping wavelength and lasing wavelength (980 and 1,030 nm), than Nd^{3+} (808 and 1060 nm, respectively), i.e., lower "quantum defect", and as a result heat losses are typically lower for Yb^{3+} materials;
- electron energy levels of ytterbium used in lasing are simpler than those of neodymium, so the former has fewer competitive pathways ("parasite channels") for energy loss;
- ytterbium exhibits wider lines and absorption bands than neodymium in electron spectra, which is better for short impulse generation and for use in adjustable lasers;
- lifetime of the highest excitation energy level is higher for ytterbium than for neodymium.

Whereas laser applications (especially in the powerful laser systems) of $CaF_2:Yb^{3+}$ single crystals have been actively studied recently [52 - 64], one of the first spectroscopic investigations of ytterbium admixtures in CaF_2 crystals was done as early as in 1969 [65].

Phase diagram of CaF_2-YbF_3 system (Figure 13) has been built according to Fedorov, Sobolev, Greis et al. data [66-69] and by using an analogy with CaF_2-YF_3 system ([70] in the area of lower temperature. It contains area of $Ca_{1-x}Yb_xF_{2+x}$ solid solution, that stretches up to 41 mol.% YbF_3 at its maximum width. Experimental measurements have confirmed that ytterbium melt-crystal distribution coefficient is about 1 for the area with the lower YbF_3 concentration in the aforementioned system. Increase of ytterbium concentration makes melting of the solid solution clearly incongruent; whereas ytterbium distribution coefficient turns to be lower than 1. Once YbF_3 content exceeds 10-15 mol.%, growth of $Ca_{1-x}Yb_xF_{2+x}$ single crystals of laser quality becomes very problematic due to concentration overcooling and instability of the crystallization front [27].

Crystal lattice parameter of $Ca_{1-x}Yb_xF_{2+x}$ solid solution cubic phase depends on the sample composition in a linear manner [66]:

$a = 5.463 + 0.112x$ [Å],

where x is a molar part of YbF_3 in CaF_2-YF_3 system. Changes in the crystal lattice parameters of $Ca_{1-x}Yb_xF_{2+x}$ solid solution phase are determined by heterovalent isomorphic substitution of calcium cations with yttrium accompanied with charge compensation via insertion of additional fluoride anions

$Ca^{2+} \Rightarrow Yb^{3+} + F_{int}^{-}$

Replacement of the relatively large calcium cations with smaller ytterbium cations is compensated by insertion of fluoride anions and leads to the smaller change of the crystal lattice parameter under variation of the sample chemical composition.

For the lower concentrations of ytterbium (ca. 0.01 mol.%), lattice defects of the opposite charge form dipole couples [69, 70] [71]; [72].

$Ca_{1-x}Yb_xF_{2+x}$ solid solution decomposes at the lower temperature with formation of Ca_2YbF_7, $Ca_9Yb_5F_{33}$, $Ca_{17}Yb_{10}F_{64}$ and $Ca_8Yb_5F_{31}$ fluorite-derived phases existing in 33-42 mol.% YbF_3 composition interval. Ca_2YbF_7, $Ca_9Yb_5F_{33}$, $Ca_{17}Yb_{10}F_{64}$ and $Ca_8Yb_5F_{31}$ phases contain Yb_6F_{37} clusters naturally and coherently built in their crystal lattices [73-77] (Figure 14), as it has been proven by ESR technique [78]. Cations in these clusters maintain the same coordination number 8 as they had it in the original fluorite crystal lattice before the replacement of the regular cubes by Thompson's antiprisms. Difference in size between Yb_6F_{37} cluster and replaced by it corresponding part of fluorite-type crystal lattice leads to the internal local strain and deformation within composite material. Difference between the cluster formal charge and formal charge of the replaced part of fluorite lattice causes addition of extra fluoride anions to the crystal matrix along with formation of Yb_6F_{37} fragment:

$$(M_6F_{32})^{20-} \Rightarrow (R_6F_{37})^{19-} + F_{int}^{-} \qquad (3)$$

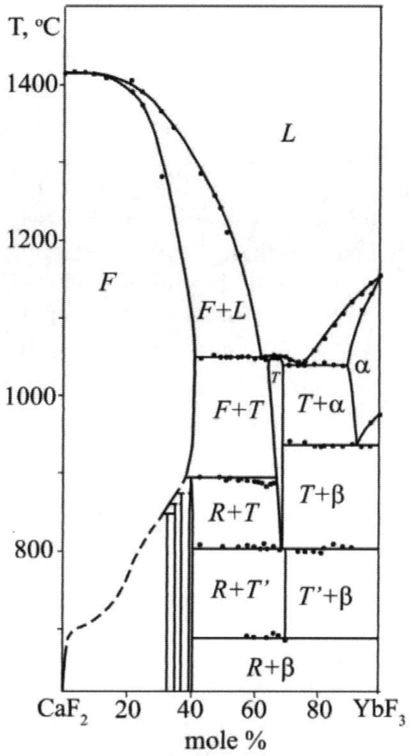

Figure 13. Phase diagram of CaF_2-YbF_3 system [66-70]: L- melt, F – $Ca_{1-x}Yb_xF_{2+x}$ solid solution, T – $Yb_{1-y}Ca_yF_{3-y}$ solid solution, T' – $Yb_7Ca_3F_{27}$, R – $Ca_8R_5F_{31}$.

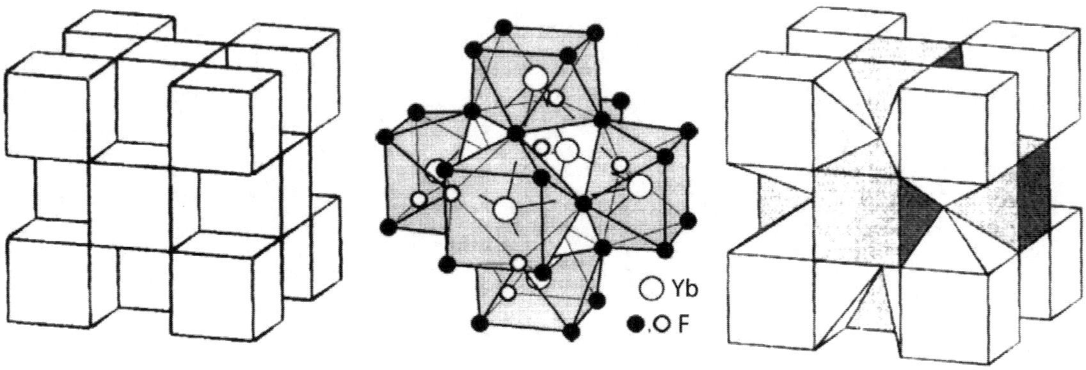

Figure 14. CaF_2 crystal lattice unit (a), Yb_6F_{37} cluster (b) and its positioning in fluorite crystal lattice (c).

Absorption bands in $Ca_{1-x}Yb_xF_{2+x}$ solid solution spectra, assigned to association of the couples of Yb^{3+} ions or combination of several complex optical centers [56, 65], more likely shall be attributed to the spectral properties of Yb_6F_{37} cluster, for these bands stay unchanged in a very wide concentration interval of ytterbium. The latter also supports suggestion regarding relatively high Yb_6F_{37} cluster stability.

Preparation of $Ca_{1-x}Yb_xF_{2+x}$ ceramics and single crystals under reducing conditions (e.g., by contacting graphite crucible surface) is accompanied with partial conversion of Yb^{3+} to Yb^{2+} ions and subsequent appearance of the corresponding absorption bands in UV segment of electron spectrum [54, 79] and possible thermal non-radiative losses under lasing [62].

It is very important to keep in mind that according to the existing CaF_2-YbF_3 phase diagram $Ca_{1-x}Yb_xF_{2+x}$ solid solution exists under non-equilibrium conditions at room temperature. Usually, transformations in these solid solutions with a few mol. % of lanthanide trifluoride are extremely slow, so no changes in the samples could have been detected over several decades. This may be related to very low values of the cations' diffusion coefficients and relatively high activation energy barriers for the formation of the centers for crystallization of the new phase. These factors, however, can cause certain problems for the ceramic materials designated for the use under high energy flow conditions. Phase transformation also can be initiated at the boundaries between twinning domains. One can describe border between twinning domains in fluorite-type crystal lattice as 2-layer cation pack instead of the regular 3-layered matrix, corresponding to the closest packing of cations in fluorite-type crystal lattice. These borders can play role of the nucleation centers for the formation of the new tysonite-type phase (low temperature orthorhombic modification of β-YF_3), but this suggestion requires further studying.

Earlier, alleged equal thermal conductivity of both laser ceramics and pure fluorite was considered as an important advantage of ceramic materials [53, 58, 60]. However, actual measurements of this parameter disproved this assumption and demonstrated that introduction of heterovalent admixture to fluorite structure leads to the dramatic decrease of the thermal conductivity, especially, at lower temperature. The reason of such essential decrease of thermal conductivity in the case of samples with heterovalent isomorphic doping vs. specimens with isovalent substitutions is that scattering of acoustic phonons occurs at defect clusters for materials with heterovalent doping instead of the single atoms in the isovalent-substituted samples (Figure 14).

Ceramics with the relatively high concentration of trifluoride dope exhibits glass-type temperature dependency of its thermal conductivity [80]. Comparison of the thermal conductivity of $Ca_{1-x}Yb_xF_{2+x}$ solid solution and ytterbium-doped Al-Y garnet [12] (Figure 15) shows that fluoride ceramics is inferior to oxide material. Decrease of ytterbium content in $Ca_{1-x}Yb_xF_{2+x}$ leads to the increase of its thermal conductivity, but relatively insufficient improvement of this property significantly limits prospective use of CaF_2:Yb^{3+} materials, including ceramics, for the powerful lasers.

Properties of various CaF_2:Yb ceramics are presented in Table 2.

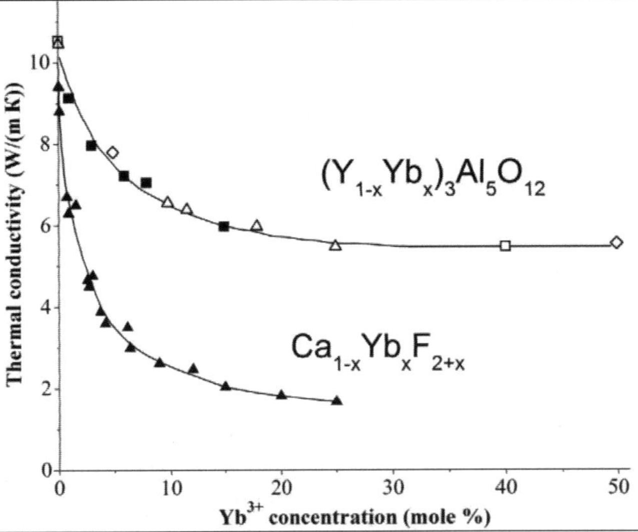

Figure 15. Thermal conductivity of $(Y_{1-x}Yb_x)Al_5O_{12}$ [12] and $Ca_{1-x}Yb_xF_{2+x}$ (P.A.Popov, [80] and private communication) solid solutions at 300 K.

Journal of Physics: Conference Series **345** (2012) 012017 doi:10.1088/1742-6596/345/1/012017

7. Synthesis of prospective materials

Crystal symmetry of laser ceramics plays a decisive role in prospective application of these materials. Both fluoride and oxide laser ceramics have cubic crystal lattices. Oxides have structure of garnet (*Ia3d* space symmetry group, or SSG) or yttrium oxide (*I*a3 SSG), whereas fluorides are represented by CaF_2, SrF_2 or their combinations (fluorite-type solid solutions) and lithium fluoride (all of fluorides belong to *Fm*3*m* SSG). These cubic crystals are very likely to form boundaries between crystalline domains that correspond to polysynthetic twinning and, therefore, produce low optical losses. From this point of view, cubic KY_3F_{10} (*Fm*3*m* SSG) could be a very good candidate for the same laser application.

Similar suggestion can be true for CaF_2:Yb,Na solid solution, too [57]. Fluorite, double-doped with sodium and ytterbium, does not produce good quality single crystals (as per phase diagram for NaF-CaF_2-YbF_3 system [81], thus ceramic CaF_2:Yb,Na looks like a very promising alternative.

Compounds with non-cubic crystal lattices do not form laser-quality ceramic materials. Because of anisotropic refraction coefficients and light scattering at the incoherent boundaries between crystalline domains, their samples exhibit essential optical losses. One could produce acceptable values of light transmittance for Y_2O_3-based Yttralox crystals, but similar result was impossible to achieve for Al_2O_3-based Lukalox ceramics. Oxosulfides of the rare earth elements also cannot be used for manufacturing of transparent laser ceramics [34]. The same is true for MgF_2 (tetragonal system, *P*42/*mnm* SSG), YF_3 (orthorhombic system, *Pnma* space group), $LiRF_4$ (tetragonal system, $I4_1$/a SSG), BaR_2F_8 (monoclinic system, C2/m SSG) fluorides, too.

As it has been already mentioned above, CaF_2-based ceramics exhibits twinning at the nano-level and, therefore, generates low optical losses because of perfectly well-structured boundaries between crystalline domains. Polysynthetic twinning is known for the phase transitions of the second type (or close to the second type), when lower-temperature modifications are slightly distorted in comparison with more symmetric higher-temperature modifications. This type of twinning with lowering crystal lattice symmetry is especially common for the phase transitions of the higher-temperature cubic phases. As an example of this phenomenon, one can mention ferroelectrics [82]. In particular, A_2BRF_6 elpasolites (A, B- alkaline metals with $r_A > r_B$, R – rare earth elements) undergo multiple phase transitions at different temperatures [83], and Rb_2NaAlF_6 transparent ceramics has been synthesized in [84] by cold-pressing of elpasolite precursor produced by sol-gel method. Another example of the same kind is neighborite $NaMgF_3$ [85]; [86]. This compound (melting point 1020°C) has an orthorhombic crystal lattice at room temperature, but at 700°C it undergoes phase transition and forms high-temperature cubic modification. Preliminary studies have shown an opportunity to synthesize transparent twinned polycrystalline samples at room temperature [19].

Ordered fluorite-type phases in MF_2-RF_3 systems [28, 66, 73] also have slightly-distorted CaF_2 structures (Fig. 13) and, therefore, they may be considered as prospective materials for transparent laser ceramics. For example, optically homogeneous crystal domains of trigonally-distorted fluorite-type $Ca_8R_5F_{31}$ have been observed in polished cross-sections of two-phase samples in CaF_2-RF_3 systems by [87].

However, one can make highly transparent ceramics from non-cubic crystalline substances only of crystalline domains will be less than the wavelength of the used electromagnetic radiation [88].

8. Conclusion

The reviewed publications show that modern fluoride ceramics technology provides materials with high optical transparency and homogeneity, so they can be widely used as laser materials.

40+ years ago Kodak scientists published their first studies in this area, and recently researchers from A. M. Prokhorov General Physics Institute (Moscow, Russia), State Optical Institute (St. Petersburg, Russia) and INKROM Company (St. Petersburg, Russia) developed new synthetic methods to manufacture fluoride laser ceramics with properties similar to the properties of corresponding single crystals, thus making fluoride laser ceramics the most prospective and easily available material for the solid-state lasers of various types and different designation.

Polycrystalline fluoride laser materials were prepared by hot-forming of single crystals instead of more conventional methods. However, use of thermal treatment of starting batches and similarity of the structure of the obtained polycrystalline samples and corresponding properties of the substances prepared by "classic" protocols makes possible application of "ceramic laser materials" term to the products of hot- deformation synthesis.

Pure CaF_2 ceramics possesses excellent mechanical and perfect optical properties. Therefore, making doped fluorites of the comparable properties becomes a priority. The co-precipitation from water medium does not produce satisfactory results, so one can consider preparation of their precursor materials from non-aqueous solutions.

9. Acknowledgements

The author wishes to express his personal thanks to V. V. Osiko, E. A. Garibin, and I. A. Mironov for the fruitful discussions, A. I. Popov and E. V. Chernova for the preparation of this manuscript.

References

[1] Vydrik G A, Solov'eva T V, and Kharitonov F Ya 1980 *Prozrachnaya keramica* (*Transparent ceramics*), Moscow: Energy (in Russian)
[2] Carnall E, Hatch S E, and Parsons W E 1966 *Mater Sci Res.* **3** 165.
[3] Swinehart C F, and Packer H 1978 US Patent 4,089,937.
[4] Volynets F K 1973 Preparation methods, structure and physico-chemical properties of optical ceramics *Optiko-mekhanicheskaya prom-st [Optical Mechanical Industry]* **9** 48-51 (in Russian)
[5] Volynets F K 1973 Optical properties and application of optical ceramics *Optiko-mekhanicheskaya prom-st [Optical Mechanical Industry]* **10** 47 (in Russian)
[6] Hatch S E, Parsons W F, and Weagley R J 1964 Hot-pressed polycrystalline $CaF_2:Dy^{2+}$ laser *Applied Physics Letters* **5** (8) 153
[7] Carnall E, Hatch S E, Parsons W F, and Weagley R J 1969 *Hot-pressed polycrystalline laser materials* US Patent 3,453,215
[8] Greskovich C, and Chernoch J P 1973 Polycrystalline ceramic lasers *J. Appl. Phys.* **44** (10) 4599
[9] Greskovich C, and Chernoch J P 1974 Improved polycrystalline ceramic lasers *J. Appl. Phys.* **45** (10) 4495
[10] Ikesue A, Kinoshita T, Kamata K, and Iosida K 1995 *J. Amer. Ceram. Soc.* **78** 1033
[11] Lu J, Ueda K, Yagi H, Yanagitani T, Akiyama Y, and Kaminskii A 2002 *J. Alloy and Compounds* **341** 220
[12] Sanghera J, Shaw B, Kim W,Villalobos G, Baker C, Frantz J, Hunt M, Sadowski B and Aggarwal I 2011 *Proc. SPIE* **7912** 79121Q-1
[13] Sanghera J, Shaw B, Kim W, Villalobos G, Baker C, Frantz J, Hunt M, Sadowski B, and Aggarwal I 2011 Ceramic laser materials *Proc. SPIE* **7912** 79121Q-1.
[14] Grass R N, and Stark WJ 2005 Flame synthesis of calcium-, barium fluoride nanoparticles and sodium chloride *Chem. Commun.* 1767
[15] Ishizawa H 2005 *Proc. 13-th Inter. Workshop on Sol-Gel Sciences Technology,* Los Angeles, CA, 289
[16] Mortier M, Bensalah A, Dantelle G, Patriarche G, and Vivien D, 2007 Rare-earth doped oxyfluoride glass-ceramics and fluoride ceramics: Synthesis and optical properties *Optical Materials* **29** 1263
[17] Aubry P, Bensalah A, Gredin P, Patriarche G, Vivien D, and Mortier M 2009 *Optical Materials,* **31** 750
[18] Fedorov P P, Osiko V V, Basiev T T, et al. Report No 4 (2006), Russian state contract No 02.435.11.2011 Development of technology for nanoceramics with low optical losses on the basis of metal fluorides doped with rare-earth elements (in Russian) 212p.

[19] Fedorov P P, Osiko V V, Basiev T T, et al. Report No 4 (2009), Russian state contract No 02.513.12.3029 "Nanoceramics on the base of compounds with high optical transmission in the mid-IR range for coherent and incoherent light sources" (in Russian) 72p.

[20] Basiev T T, Doroshenko M E, Fedorov P P, Konuyshkin V A, Kouznetsov S V, Osiko V V $LiF{:}F_2^-$ color center ceramic laser material *3rd Laser Ceramics Symposium: Int. Symposium on Transparent Ceramics for Photonic Applications* 2007 October 8-10 Paris Abstract O-L-1

[21] Basiev T T, Doroshenko M E, Konuyshkin V A, Osiko V V, Ivanov L I, and Simakov S V 2007 Lasing in diode-pumped fluoride nanostructure $F_2^-{:}LiF$ color centers ceramics *Quantum electronics* **37**(11) 989

[22] Basiev T T, Doroshenko M E, Kouznetsov S V, Konuyshkin V A, Osiko V V, Fedorov P P 2007 *Ceramic laser microstructured material with twinned nanostructure and method of manufacture* RU Patent No 2358045

[23] Basiev T T, Doroshenko M E, Fedorov P P, Konuyshkin V A, Kouznetsov S V, Osiko V V, Akchurin M Sh (2008), "Efficient laser based on CaF_2-SrF_2-YbF_3 nanoceramics", *Optical Letters*, 33(5), 521-523.

[24] Basiev T T, Doroshenko M E, Konyushkin V A, Osiko V V, Fedorov P P, Demidenko V A, Dukelrskii K V, Mironov I A, and Smirnov A N 2008 Fluoride optical nanoceramics *Russian Chemical Bulletin* International Edition **57** (5) 877

[25] Alimov O K, Basiev T T, Doroshenko M E, Fedorov P P, Konuyshkin V A, Kouznetsov S V, Nakladov A N, Osiko V V, Shlyakhova O V Spectroscopic and Laser Properties of Yb^{3+} ions in BaF_2-SrF_2-CaF_2 Single Crystals and Nanoceramics 2008 July 7-11 *The 15th Int. Conf. on Luminescence and Optical Spectroscopy of Codence Matter.* Lyon 147

[26] Fedorov P P, Osiko V V 2005 Crystal Growth of Fluorides In: Bulk *Crystal Growth of Electronic, Optical and Optoelectronic Materials*. Ed. P.Capper. Wiley Series in Materials for Electronic and Optoelectronic Applications. John Wiley & Son, Ltd. 339-356

[27] Kuznetsov S V, and Fedorov P P 2008 Morphological Stability of Solid-Liquid Interface during Melt Crystallization of Solid Solutions $M_{1-x}R_xF_{2+x}$ *Inorganic Materials* **44** (13) 1434 (Supplement)

[28] Sobolev B P 2000 *The Rare Earth Trifluorides* p.1 *The High-temperature Chemistry of the Rare Earth Trifluorides* Institute D'Estudis Catalans Barcelona 520p.

[29] Fedorov P P, Kuznetsov S V, Voronov V V, Yarotskaya I V, Arbenina V V 2008 Soft chemical Synthesis of $NaYF_4$ Nanopowders *Russian J Inorg Chemistry* **53**(11) 1681

[30] Kuznetzov S V, Osiko V V, Tkatchenko E A, Fedorov P P 2006 *Russian Chem. Rev.* **75**(12) 1065

[31] Bensalah A, Mortier M, Patriarche G, Gredin P, and Vivien D 2006 *J. Solid. State. Chem.* **179** 2636-44

[32] Kouznetsov S V, Basiev T T, Voronov V V, Lavristchev S V, Osiko V V, Tkatchenko E A, Fedorov P P, Yarotzkay I V 2006 Preparation of nanoparticles of $M_{1-x}R_xF_{2+x}$ from water solutions. Proceedings of ISIF-2006. *The Second International Siberian Worksop INTERSIBFLUORINE- «Advanced Inorganic Fluorides»* Tomsk 2006 June 11-16, 124

[33] Kouznetsov S V, Yarotzkay I V, Fedorov P P, Voronov V V, Lavristchev S V, Basiev T T, and Osiko V V 2007 Preparation of Nanopowdered $M_{1-x}R_xF_{2+x}$ (M=Ca, Sr, Ba; R=Ce, Nd, Er, Yb) Solid Solution *Russian J. Inorg. Chem.* **52** 315

[34] Fedorov P P, Osiko V V, Basiev T T, Orlovskii Yu V, Dykel'skii K V, Mironov I A, Demidenko V A, Smirnov A N 2007 Optical Fluoride and Oxysulphide Ceramics: Preparation and Characterization. In: *Developments in Ceramic Materials Research,* NOVA Science Publishers Inc. 53-95

[35] Kuznetsov S V, Fedorov P P, Voronov V V, Samarina K S, Ermakov R P, and Osiko V V 2010 Synthesis of $Ba_4R_3F_{17}$ (R stands for Rare-Earth Elements) Powders and Transparent Compacts on Their Base *Russian J. Inorg. Chem.* **55**(4) 484

[36] Basiev T T, Bolyasnikova L S, Demidenko V A, Fedorov P P, Kouznetsov S V, Mironov I A,

Ovsyannikova O P, Osiko V V, and Akchurin M Sh 2006 Fluoride nanoceramics *National Russian Conference on Crystal Growth 12*. Moscow 2006 October 23-27 Abstract 433 (in russian)

[37] Lyberis A, Patriarche G, Gredin P, Vivien D, and Mortier M 2011 Origin of light scattering in ytterbium doped calcium fluoride transparent ceramic for high power lasers *J. European Ceramic Society* **31** 1619

[38] Akchurin M Sh, Gainutdinov R V, Garibin E A, Golovin Yu I, Demidenko A A, Dukel'skii K V, Kuznetsov S V, Mironov I A, Osiko V V, Smirnov A N, Tabachkova N Yu, Tyurin A I, Fedorov P P, and Shindyapin V V 2011 Nanostructure of Optical Fluoride Ceramics *Inorganic Materials*: *Applied Research* **2**(2) 7

[39] Akchurin M Sh, Gainutdinov R V, Smolyanskii P L, and Fedorov P P 2006 Anomalously High Fracture Toughness of Polycrystalline Optical Fluorite from the Suran Deposit (South Urals) *Doklady Physics* **51**(1) 10

[40] Popov P A, Dykel'skii K V, Mironov I A, Demidenko V A, Smirnov A N, Smolyanskii P L, Fedorov P P, Osiko V V, and Basiev T T 2007 Thermal conductivity of CaF_2 optical ceramics, *Doklady Physics* **52**, 7

[41] Palashov O V, Khazanov E A, Mukhin I B, Mironov I A, Smirnov A N, Dukel'skii K V, Fedorov P P, Osiko V V, and Basiev T T 2007 Comparison of the optical parameters of a CaF_2 single crystal and optical ceramics *Quantum Electronics* **37**(1) 27

[42] Palashov O V, Khazanov E A, Mukhin I B, Smirnov A N, Mironov I A, Dukel'skii K V, Garibin E A, Fedorov P P, Kuznetsov S V, Osiko V V, Basiev T T, and Gainutdinov R V 2009 Optical absorption in CaF_2 nanoceramics *Quantum Electronics* **39** (10) 943

[43] Khazanov E N, Taranov A V, Fedorov P P, Kuznetsov S V, Basiev T T, Mironov I A, Smirnov A N, Dukel'skii K V, and Garibin E A 2009 *Doklady Physics* **54**(1) 14

[44] Bol'shchikov F A, Garibin E A, Gusev P E, Demidenko A A, Kruglova M V, Krutov M A, Lyapin A A, Mironov I A, Osiko V V, Reiterov V M, Ryabochkina P A, Sakharov N V, Smirnov A N, Ushakov S N, Fedorov P P 2011 Nanostructured Tm: CaF_2 ceramics: potential gain media for two-micron lasers *Quantum Electronics* **41**(3) 193

[45] Shcheulin A S, Ryskin A I, Angervaks A E, Fedorov P P, Osiko V V, Demidenko A A, Garibin E A, Smirnov A N, Dukel'skii K V, and Mironov I A 2011 *Optics and Spectroscopy* **110** (4) 604

[46] Batygov Kh S, Bolyasnikova L S, Garibin A E, Demidenko V A, Doroshenko M E, Dykelskii K V, Lugingina A A, Osiko V V, and Fedorov P P 2008 *Doklady Physics* **53** (9) 485

[47] Rodnyi P A, Gain S D, Mironov I A, Garibin E A, and Demidenko A A, Seliverstov D M, Gusev Yu I, Fedorov P P, and , Kuznetsov S V 2010 Spectral-kinetic characteristics of crystals and nanoceramics based on BaF_2 and BaF_2: Ce *Physics of the solid state* **52**(9) 1910

[48] Basiev T T, Doroshenko M E, Fedorov P P, Konyushkin V A, Kouznetsov S V, Nakladov A N, Osiko V V 2009 Laser properties of Nd^{3+} and Yb^{3+} ions in CaF_2 Crystals and Ceramics *5th Laser Ceramics Symposium: International Symposium on Transparent Ceramics for Photonic Applications* Bilbao, Spain, 2009 December 9-11 Abstracts

[49] Basiev T T, Doroshenko M E, Konyushkin V A, and Osiko V V 2010 $SrF_2{:}Nd^{3+}$ laser fluoride ceramics *Optics Letters* **35**(23) 4009

[50] Basiev T T, Konyushkin V A, Konyushkin D V, Doroshenko M E, Huber G, Reishert F, Hansen N-O, Fechner M 2011 New Visible $SrF_2{:}Pr^{3+}$ Ceramic Laser at 639 nm *ASSP 2011 Postdeadline Paper AMF* 1

[51] Krupke W F 2000 Ytterbium solid-state lasers. The first decade *IEEE J. of Quantum Electronics* **6** (6), 1287

[52] Lucca A, Dbourg G, Jacquemet M, Druon F, Balembois F, and Georges P 2004 High-power diode-pumped $Yb^{3+}{:}CaF_2$ femtosecond laser *Optics Letters* **29** (23) 2767

[53] Ito M, Goutaudier Ch, Guyot Y, Lebbou Kh, Fukuda T, and Boulon G 2004 Crystal growth, Yb^{3+} spectroscopy, concentration quenching analysis and potentiality of laser emission in

$Ca_{1-x}Yb_xF_{2+x}$ *J. Phys.: Condens. Matter.* **16** 1501

[54] Kaczmzrek S M, Tsuboi T, Ito M, Boulon G, and Leniec G 2005 Optical study of Ym^{3+}/Yb^{2+} conversion in CaF_2 crystals *J. Phys. Condens. Matter.* **17** (25) 3771

[55] Bensalah A, Ito M, Guyot Y, Goutaudier C, Jouini A, Brenier A, Sato H, Fukuda T, and Boulon G 2007 Spectroscopic properties and quenching processes of Yb^{3+} in fluoride single crystals for laser applications *J. of Lumin.* **122-123** 444-446

[56] Petit V, Camy P, Doualan J-L, and Moncorge R, 2007 Refined analysis of the luminescent centers in the Yb^{3+}:CaF_2 laser crystal *J. of Lumin.* **122–123** 5-7

[57] Hraiech S, Jouini A, Kim K J, Guyot Y, Goutaudier C, Yoshikawa A, Trabelsi-Ayadi M, and Boulon G 2007 Breakage of Yb^{3+} pairs by Na^+ in Yb^{3+}-doped CaF_2 laser host *Annales de Physique* **32** (2-3) 59-61

[58] Jouini A, Brenier A, Guyot Y, Boulon G, Sato H, Yoshikawa A, Fukuda K., and Fukuda T 2008 Spectroscopic and Laser Properties of the Near-Infrared Tunable Laser Material Yb^{3+}-Doped CaF_2 Crystal *Crystal Growth & Design* **8** (3) 808-811

[59] Petit V, Voretti P, Camy P, Doualan J-L, and Moncorge R 2008 Active waveguides produced in Yb^{3+}:CaF_2 by H^+ implantation for laser applications *J. Alloys and Compounds.* **451** 68-70

[60] Siebold M, Bock S, Schramm U, Xu B, Doualan J L, Camy P, and Moncorge R 2009 Yb:CaF_2– a new old laser crystal *Appl. Phys. B* **97** 327-338

[61] Hraiech S, Jouini A, Kim K J, Guyot Y, Yoshikawa A, and Boulon G 2010 Role of monovalent alkali ions in the Yb^{3+} centers of CaF_2 laser crystals *Radiation Measurements* **45** 323-327

[62] Druon F, Ricaud S, Papadopoulos D N, Pellegrina A, Camy P, Doualan J L, Moncorge R, Courjaud A, Mottay E, and Georges P 2011 On Yb:CaF_2 and Yb:SrF_2: review of spectroscopic and thermal properties and their impact on femtosecond and high power laser performance *Optical Materials Express* **1** (3) 489-502

[63] Ricaud S, Papadopoulos D N, Pellegrina A, Balembois F, Georges P, Courjaud A, Camy P, Doualan J L, Moncorge R, and Druon F 2011 High-power diode-pumped cryogenically cooled Yb:CaF_2 laser with extremely low quantum defect *Optics Letters* **36**(9) 1602-1604

[64] Pena A, Camy P, Benayad A, Doualan J-L, Maurel C, Olivier M, Nazabal V, Moncorge R 2011 Yb:CaF_2 grown by liquid phase epitaxy *Optical Materials* **33** 1616-1620

[65] Voron'ko Yu K, Osiko V V, and Shcherbakov I A 1969 Optical centers and the interaction of ions Yb^{3+} in cubic fluorite crystals *Sov. Phys.* JETP. **29**(1) 86

[66] Sobolev B P, Fedorov P P 1978 Phase diagramms of the CaF_2 - $(Y,Ln)F_3$ systems. I. Experimental *J. Less-Common. Metals.* **60** 33-46

[67] Lechtenbohmer C, Greis O 1978 Phasenuntersuchungen am system CaF_2-YbF_3 *J. Less-Common. Metals.* **61** (2) 177

[68] Bevan D J M, and Greis O 1978 Fluorine-deficient tysonite-type solid solutions $(Ca, Ln)F_{3-d}$ and related superstructure phases $(Ca_{13}Ln_7)F_{27}$ *Rev. Chim. Miner.* **15** (4) 346

[69] Greis O, and Kieser M 1981 Preparation and characterization of the solid solutions $(Ca,R)F_{2.33}$ with R = Y, La Lu and corresponding superstructure phases Ca_2RF_7 with R = Er- Lu and Y *Z. anorgan. und allgem. Chem.* **479** 165-170

[70] Fedorov P P 2010 Third law of thermodynamics as applied to phase diagrams *Russian J. Inorg. Chem.* **55** (11) 1722-1739

[71] Corish J, Catlow C R A, Jacobs P W M, and Ong S H 1982 Defect aggregation in anion-excess fluorites. Dopant monomers and dimmers *Phys. Rev. (B)* **25** (10) 6425-6438

[72] Murin I V 1992 Relaxation methods for the study of ion transport in halide systems *Solid State Ionics* **53-56** 837-842

[73] Greis O, and Haschke J M 1982 Rare earth fluorides *Handbook on the Physics and Chemistry of Rare Earth.* Ed. by K A Gscheidner & Le Roy Eyring, North-Holland, Amsterdam, N.-Y., Oxford. 5 Ch. 45, 387-460

[74] Bevan D J M, Greis O, and Strahle J 1980 A New Structural Principle in Anion-Excess Fluorite-Related Superlattices *Acta Cryst.* **36** (6) 889-890

[75] Bevan D J M, McCall M J, Ness S E, and Taylor M R 1988 Cuboctahedral anion clusters in fluorite-related superstructures: the structure of Ca_2YbF_7 Part 2 *Eur. J. Solid. State Inorg. Chem.* **25** (5-6) 517-526

[76] Bevan D J M, Ness S E, and Taylor M R 1988 On the crystal chemistry of Ca_2YbF_7 and other closely-related structures with cuboctahedral anion clusters *Eur. J. Solid. State Inorg. Chem.* **25** (5-6) 527-534

[77] Fedorov P P 1991 Association of point defects in non stoichiometric $M_{1-x}R_xF_{2+x}$ fluorite-type solid solutions *Butll. Soc. Cat. Cien.* **12** (2) 349-381

[78] Kazanskii S A, Ryskin A I, Nikiforov A E, Zaharov A Yu, Ougrumov M Yu, and Shakurov G S 2005 EPR spectra and crystal field of hexamer rare-earth clusters in fluorites *Physical Review (B)* **72** 014127

[79] Nicoara I, Pecingina-Garjoaba N, and Bunoiu O 2008 Concentration distribution of Yb^{2+} and Yb^{3+} ions in YbF_3:CaF_2 crystals *Crystal Growth* **310** (7-9) 1476-1481

[80] Popov P A, Fedorov P P, Kuznetsov S V, Konyushkin V A, Osiko V V, Basiev T T 2008 Thermal conductivity of single crystals of $Ca_{1-x}Yb_xF_{2+x}$ solid solutions *Doklady Physics* **53** (4) 198

[81] Fedorov P P, and Rappo A V 2008 NaF-CaF_2-YbF_3 phase diagram *Russian J. Inorg. Chem.* **53** (7) 1126-1129

[82] Zheludev I S 1971 *Physics of Crystalline Dielectrics* **2**, "Electrical Properties", Plenum Press.

[83] Flerov I N, Gorev M V, Aleksandrov K S, Tressaud A, Grannec J, and Couzi M 1998 Phase transitions in elpasolites (ordered perovskites) *Mat. Sci. & Engin.* **R24**(3) 81-151

[84] Ahrens M, Schuchke K, Redmer S, Kemnitz E 2007 Transparent ceramics from sol-gel derived elpasolites by cold pressing *Solid State Sciences* **9** 833-837

[85] Ocafrain A, Chaminade J P, Viraphong O, Cavagnat R, Couzi M, and Pouchard M 1996 Growth by the heat exchanger method and characterization of neighborite, $NaMgF_3$ *J. Crystal. Growth.* **166** 414-418

[86] Yoshiasa A, Sakamoto D, Okudera H, Ohkawa M, and Ota K 2003 Phase relation of $Na_{1-x}K_xMgF_3$ ($0 \leq x \leq 1$) perovskite-type solid solutions *Mater. Res. Bull.* **38** 421-427

[87] Fedorov P P, Izotova O E, Alexandrov V B, Sobolev B P 1974 New phases with fluorite-derived structure in CaF_2 - (Y, Ln)F_3 systems *J. Solid State Chem.* **9** (4) 368-374

[88] Ikesue A, and Aung Y L Ceramic laser materials 2008 *Nature Photonics* **2** 721

Journal of Physics: Conference Series **345** (2012) 012018 doi:10.1088/1742-6596/345/1/012018

New methods of highly efficient controlled generation of radiation by liquid crystal nanostructures in a wide spectral range

S.N. Bagayev, V.M. Klementyev, B.N. Nyushkov, V.S. Pivtsov, S.I. Trashkeev*
Institute for laser physics, Siberian branch of the Russian academy of science, 13/3, Academic Lavrentyev's prospect, Novosibirsk 630090, Russia
E-mail: sitrskv@mail.ru

Abstract. We report the recent results of research focused on a new kind of soft matter – the liquid-crystal nanocomposites with controllable mechanical and nonlinear optical properties. These are promising media for implementation of ultra-compact photonic devices and efficient sources of coherent radiation in a wide spectral range. We overview the technology of preparation of nematic-liquid-crystal media saturated with disclination defects. The defects were formed in different ways: by embedding nanoparticles and molecular objects, by exposure to alpha-particle flux. The defect locations were controlled by applying an electric field. We also present and discuss the recently discovered features of nematic-liquid-crystal media: a thermal orientation effect leading to the fifth-order optical nonlinearity, enormous second-order susceptibility revealed by measurements, and structural changes upon exposure to laser radiation. We report on efficient generation of harmonics, sum and difference optical frequencies in nematic-liquid-crystal media. In addition, transformation of laser radiation spectra to spectral supercontinua, and filamentation of laser beams were also observed in nematic-liquid-crystal media. We conclude that most nonlinear optical effects result from changes of the orientational order in the examined nematic liquid crystals. These changes lead to the symmetry breaking and disclination appearances.

1. Introduction

Nowadays a special kind of physical medium referred to as soft matter (or partially ordered media) is under intensive study in the physics of condensed state of matter [1]. Distinctive feature of soft matter is that the properties of such matter can undergo significant changes under relatively weak external action. The main focus of the mentioned study is exploration of intermolecular structures and forces, viscoelastic interactions, chemical bonds, and other features which govern mainly mechanical and linear optical properties of soft matter. Nonlinear optical properties of such matter are also varied. Their study may be both of scientific interest and leading to new practical applications.

Liquid crystals (LCs) are perhaps the most known and widely-used representatives of soft matter systems [2]. Besides nonlinear optical properties of LCs themselves, of particular interest are the properties manifested by LC-based composites with nanoparticles. Interest in LC-nanocomposites is aroused so far as creation of ordered defect structures [3] in such media opens up possibilities for design and implementation of a novel type of multidimentional photonic crystals with controllable nonlinear optical properties.

It is well known that, liquid crystals are viscous liquids with a structure determined by order and shape of constituent molecules [1,2]. Such liquids feature properties peculiar to crystals, in particular,

– anisotropy. Orientation vector of dipole molecules – *director* – determines the direction of an optical axis in LCs. The director is strongly affected by external factors, namely, electromagnetic fields (Freedericksz effect [1,2]), and thermal flows (thermal orientation effect [4]).

Orientation structure of LCs is not necessarily homogeneous. It may have singularities with ambiguous orientation of the director [3]. Orientation defects (in particular, disclinations) can be caused both by external actions and by internal factors. The volume form of a LC itself is one of the structuring factors. The defects can be produced also by particles embedded in LC structure. Of particular interest are nanoparticles, molecular particles, and alpha-particles that interact with LCs [5,6]. A foreign particle can make the center (singular point) of disclination the same as the volume form do. The state of LCs in the neighbourhood of singularities, the birth and interaction of disclinations, their impact on optical properties of LCs are not yet exhaustively studied and explained. However, it is precisely this fact that the mentioned particular state of LCs saturated with defects features extraordinary physical properties leading to a variety of effects useful for new practical applications. In what follows we study such effects and propose their application in photonics. .

2. Nematic liquid crystals with particles-induced defect structures

In First, the effects given by insertion of nano-, molecular-, and alpha- particles into nematic liquid crystals (NLCs) subjected to an electrostatic field were studied experimentally and theoretically. Inserted particles form charged defects in the orientation structure of NLCs. The radius of these defects exceeds many times the size of the particles themselves. This phenomenon can be used as an efficient method to produce NLC orientation structures saturated with charged defects. The figure 1 shows the experimental setup used to observe modification of NLC orientation structures upon insertion of particles and application of electric field. The areas of complex optical anisotropy appear around singular points created by inserted particles. The resulting optical texture of the NLC orientation structure was visualized by means of polarizing microscope. The figure 2 and figure 3 show the images of the textures obtained with the use of NLCs doped by gold nanoparticles (~20 nm in diameter) and molecular particles (cromolyn), respectively. The textures appeared when an external spatially uniform electric field was applied. This electric field moves charged nanoparticles, initially situated on the substrates, into the volume of NLC. In this experiment the thickness of the NLC layers was ~ 100 microns. The concentration of the particles in the NLC layers was $10^4 \div 10^5$ mm^{-2}. The dc voltage applied to the electrodes was ~3 V.

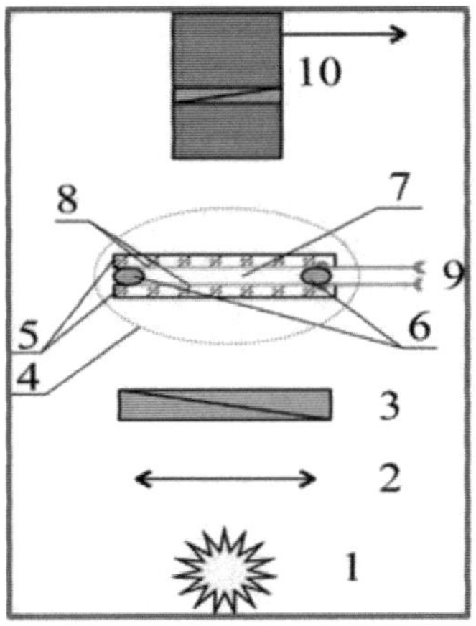

Figure 1. Experimental setup used for observation of nanoparticle-induced singularities in the NLC orientation structure: 1 – light-emitting diode, 2 – condenser, 3 – polaroid, 4- NLC sample (5 – glass plates with electrodes formed by ITO-coating, 6 – spacing, 7 – NLC volume, 8 – electrodes formed by ITO-coating, 9 – leads of electrodes), 10 – polarizing microscope with a CCD – camera.

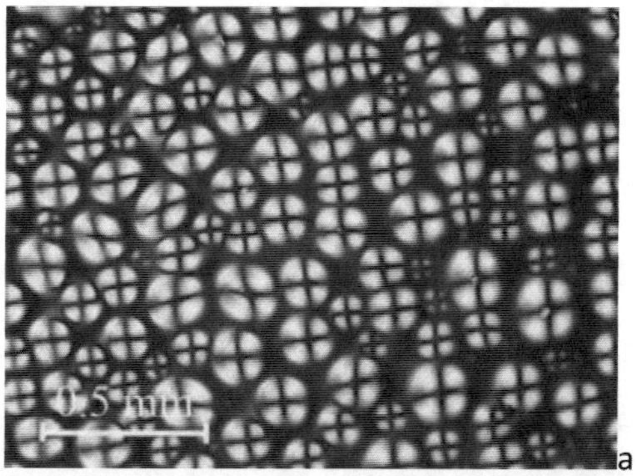

Figure 2. The optical texture of the NLC orientation structure affected by nanoparticles.

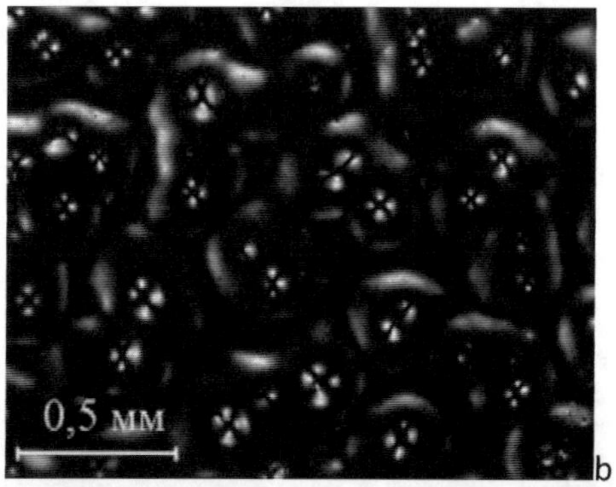

Figure 3. The optical texture of the NLC orientation structure affected by foreign molecular particles.

When NLC samples with nanoparticles were placed in a spatially periodic electrostatic field, we observed ordering optical textures, as shown in Figure 4. To produce such a regular array of nanoparticle-induced singularities we made use of glass substrates with special arrangement of electrodes formed by ITO-coatings (Figure 5). In these structures the spatial period (the lattice spacing) is fixed, but the refractive index and its gradient in the cells formed by singularities can be tuned by varying the strength of external electric field. Scaling down the dimensions of electrodes and applying independent voltage control to them, one can create a completely controllable phase grating. It is opens up opportunities for the development of a new class of photonic devices – 2-D and 3-D electronically controlled, variable photonic crystals.

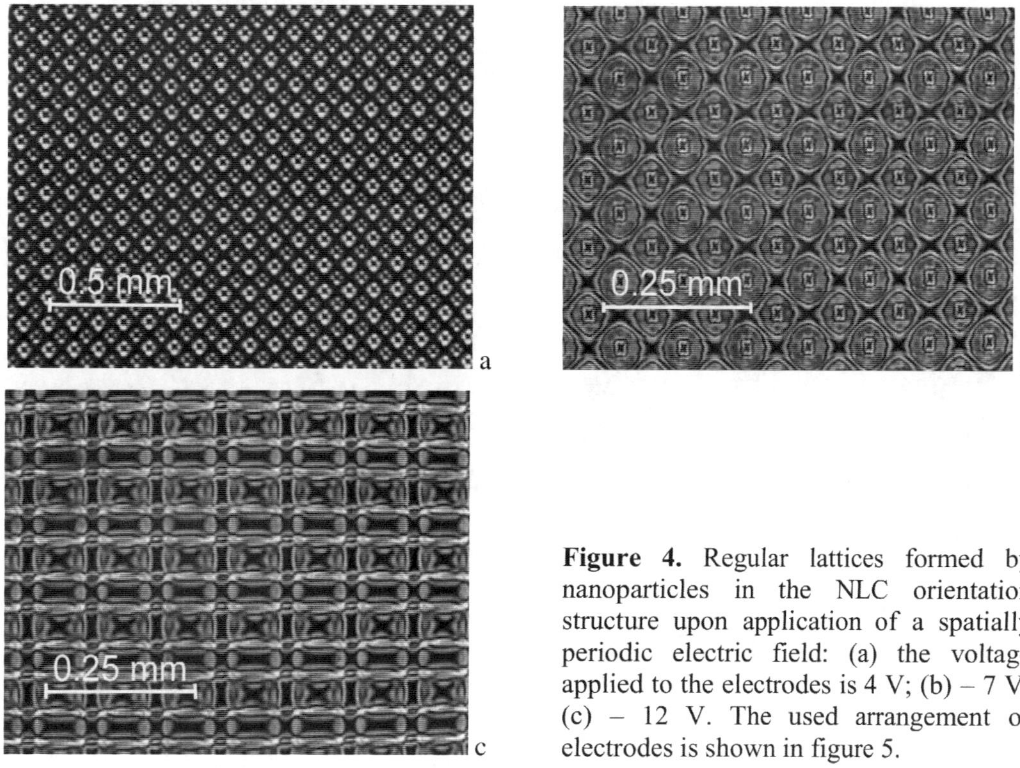

Figure 4. Regular lattices formed by nanoparticles in the NLC orientation structure upon application of a spatially periodic electric field: (a) the voltage applied to the electrodes is 4 V; (b) – 7 V; (c) – 12 V. The used arrangement of electrodes is shown in figure 5.

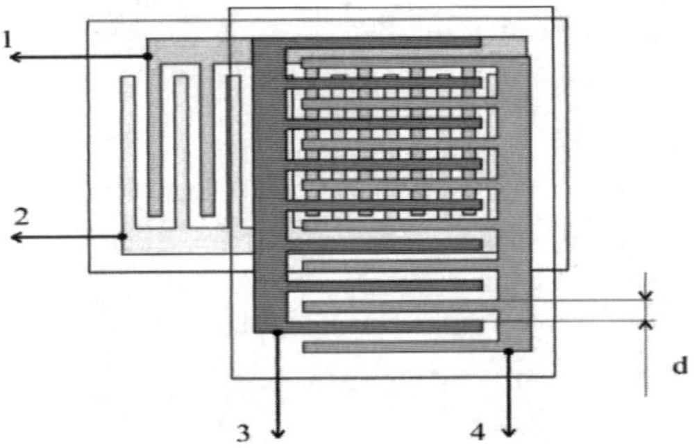

Figure 5. The arrangement of electrodes (formed by ITO-coating on the glass plates) which was used to package nanoparticles into a regular lattice and, thus, to obtain the spatially periodic NLC orientation structures shown in figure 4.

A steady-state configuration of the director orientation in the vicinity of a nanoparticle-induced singular point, and the resulting optical pattern in light transmitted through crossed polarizers were also found theoretically [6]. Examples of the calculated graphs are depicted in figure 6 (a) and figure 6 (b). The calculations yielded almost all defect structures observed in the experiments with single nanoparticles (one nanoparticle in a large volume of NLC). Figure 6 (c) presents the microphotograph of an experimentally observed optical pattern of the defect arising from a single nanoparticle in NLC. The dense embedment of interacting nanoparticles in NLCs is not yet studied theoretically.

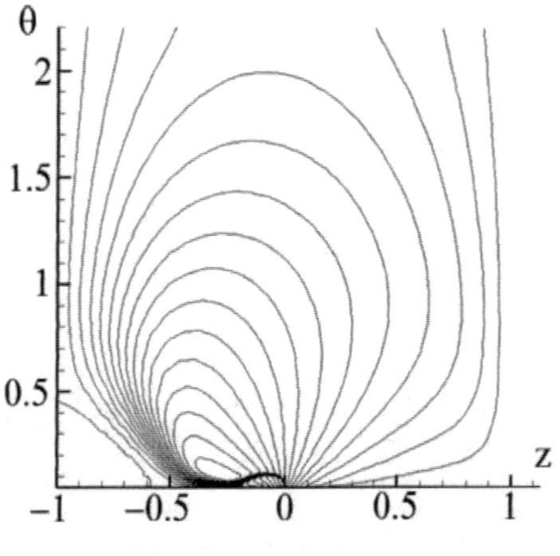

a

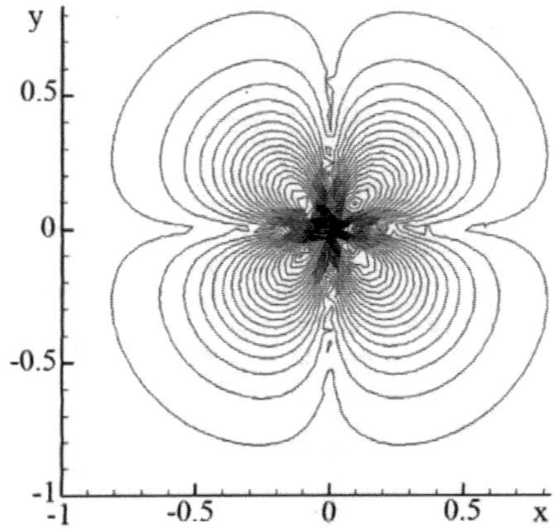

b

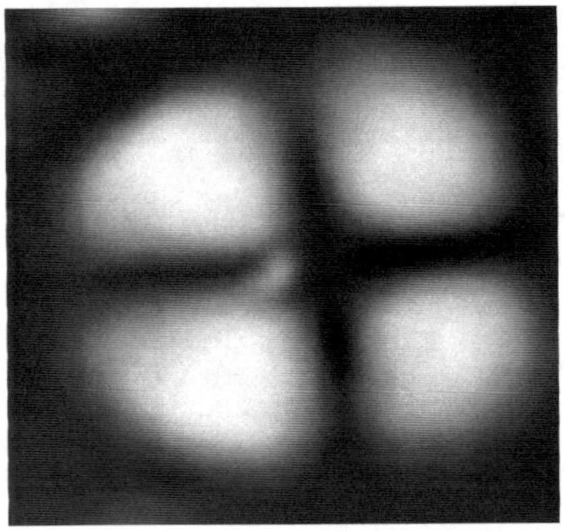
c

Figure 6. Calculated and experimental data for an orientation defect induced by a single nanoparticle: (a) – polar angle θ of the director orientation in a radial-axial plane; (b) – the intensity distribution in light transmitted through crossed polarizers; (c) – the image acquired by means of a polarizing microscope.

In addition we visualized the defect structures formed in NLCs by alpha-particles. To this effect the NLC sample was irradiated by alpha-particles from radioactive isotopes placed under the glass substrate. The particle energy was 4.5 MeV, the intensity of the particle flux was 10 decays•(s^{-1}•cm^{-2}). Because of non-transparency of the used radioactivity source we performed observation with incident light only. When a spatially uniform electric field was applied to the NLC sample being irradiated by alpha-particles, we observed (by means of a polarizing microscope) moving bright spots with irregular shapes (figure 7). The alpha-particles produce local dipole fields [5] which, being subjected to an external electric field, deform the NLC orientation structure and allow visualization of the particle-induced defects. The observed effect can be used for the development of a new type of sensitive alpha-particles detectors that will feature a visualization option (for the purpose of ionizing radiation characterization).

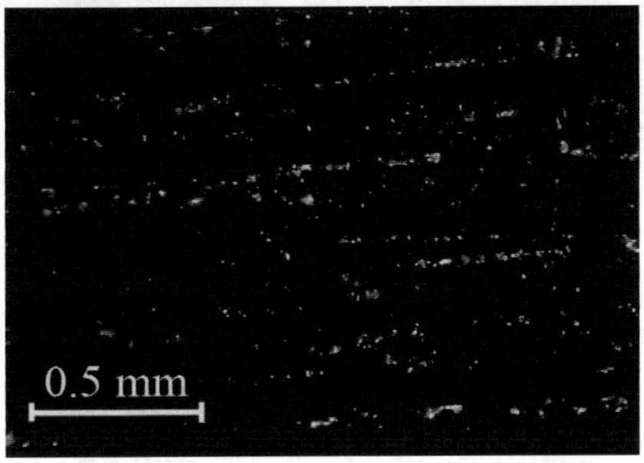

Figure 7. A shot of a dynamic optical texture visualizing NLC orientation deformations induced by an alpha-particle flux.

A steady-state configuration of the director orientation in the vicinity of a nanoparticle-induced singular point, and the resulting optical pattern in light transmitted through crossed polarizers were also found theoretically [6]. Examples of the calculated graphs are depicted in Figure 7. The calculations yielded almost all defect structures observed in the experiments with single nanoparticles (one nanoparticle in a large volume of NLC). Figure presents the photo of an experimentally observed optical pattern arising from a single nanoparticle in NLC. The dense embedment of interacting nanoparticles in NLCs is not yet studied theoretically.

3. Thermal orientation effect in NLCs

Thermal flow influence on the orientation structure of NLCs was detected experimentally. This effect was discovered in NLCs which are centrally symmetrical. Therefore it cannot be explained as the well-known thermo-mechanical effect attributed to cholesteric liquid crystals and deformed NLCs with broken symmetry [4]. A theoretical model was proposed for explanation. This model reveals existence of NLC's orientation thermoelasticity governed by quadratic dependence on the temperature gradient [9]. The detected thermal orientation effect (unavoidable in the presence of radiation absorption) along with the well-known light-induced Fredericks effect [2] cause the central symmetry breakdown and formation of disclinations. Figure 8 demonstrates a calculated orientation structure and the photo of a corresponding optical texture in transmitted light which were obtained under impact of the both effects. In what follows we demonstrate experimentally that the complex impact of these effects on the NLC structure (first of all, central symmetry breaking) allows second-harmonic, difference-frequency, and supercontinuum generation in NLCs.

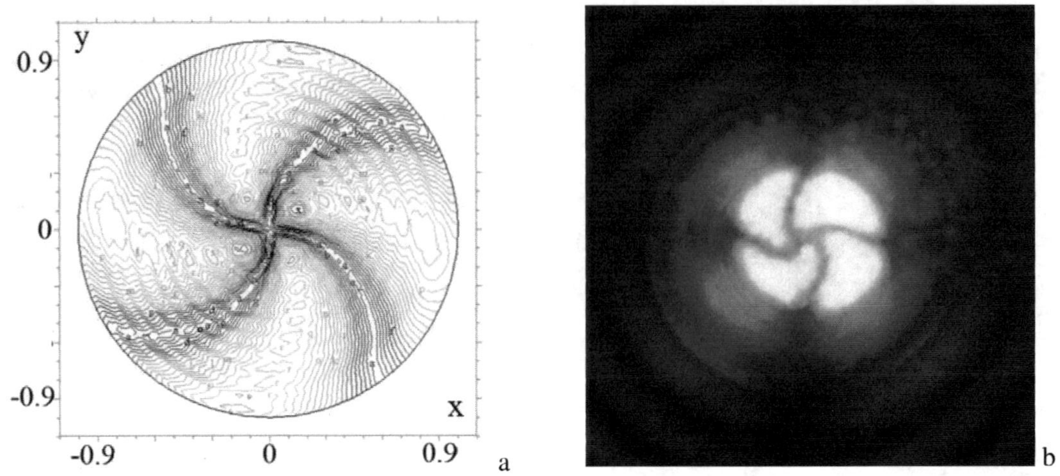

Figure 8. The intensity distribution in light transmitted through crossed polarizers calculated (a) and photographed (b) for the case of combined action of the thermal orientation effect and the light-induced Fredericks effect in NLC.

4. Laser frequency conversion in NLCs

Various transformations of laser radiation spectra in NLCs and accompanying modifications of the NLC orientation structure were studied experimentally. Thus, highly efficient difference frequency generation was obtained, for the first time, in a nematic liquid crystal excited by several visible lines from a cw argon laser with a total power of 0.08—1.5 W [7]. The maximum conversion efficiency was ≈1% and the measured quadratic susceptibility was $\approx 2\times10^{-6}$ m/V. This value is about 6 orders of magnitude higher than the maximal known susceptibility attributed to an organic crystal NPP. The experimental data related to the parameters of generated difference-frequency radiation are depicted in figure 9. The observed laser frequency conversion in NLC has specific features compared to generation at difference frequencies in solid crystals. Therefore we conclude that approximation of quadratic nonlinearity in NLCs requires more precise definition.

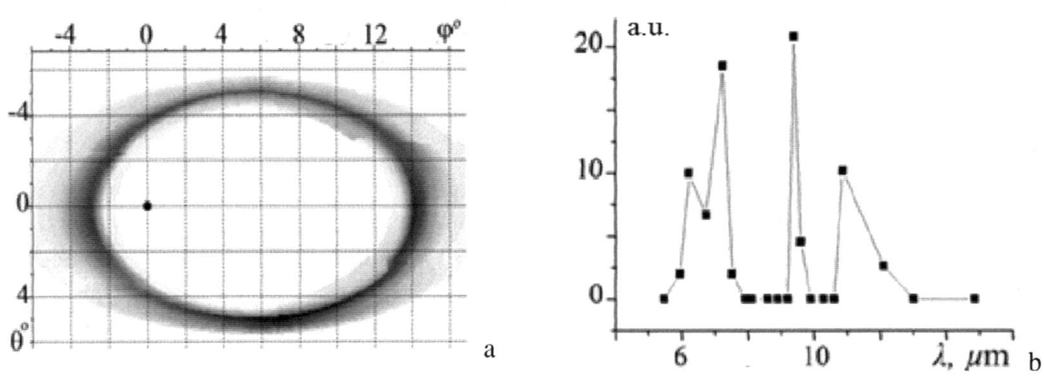

Figure 9. The measured parameters of optical radiation obtained through difference frequency generation in NLC: (a) – indicatrix of the radiation, (b) – optical spectrum.

To study other types of laser radiation nonlinear transformation in NLCs, we made use of two fiber-coupled cw diode lasers with wavelengths of 980 nm and 1480 nm, respectively. The experiments were arranged in two different ways. In the first arrangement, the tip of an single-mode optical fiber delivering laser radiation was immersed in a narrow slot which was formed by glass plates and filled with NLC. The plate spacing was determined by the fiber tip diameter (125 microns).

Two induced orientation structures were observed in the NLC region being exposed to the laser radiation at 1480 nm: a 2-mm-long 10-μm-narrow beam with a divergence of less than 1^0 (depicted in figure 10), and nearly symmetric radial structures centered at the fiber tip (depicted in figure 11). These structures are attributed to the linear and point defects. As seen in figure 10, the self-confined beam in the NLC orientation structure has regular transversal dashes. This filamentation-like effect can be caused by combination of the thermal orientation effect due to absorption [4, 9] and the nonlinear orientation self-focusing due to light-induced Fredericks effect [2].

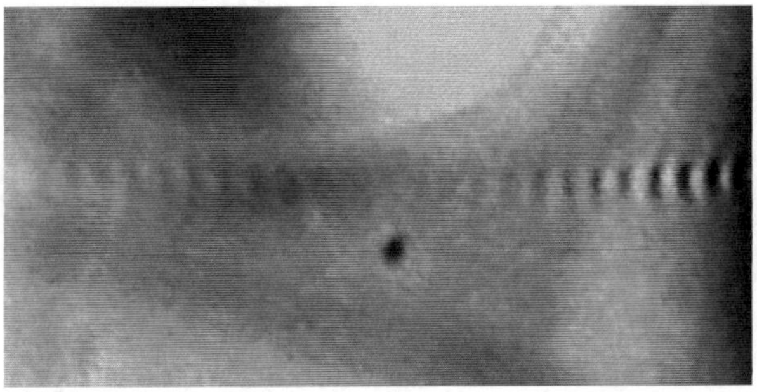

Figure 10. The particular orientation deformation induced by a laser beam in a homogeneous NLC structure, which features filamentation-like appearance. The image was obtained in transmitted light with the use of crossed polarizers and a microscope. The laser beam propagates from right (from an optical fiber tip) to left.

The radial structures (figure 11) appeared when a heat-spreading mirror-like plate was used in the above experimental arrangement instead of the bottom glass plate. In addition, an electrostatic field was applied to the NLC sample in that case. The voltage between electrodes was set about 4.5 V. The photos in figure 11 indicate that disclinations of different strengths can be obtained depending on the radiation intensity. Upon reaching a threshold intensity of the pump radiation we observed appearance of a visible irradiating spot (figure 11 (c)). This spot can be attributed, as we discovered later, to the complex laser frequency conversion in the NLC

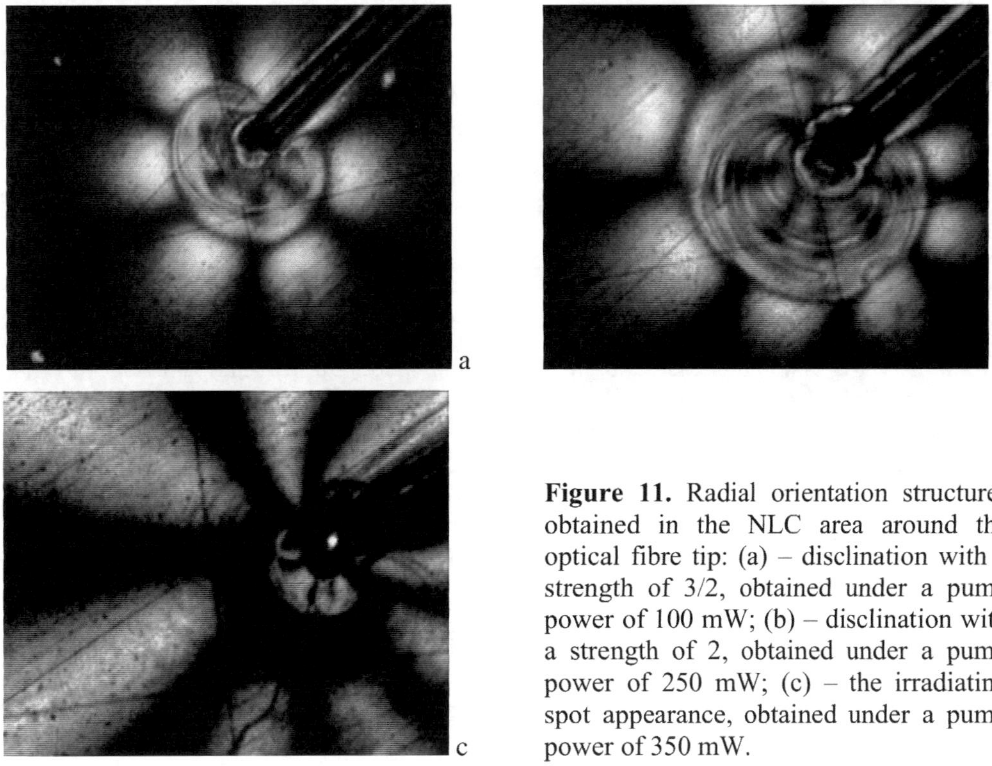

a b c

Figure 11. Radial orientation structures obtained in the NLC area around the optical fibre tip: (a) – disclination with a strength of 3/2, obtained under a pump power of 100 mW; (b) – disclination with a strength of 2, obtained under a pump power of 250 mW; (c) – the irradiating spot appearance, obtained under a pump power of 350 mW.

To examine the optical spectra of laser radiation transformed in NLC we used another experimental arrangement – a small drop of NLC was placed directly onto the tip of a radiation-delivery fiber. Examination of the optical spectra has revealed the following transformations. An octave spanning spectral supercontinuum with an apparent maximum forms in a spectral range from ~500 nm to ~1200 nm, when the NLC drop is pumped at 1480 nm (figure 12). The wavelength of the supercontinuum maximum and the supercontinuum width depend on the pump power. Narrow peaks of the second (~493 nm) and third (~740 nm) harmonics manifested themselves against the wide spectral profile of the supercontinuum. These spectral peaks appear in different points of the radiation indicatrix. Therefore it is hardly possible to register these harmonics simultaneously by a spectrometer. This phenomenon has a simple explanation – very different conditions of phase synchronisms. At a low level of pump power (30 to 100 mW) the generated radiation and the formed orientation structure have stationary properties. At a high pump power level (>250 mW) the intensity of the generated supercontinum oscillates continuously, and the harmonics disappear (become indistinguishable against supercontinuum).

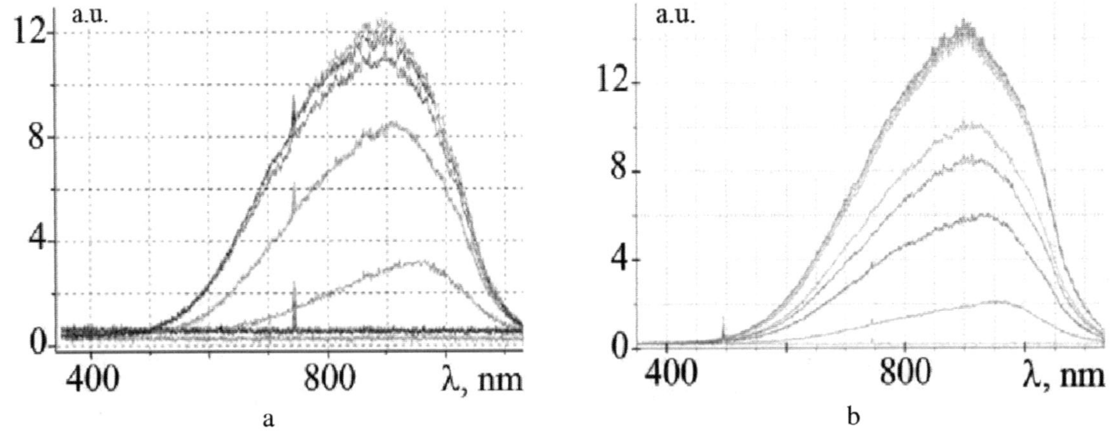

Figure 12. The spectral supercontinua generated in a drop of NLC pumped by a fibre-coupled diode laser at 1480 nm. The diversely coloured traces were acquired at different moments of time to illustrate the intensity oscillations. The spectrum traces (a) and the spectrum traces (b) were acquired from different points of the radiation indicatrix to manifest the contribution of the second and third harmonics, respectively.

When pumping was done at a wavelength of 980 nm, the third harmonic radiation belonged to the ultraviolet (λ_{th}~326 nm) as seen in figure 13 (a). The second harmonic (~490 nm) at the same point of the radiation indicatrix is much weaker and has a particular spectral profile shown in figure 13 (b). The observed spectral profiles underwent some stochastic transient changes upon switching the pump mode and adjusting the power level. Generation characteristics of the supercontinua and harmonics have threshold and hysteresis dependence on the pump power. It was noticed that when we used a multimode radiation-delivery fiber (pumping at 980 nm), the spectral profiles manifested much more irregularities as compared with the spectra obtained using a single-mode delivery fiber (pumping at 1480 nm). In both cases generation characteristics of the supercontinua and harmonics had threshold and hysteresis dependence on the pump power. It is also important to notice that exposure to the highest available radiation intensity of 400 kW/cm^2 did not lead to overheating of the NLC.

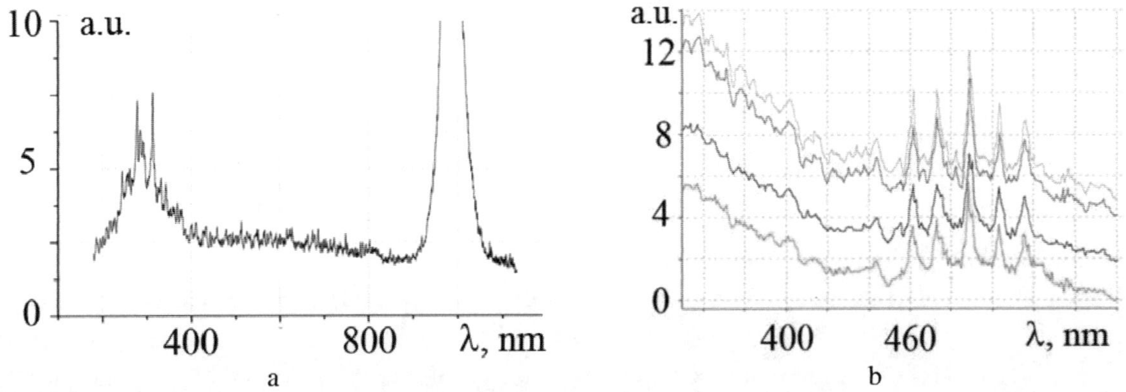

Figure 13. Optical spectrum of radiation generated in a drop of NLC pumped by a fibre-coupled diode laser at 980 nm: (a) – a full range spectral trace which manifests the dominating contribution of the third harmonic (~326 nm) in the selected point of the radiation indicatrix; (b) – magnification of the optical spectrum in the vicinity of the second harmonic (the diversely coloured traces were acquired at different moments of time to illustrate the intensity oscillations).

5. Conclusion

A concept of non-mechanical transportation of nanoparticles and their packaging into a regular lattice has been implemented using the possibility to control the orientation of NLC molecules by means of applied quasi-static electric field. These particles induce a set of spatially periodic singularities in the orientation structure of NLC, and thus yield an optical medium with a spatially periodic anisotropy – a photonic crystal. The key parameters of such a phtonic crystal can be controlled by applied electric fields. It opens up possibility of development of new types of photonics devices for laser technology, telecommunications and optical computing. The demonstrated method of alpha-particles flux visualization by means of NLCs turned to be very sensitive, and, therefore, can be used for characterization of ionizing radiation beams in many practical applications. Enormous quadratic and cubic optical nonlinearities of NLCs subjected to thermal orientation effect along with light-induced Fredericks effect lead to efficient conversion of laser frequencies. Feasibility of laser radiation nonlinear transformation in a volume of NLC was demonstrated experimentally. The most effective transformations are the difference frequency generation in the mid-infrared and the spectral supercontinua generation in the visible and near-infrared ranges. The generation of the second and third harmonics was also achieved.

The obtained research results allow us to assume that nonlinear optical properties of NLCs make it possible to implement a new type of coherent radiation sources which can generate across the whole optical spectrum, from the ultraviolet to the far-infrared.

References

[1] Kléman M, Lavrentovich O D 2003 *Soft matter physics* (Springer)
[2] Iam-Choon Khoo 2007 *Liquid crystals* (John Wiley & Sons)
[3] Kurik M V, Lavrentovich O D 1988 *Sov. Phys. Usp.* **31** 196
[4] Demenev E I, Pozdnyakov G A and Trashkeev S I 2009 *Technical Physics Letters* **35** 76
[5] Pozdnyakov G A and Trashkeev S I 2007 *Technical Physics Letters* **33** 70
[6] Trashkeev S I, G. N. Grachev and Pozdnyakov G A 2007 *Technical Physics* **77** 85
[7] Trashkeev S I, Klement'ev V M and Pozdnyakov G A 2008 *Quantum Electronics* **38** 373
[8] Trashkeev S I, Klement'ev V M and Statsenko P A 2006 *Liquid Crystals* **33** 417
[9] Trashkeev S I and Britvin A V 2011 *Technical Physics* **56** 747

Interaction of nanodiamonds materials with influenza viruses

V T Ivanova[1], M V Ivanova[1], B V Spitsyn[2], K O Garina[1], S V Trushakova[1], A A Manykin[1], A P Korzhenevsky[3] and E I Burseva[1]

The D.I. Ivanovsky Institute of Virology of the Ministry of Health and Social Development of the Russian Federation, Gamaleya st, 16, Moscow 123098, Russia,[2] A.N. Frumkin Institute of Physical Chemistry and Electrochemistry, RAS, Leninsky prospect, 31, Moscow 119071, Russia, [3]Joint-Stock Scientific and Production Company "SINTA", 220136, 2 per. Montazhnikov, 4, Minsk, Republic of Belarus

E-mail: valivanova1946@mail.ru

Abstract. The perspectives of the application of modern materials contained nanodiamonds (ND) are considered in this study. The interaction between detonation particulate ND, soot and influenza A and B viruses, fragments of cDNA were analyzed at the normal conditions. It was shown that these sorbents can interact with the following viruses: reference epidemic strains of influenza A(H1N1), A(H1N1)v, A(H3N2) and B viruses circulated in the word in 2000-2010. The allantoises, concentrated viruses, cDNA can be absorbed by ND sorbents and getting removed from water solutions within 20 min. ND sorbents can be used for the preparation of antivirus filters for water solution and for future diagnostic systems in virology.

Introduction

The contamination of the environment by biological and chemical waste products becomes a world-scale problem, which requires a different ways of solving. Purification of the water, containing micropatogens, is one of the most difficult task, especially related to viruses, sized in nm. Charcoal has been used as adsorbent for many years. Searching new materials with higher sorption efficiency becomes very actual. It was shown that modified graphite after treatment in water at 95°C [1], the carbon nanotubes [2] belonging to the group of carbon materials which can interact with chemical matters, drugs and viruses and remove them from air and water solutions. During the investigation it was discovered that all of them have both advantages and deficiencies. The detonation nanodiamonds were discovered in 1960s in Russia as the results of the explosion of mixture of trotyl and hexogen in optimal proportion 3:5 at pressure 16-23 GPa and T above 3000K [3,4]. First investigation period of ND was devoted mainly to physical-chemical properties and leaded to the practical results - the creation of polished, lubricant additive materials and others. Recently the studies devoted to interaction of ND with proteins, enzymes resulted to the proposal of application of ND for drug delivery to cancer cells, in anti cancer therapy [4,5].The interaction of ND with micropatogens was almost not investigated yet. According to [6], it was found that ND could interact with bacterium *E – Coli.* This study continued the serial investigations devoted to the interactions of modern materials with viruses, in particular between ND nanomaterials and influenza viruses. The aim of this work is manufacturing sorbents for antivirus filters and virus control system based on ND nanomaterials.

Experimental

Preparation of nanodiamonds and soot

The nanodiamonds (ND) and soot were manufactured by detonation method [3]. The electron microscopy analysis showed that both nanodiamonds and soot contain the clusters with different forms and sized (Fig.1). The content of different impurity atoms ND and soot in concentration more then 0.01% atom presented in table. It was studied by standard C-H-N chemical analysis.

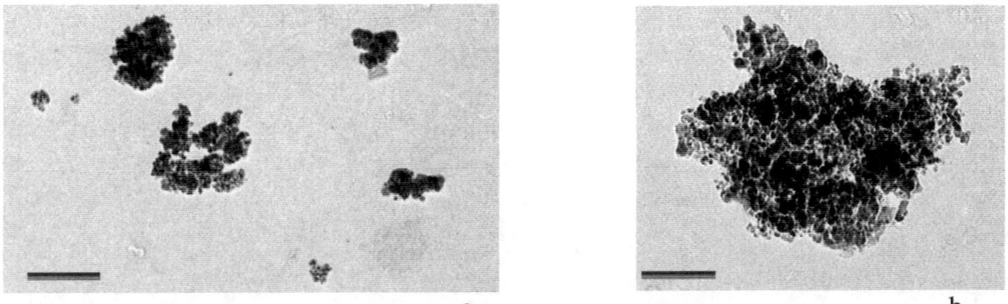

Figure 1. The structure of nanodiamonds - a and soot - b, by tansmission electron microscopy.

Virus cultivation and registration

The reference strains A/New Caledonia/20/99(H1N1), A/Perth/16/09(H3N2), B/Sichuan/379/99, B/Florida/04/06 were chosen for study. The viruses were grown in 9-10 embryonated chicken eggs. The viruses were concentrated by ultracentrifugation and diluted in physiological solution. The hemagglutination test with human erythrocytes was used to determined viruses in solutions.

Interaction of viruses with sorbents

The method of investigation virus-sorbent interaction was developed by Ivanova et al and described in_[1,2]. It includes several stages: the intensive contact of viruses with sorbent on shaker during 15min at temperature 22°C, low speed centrifugation (2000 circle /min), the determination of viruses in solutions before sorption and in supernatant after sorption . In the case of influenza viruses solution volume was 200-400 μl, sorbent mass - 4 mg. The hemagglutination test with human erythrocytes was used for virus detection in solutions. The experiments were carried out in the follow solutions: H_20 after previously special treatment by heating to destroy other pathogens, 0.15 M NaCl in H_20, allantoises liquid of chicken embryos.

Desorption of viruses from sorbents

The desorption of viruses into solution was also investigated. The complex virus + sorbent put into 0.15M NaCl pH= 7,2 for 24, 48 h at 20ºC. Then the complex was precipitated by centrifugation at 2000 circle/min during 5 minutes. Supernatant was investigated for presents of viruses in hemagglutination test with human erythrocytes.

cDNA preparation and registration

The fragments of cDNA were obtained from influenza A viruses in Real Time polymerase chain reaction (PCR) using Russian test - system "Influenza A virus" (firm "DNA- technology", Russia). The cDNA samples (80-100 μl) connected with sorbent (mass 2mg) by the same method , which we used for viruses. The samples before and after sorption were analyzed by electrophoresis in 2% agarose gel during 1.5 h., voltage = 80V. The volume of the probe injected into gel was 7 μl. The positive control was the control sample included in the test -system. The gel was photographed and analyzed by transluminator, detection wavelength λ=254 nm.

Results and discussion

The first biological objects were influenza viruses which caused the epidemics in the world during 1998- 2010 according data of World Health Organization. The surface viral proteins - hemagglutinin and neuraminidase had different structure and according antigenic properties. All these viruses were concentrated to remove majority of allantoises protein and to increase the initial virus concentration. It was established that both sorbents interacted with all viruses in solutions. The hemagglutination titers of virus solutions decreased in 8-1000 times after contact with sorbents (Fig.2). The adsorption did not depend on antigenic properties of viruses. The decreasing of this value depended on the virus titers before sorption and the sorbent mass. The soot was approximately 2,5 times more effective than nanodiamonds. The experiments for desorption of viruses from sorbents were carried out to determine the stability of the binds of viruses with sorbents. It was shown that viruses absorbed at ND did not desorb at all from sorbent or desorption was very small. The results for influenza virus

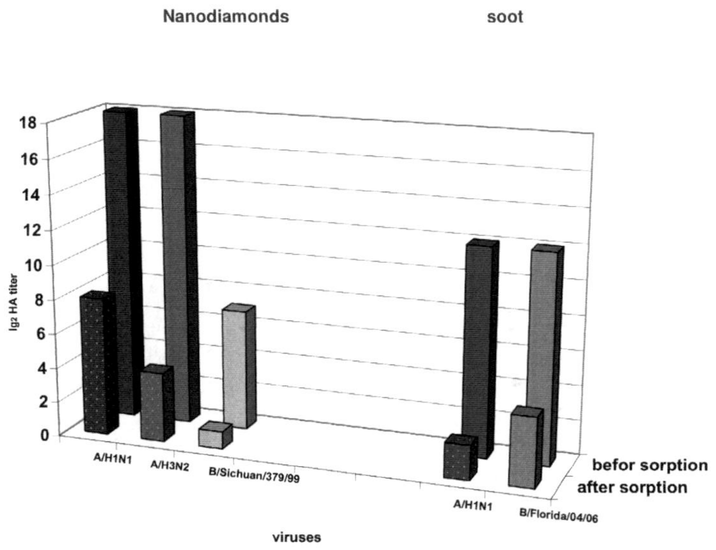

Figure 2 . Sorption of influenza A and B viruses on nanodiamonds and soot

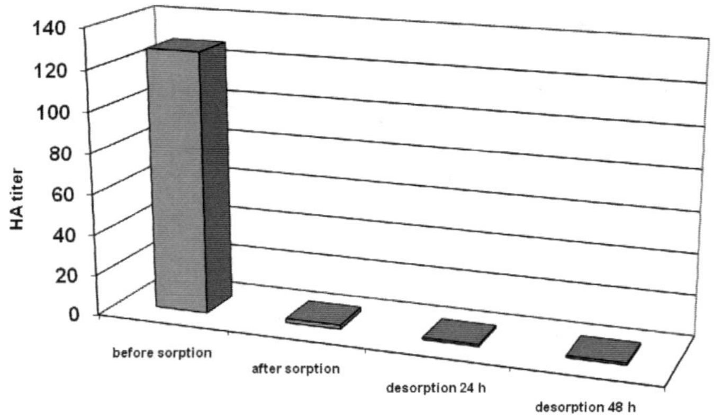

Figure 3. Sorption and desorption of influenza viruses B/Florida/04/06 during 24h and 48h from nanodiamonds.

B/Florida/04/06 was present at Fig. 3. Viruses titer in solution before sorption was 128 HAE, after sorption - <2 HAE and after desorption 24-48 h was <2 HAE. It shows the stability of the bind between viruses and sorbents at the chosen conditions in solutions.

The next objects for investigation were the fragments of cDNA. After preparation in PCR, the samples were added to tubes with sorbents for contact during 20 min at 22°C, then the complexes cDNA with sorbent were pellet by centrifugation. The supernatant was analyzed by electrophoresis. Both sorbents interacted with fragments of cDNA with size 190 b.p. Fragments of cDNA sized more 560 b.p. were removed by ND partly, by soot –completely (Fig.4).

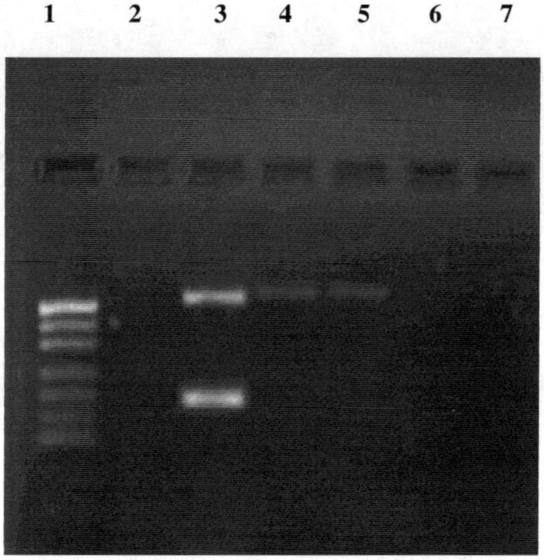

Figure 4. Electrophoreses of cDNA after interaction with nanodiamonds and soot.
Pattern 1- plasmid PUC19 (molecular size marker 34-501 b.p.), pattern 3- fragments of cDNA before sorption, patterns 4,5 - after sorption on ND, patterns 6,7- after sorption on soot.

Early it was shown the adsorption of liner DNA by ND and the absence the interaction ND with ring DNA [7]. Our data showed that the size of fragment cDNA also important for effective interaction for ND in instead of soot in size DNA 190-560 b.p.

The interaction between proteins and cDNA with ND materials and its absence in the case of diamonds can be explained by the differences in their structure and chemical purity [3,4]. Instead of diamonds ND and soot had on their surfaces besides carbon such atoms as N, O, S, H. The percent of these elements varied from 0,03 till 5,69 % atom in the case sulphur and oxygen accordingly (Table). They could form the binds with functional amino acid groups of viral proteins, nonviral proteins or enzymes [4].

After the analysis of our data we concluded that viral proteins and cDNA interact with ND sorbents. We observed the differences in absorbed properties of ND and soot. The soot absorbs these biological objects better then ND. This phenomenon could be explaining the structure and its specific surfaces of sorbents. The specific surface of soot is 450 m^2/g, it is much more than the same surface of nanodiamonds – 300 m^2/g. In addition the soot has a difference in the properties of surface carbons. It has more amorpheous carbon [3]. Therefore more intensive soot interaction with viruses and cDNA (compare to ND) correlates with its physical-chemical properties.

Table. Chemical composition of nanodiamonds and soot used

element	Nanodiamonds		Soot	
	% atom	%mass	% atom	%mass
C	91,7954	89,1718	93,5983	90,6255
N	2,0570	2,3310	1,3851	1,5644
O	5,6998	7,3763	3,8191	4,9263
Na	0,1095	0,2037	0,1419	0,2631
Al	0,0712	0,1554	0,6012	1,3071
Si	0,0982	0,2228	0,1067	0,2414
S	0,0314	0,0814	0,0309	0,0800
Cl	0,0190	0,0544	0,0669	0,1912
K	0,0298	0,0941	0,0429	0,1354
Ca	0,0119	0,0386	0,0589	0,1904
Fe	0,0223	89,1718	0,0391	0,1759

Conclusion

Our results showed that nanodiamond sorbents can interact with influenza viruses independently on antigen properties of surface proteins and we suppose that interaction of these sorbents with other viruses might also has place (for example could interact with other envelope viruses besides influenza viruses). The ND materials absorb viruses from solution more effectively compare to carbon nanotubes [2]. The ability of ND materials to absorb the viruses and fragments of cDNA can be used for the creation of the antimicrobial filters for water decontamination from micropathogens and the development of new diagnostic systems. Our data and early established the absence of toxic properties of ND and ability to keep the enzyme activities of proteins connected with ND [4] extend possibilities to use ND material as carrier for virus proteins.

Acknowledgments
The work was partially supported by Influenza division of the Center of disease control and prevention, USA, Atlanta, Georgia, Grant number 1U51 | P000527-01.

References
[1] Ivanova V T, Kurochkina YA E, Buravsev V N, Nikolaev A V , Timofeeva A V, Baratova L A, 2008, *Problems of virology* 53 2 40-43
[2] Ivanova V T , Katrukha G S, Timofeeva A V, Ilyna M V, Kurochkina Y E, Baratova L A, Sapurina I Yu, Ivanov V.F. 2011, *J. Phys.: Conf. Ser.* 291 012004
[3] Spitsyn B V, Davidson J L et al. 2006, *Diamond and Related Materials* 15 298
[4] Schrand A M, Hens S A C , Shenderova O A, 2009, *Critical reviews in solid and material sciences* 34 1-2, 18
[5] Adnan A, Lam R, Chen H et al. 2011, *Molecular Pharmaceutics* . 8 2 368
[6] Liu Y. L. and Sun K. W. 2010.*Nanoscale Res Lett*. June, 5 6 1045
[7] Purtov K V, Duratova L P , Puzyr A P, Bondar V S. 2008, *Nanotechnology* 325101 19 1

Journal of Physics: Conference Series **345** (2012) 012024 doi:10.1088/1742-6596/345/1/012024

Nanosize Powders of Transition Metals Binary Systems

Yu.A. Zaharov[1,2], V.M. Pugachev[1], V.G.Dodonov[1], A.N. Popova[1,2], R.P. Kolmykov[1], G.A. Rostovtsev[1], O.V. Vasiljeva[1], E.N. Zyuzyukina[1], A.V. Ivanov[1], I.P. Prosvirin[3]

[1] Kemerovo State University, 650000, Russia, Kemerovo, Krasnaya st., 6

[2] Institute of Coal-chemistry and Material science of the RAS SB, 650000, Russia, Kemerovo, Sovetskiy av., 18

[3] Institute of Catalysis of the RAS SB, 630090, Russia, Novosibirsk, ak. Lavrentieva av., 5

E-mail: zaharov@kemsu.ru

Abstract. The review of the results obtained by the authors in the field of synthesis of Fe-Co, Fe-Ni, Co-Ni and Cu-Ni nano-size binary systems (NBS) by the liquid-phase reducing of metal chlorides by hydrazine (pH>10) and sodium tetrahydroborate ($NaBH_4$) (pH = 7-8) by the process conditions variation (temperature of reaction medium, concentration and input order of reagents) as well as NBS properties investigation is presented.

1. Introduction

The review of the results obtained by the authors in the field of synthesis of Fe-Co, Fe-Ni, Co-Ni and Cu-Ni nanosize binary systems (NBS) by the liquid-phase reducing of metal chlorides by hydrazine (pH>10) and sodium tetrahydroborate ($NaBH_4$) (pH = 7÷8) by the process conditions variation (temperature of reaction medium, concentration and input order of reagents) as well as NBS properties investigation is presented.

2. Experiment and results

The basic stages of NBS synthesis were established [1-4] by the methods of X-ray structural (RSA), phase (RPA), fluorescent (RFA) and small angle X-ray scattering (SAXS) analyses.

The reducing by hydrazine was carried out by adding alkali in reaction medium before or simultaneously with hydrazine. Thus at first the fast reaction of sedimentation proceeds (1, see below) with the formation of mixed hydroxides. The hydroxides of the iron group metals as known belongs to a highly symmetric hexagonal brucite structure type, therefore it is natural to expect their co-crystallization under the NBS synthesis conditions at any components ratios. It was confirmed experimentally that the dependence of lattice parameter on composition has monotonous almost linear character (Vegard's law) (see, figure 1).

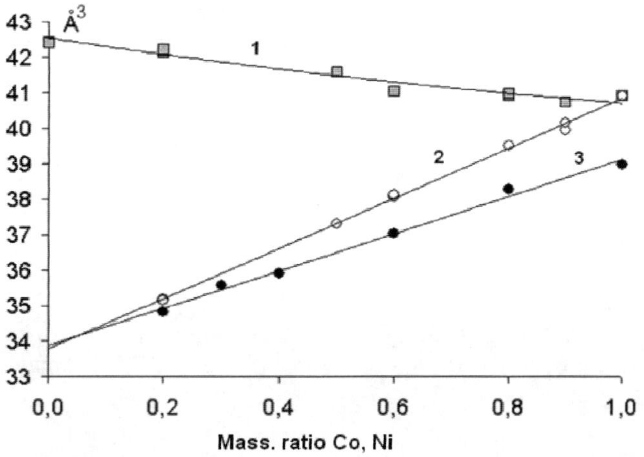

Figure 1. Unit cell volumes of mixed hydroxide $[Co,Fe](OH)_2$ (1) and oxide-hydroxide phases of Fe-Co (2), Fe-Ni (3).

After sedimentation the basic Red-Ox process of interaction of cations occurring by hydroxides dissociation (counterreaction to 1) with hydrazine proceeds. In the case of simultaneous adding of alkali with hydrazine the process goes somewhat simultaneously with more fast one of hydroxides formation. On scheme below the basic reactions at mole ratio of metals 1:1 (for example) are given:

1) $M_1^{2+} + M_2^{2+} + 4\ OH^- = M_1(OH)_2 \cdot M_2(OH)_2$

2) $M_1^{2+} + M_2^{+} + N_2H_4 + 4\ OH^- = M_1 \cdot M_2 + N_2 + 4\ H_2O$

Copper hydroxide has more complex structure, its orthorhombic unit cell looks like pair of deformed brucite cells. Therefore mutual solubility of copper and nickel hydroxides is limited. At the temperature 298 K and the content of copper less than 30 mass % (here and further in the text the composition corresponds to metal only) the high-dispersed X-ray-amorphous mixed hydroxide with the brucite type structure is formed. This mixed hydroxide gradually crystallize if obtained suspension heats during 15-30 min at 358 K. If the contents of copper is from 70 to 90 mass % the moderately crystallized orthorhombic phase with the structure of copper hydroxide is formed.

The formation of mixed hydroxides in the given composition regions is confirmed by the fact that the obtained products are stable with heating up to 358 K for a long time whereas the pure copper hydroxide is quickly decomposed by dehydration forming the phase of black CuO even at room temperature (about 298 K). It should be noted that the aging was carried out at the conditions of alkali medium (pH $\approx$ 9) that promotes dehydration.

If the contents of copper ranges from 40 to 60 mass % the X-ray-amorphous product is also formed. This product also becomes black with time and the reflexes of CuO on the X-ray diffraction picture arise. Obviously it indicates that in this range of concentrations an oversaturated by copper metastable solid solution with brucite structure is formed, which is gradual-ly decompose giving off copper oxide and water.

The X-ray investigation ex situ of nickel and copper hydroxides coprecipitated at the temperature 348-358 K corresponding to synthesis conditions showed that in such conditions only mixed hydroxides containing not more than 20 mass % of copper are stable. It is possible that at high temperature and alkalinity of reaction medium the high-dispersed mixed hydroxide is less stable. It is also possible, that

Journal of Physics: Conference Series **345** (2012) 012024 doi:10.1088/1742-6596/345/1/012024

the priority (for copper) reaction in such conditions is copper oxide formation. The high-speed synchrotron radiation diffraction measurements in situ, planned in future, might give more clear answer to that question. It may also allow to establish what actually is subjected to reducing – mixed hydroxide or its mixture with copper oxide.

In the systems containing iron the coprecipitated hydroxides may turn into oxide-hydroxide phases (OHP) due to oxidation of Fe^{2+}. So the hydroxide $Fe(OH)_2$ as a component of mixed hydroxide turn according to RSA data into FeOOH form. As a result the stoichiometry similar to a brucite structure type remains. Thus the oxidation process is not due to introducing of oxygen into the substance but due to withdrawal of proton taking away an electron of iron atom. X-ray diffraction methods showed that both atomic and superatomic structure of the substance by this process remain almost unchangeable. The composition of the OHP phase formed may be expressed as: $[Fe,M](OH)_2 \cdot FeOOH$.

The higher charge and smaller size of Fe^{3+} cations promotes their partial transition into the tetrahedral voids which in normal brucite structure are occupied by protons. This makes the OHP structure similar to the spinel one. The long heating or aging of OHP in water leads to dehydration and transformation of OHP in spinel. It is known that the substances with spinel type structure are rather persistent and more inert chemically in comparison with mixed hydroxides. As a result the final product in case of iron-containing systems may contain impurities of spinel and OHP phases.

On the other hand the formation of spinel or OHP thin film covering on a surface of metal particles may provide a protection against their further oxidation [5, 6] that is especially important in case of systems rich by iron. For example, it was established that the suppression of oxidation process (synthesis in an atmosphere of nitrogen) allows to obtain pure (without OHP and spinel impurities) but rather pyrophoric product. On the contrary, if synthesis is in condition of high aeration, the content of OHP and spinel impurities in final product increases. For example, the corresponding reactions for mole ratio of metals 2:1 are given below:

3) $4M(OH) \cdot 2Fe(OH)_2 + O_2 = 4M(OH)_2 \cdot 2FeOOH + 2H_2O$

4) $M(OH)_2 \cdot 2FeOOH = MO \cdot Fe_2O_3 + 2H_2O$

It was shown [7, 8] that the formation of mixed oxides as a intermediate products is a favorable factor for the formation by the reducing process of mixed metal phases – solid solutions.
The scheme of reducing by sodium tetrahydroborate near pH=7 is much more compli-cated. The results of products studying obtained by us using RPA, RFA, derivato-mass-spectrometry and X-ray photoelectron spectroscopy (RPES) at the temperatures 550-650 K in view of results [9] allows to present it in the first approximation in a form:

$$5)\; M_1^{2+} + M_2^{2+} \xrightarrow{THB} M_1 \cdot M_2^{nano} + (M_1 \cdot M_2)_m B_n + [M_1 \cdot M_2]_x (B_y O_z)$$

$$\xrightarrow[-(BO_2^-,\, BO^-)]{550\text{-}650\,K} M_1 \cdot M_2^{nano} + B + B_2O_3$$

The products obtained are usually X-ray-amorphous. By the heating up to the temperatures 550-650 K the crystallization of metal phase remaining in nano-size conditions and the quantitative growth of the latter owing to the thermal decomposition of borides, borate compounds and complicated complexes with boron (III) oxide formation, segregation of boron on the particles surface and apparently B_2O_3 formation are observed.

The combination of the methods of small-angle X-ray scattering (SAXS), scanning electron microscopy, specific surface determination (BET) and porosity allows to establish that all examined NBS synthesized according to scheme (1-4) are characterized by qualitatively similar morphology of the particles: nanocrystallites by the size of 6-30 nm (for different NBS synthesized by various conditions) with narrow size distributions (3-6 nm on a half-height of maximum according to SAXS data) forms compact agglomerates undestroyable by mechanical charge and ultrasound action with zero open porosity but high closed porosity, lager units (60-250 nm for various systems with different shape depending on NBS composition and conditions of synthesis) in its turn give friable globules of micron sizes easy destroyable by ultrasound treatment up to smaller agglomerates. The spatial organization of the particles may also be controlled by realization of synthesis in a constant magnetic field (1-5 kOe). The morphology of Cu-Ni nanoparticles looks simpler: the material consists of friable agglomerates which are formed directly from nanocrystallites by the size of 10-40 nm. The typical shapes of the NBS particles and mass distribution functions computed from SAXS data are given on figures 2, 3.

Fe-Co, 30% Co Fe-Ni, 50% Ni Co-Ni, 40% Co

Figure 2. Typical SEM micrograph of nanosystems (Fe-Co, Fe-Ni and Co-Ni).

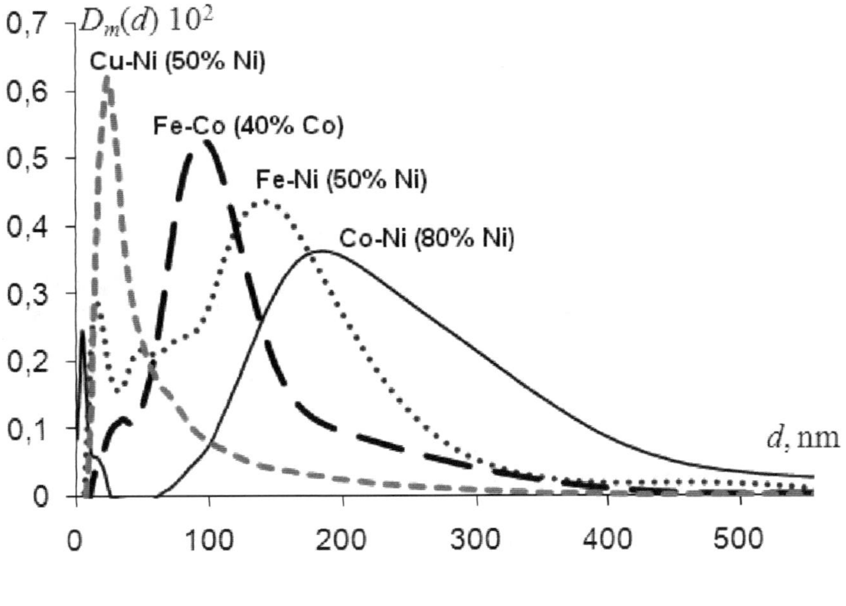

Figure 3. Mass-functions of particle size distribution according to SAXS data (for solid solutions Cu-Ni, Fe-Co, Fe-Ni, Co-Ni).

The phase portraits of studied NBS as cross-sections of phase conditions are generally correspond to the known phase diagrams of massive (macro-size) alloys provided conception on specific for each nanosystem raised temperature connected with their energy saturation. The phase composition of each NBS depending on chemical composition does not correspond to a fixed temperature above the temperature of synthesis and RPA, but corresponds to a certain temperature range, the borders of which are difficultly defined rigorously (figure 4). This is probably due to the fact that these NBS consist of particles of different sizes with different energy saturation.

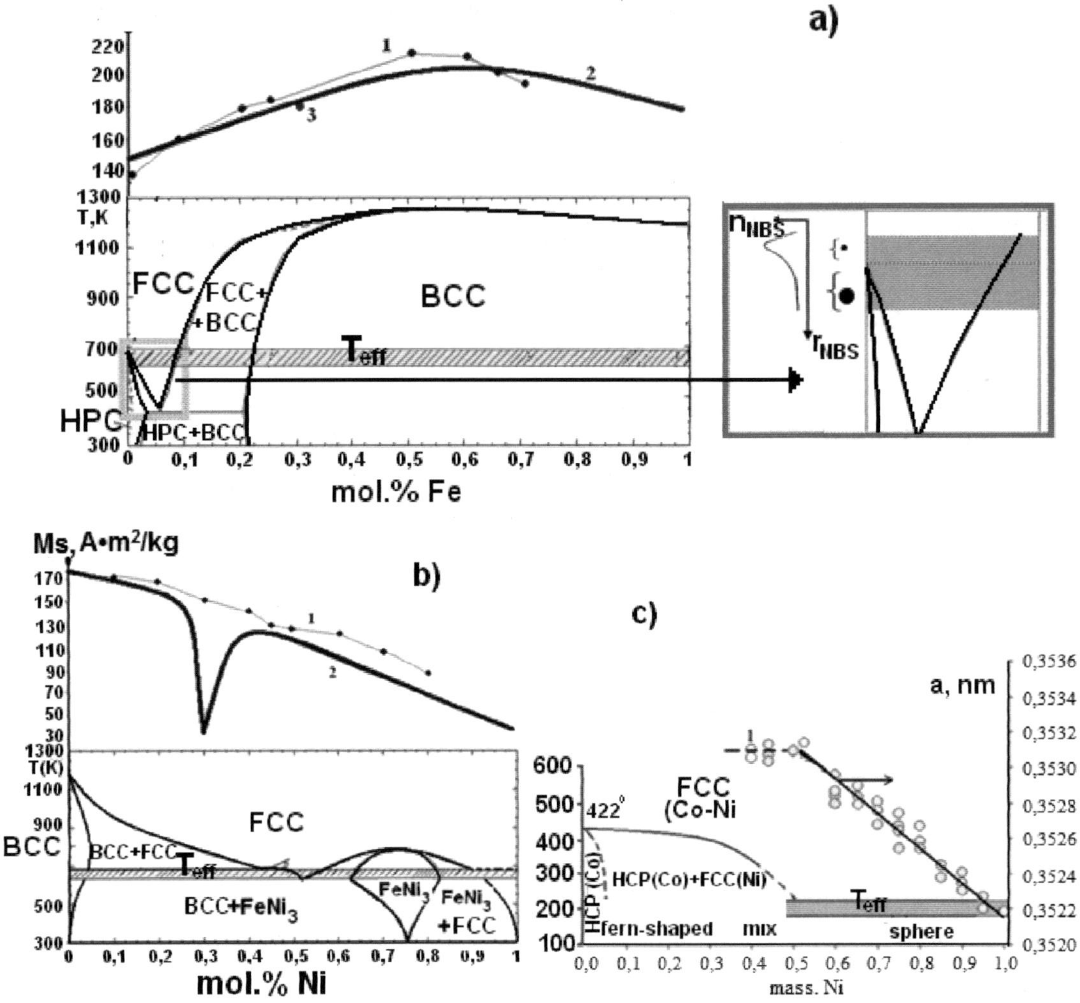

Figure 4. Part of phase diagrams (–) and phase portraits at T_{eff} (▒▒▒)of nanosystems: a) Fe-Co; b) Fe-Ni; c) Co-Ni (1 – the lattice parameters). Upper figures (a, b) – evaluation of saturation magnetization versus the system composition.

It is established [10] that for pure cobalt in the particle size range of 20-40 nm the FCC-phase is more preferred at smaller particle sizes and HCP-phase at larger ones. It is also found by X-ray diffraction methods (including SAXS) that the sizes of the crystallites (nanoparticles) for pure cobalt and examined NBS correspond to a specified range (figure 3).

According to the idea of the phase portraits shifting up to higher temperatures, both nor-mal (hexagonal - HCP) and high-temperature (cubic - FCC) phases are present in the synthe-sized systems containing large amounts of cobalt (as well as pure cobalt). The almost complete disappearance of the hexagonal phase in the Fe-Co system already at 2 wt.% of iron (figure 7) allows us to determine satisfactorily the lower limit of the range of effective temperatures (figure 4) in this range of composition. Besides the relative content of FCC phase considerably increases with particle size (SAXS data) decreasing, achieved by higher concentrations of the reducing agent (figure 7). The HCP and FCC phases are also formed in the Co-Ni system at low concentrations of nickel.

Another important consequence of the shift of phase portraits up to higher temperature is the changing of limits of components mutual solubility, which is especially evident in the Ni-Cu system. In spite of phase diagram the formation of limited solid solutions of the nickel is observed at temperatures of synthesis and storage 80 °C and 25 °C respectively.

This allows us to determine qualitatively the range of effective temperatures for the sys-tem in the region of these components ratio (figures 5, 6).

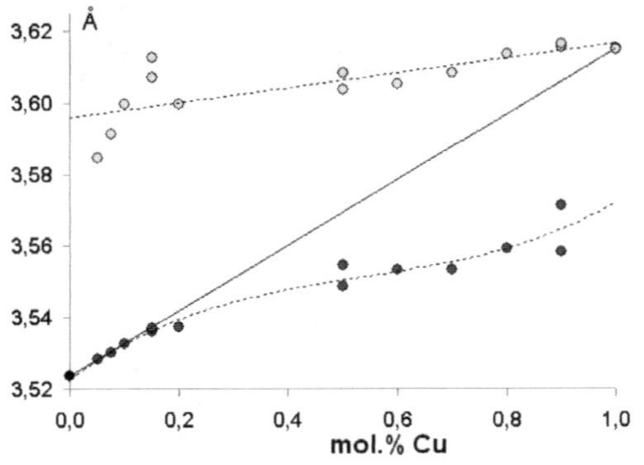

Figure 5. The lattice parameter of FCC-phase Ni-Cu.

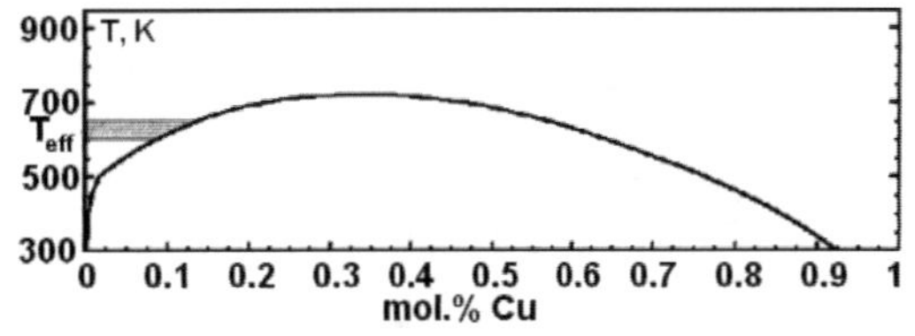

Figure 6. Part of phase diagrams (–) and phase portraits (▒) of Ni-Cu nanosystems (according to figure 5).

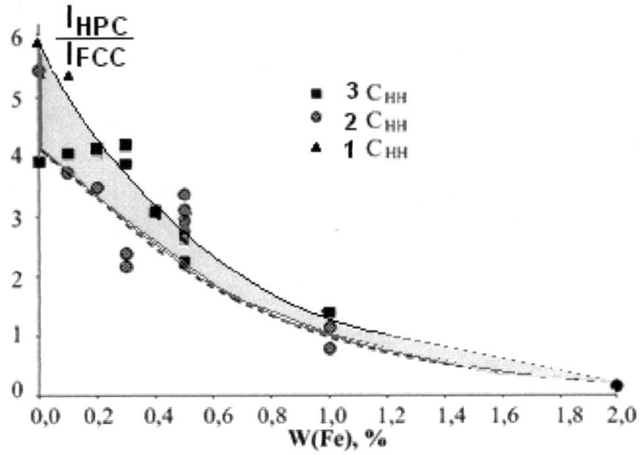

Figure 7. The intensity ratio of cobalt reflections: 101 (HPC-phase) versus 200 (FCC-phase) at the stoichiometric concentration of hydrazine hydrate (▲), in 2 (●) and 3 (■) times greater than its.

It should be noted however that higher mutual solubility of the components can also be connected with the non-equilibrium character of the metal particles process formation.

In contrast, another feature of the phase portraits of this NBS is that the solid solutions are formed with reluctance in thermodynamically allowed region. In this case it is probably connected with a significant difference in redox potentials of nickel and copper. In the region of relatively high contents of copper at $E_{Cu}^{Red/Ox} > E_{Ni}^{Red/Ox}$ the reducing of Cu^{2+} cations goes significantly faster and the complicated layer non-equilibrium structures instead of solid solutions (allowed by phase diagram) are formed (figure 5).

The process of NBS oxidation also has some features:

- the kinetics of the process for all investigated systems in the region 50-150 °C (automatic gas micro-burette, atmosphere of oxygen) is three-stage: the fast initial stage with low activation energy of oxide films formation (according to RPES data) by the thickness of several nanometers, the second stage with the usual "parabolic" kinetic dependence (for sphere-like aggregates) and E_a=26÷40 kcal/mol for different NBS and then the third stage – not trivial process with the almost constant rate of oxidation – up to observed oxidation degree (≈7%);
- the phase composition of NBS essentially influences speed of oxidation (according to the obtained phase portraits);
- in the NBS containing the metal of variable valence the consecutive oxidation reaction takes place: $Me^0 \rightarrow Me^+ \rightarrow Me^{2+}$, and the relative rates of the stages ambiguously depends on particles surface etching;

the passivation of the particles by specially formed oxide-spinel (in case of Fe-Co and Fe-Ni systems) or enriched by nickel (for Cu-Ni system) thin surface films is observed; the partial auto-oxidation of NBS in high vacuum (≈ 10^{-9} mm hg) or in the inert atmosphere is observed at raised temperatures (400 °C and more) as a general phenomenon, as we guess it may be owing to O_2 and H_2O accumulated in closed nano- and micropores.

It is established (SQUID-magnetometer together with ITC SB RAS) that the synthesized pure (RPA) solid solutions of Fe-Co, Fe-Ni and Co-Ni systems with narrow crystallite size distributions and sphere-like aggregates at optimum compounds (40-50 weight % Co, ≈90 weight % Fe) in the region 5÷350 °K have the magnetization saturation values on 10÷15% greater than the ones achieved for massive alloys and produced by other ways NBS of similar compounds [11, 12].

3. Conclusions

For the first time the comprehensive research of both NBS synthesis by the reducing of metal aqua-chlorides and its properties is carried out. Besides, it is shown general scheme of synthesis. Additionally, it is presented phase portraits of investigated nanosystems.

References

[1] Yu. A. Zaharov, A. N. Popova, R. P. Kolmikov, V. M. Pugachev, V. G. Dodonov // 2008 *Perspective Materials,* **6,** 249-254

[2] Yu. A. Zaharov, A. N. Popova, R. P. Kolmikov, V. M. Pugachev, V. G. Dodonov // 2010 *Perspective Materials*, **3,** 93-104

[3] Yu. A. Zaharov, R. P. Kolmikov, V. M. Pugachev, V. G. Dodonov // 2010 *Russian national conference on nanomaterials «Nanotechnologies of Functional Materials»* (St.-Petersburg: St-PbSU), 2010, 214-215

[4] Yu. A. Zaharov, V. M. Pugachev, V. G. Dodonov, R. P. Kolmikov // 2011 *Perspective Materials,* **4,** 156 –164

[5] V. M. Pugachev, Yu. A. Zaharov, A. N. Popova, et al. // 2010 *«The 1st conference «Usov's Proceedings»* (Novosibirsk: GEO), 2010, 41-46

[6] Yu. A. Zaharov, A. N. Popova, V. M. Pugachev, V. G. Dodonov // 2010 *Book of articles and reports of scientific and technological sections. III International forum on nanotechnologies «Rosnano 2010»* (Moscow)

[7] Yu. A. Zaharov, A. N. Popova, V. M. Pugachev, et al. // 2009 *Book of articles and reports of scientific and technological sections. II International forum on nanotechnologies «Rosnano 2009»* (Moscow), 2009, 364 - 367

[8] Yu. A. Zaharov, R. V. M. Pugachev, V. G. Dodonov, A. N. Popova // 2010 *«The 4-th conference* «Sviridov's *Proceeding*s» (Minsk: BSU), 2010, 41-46

[8] Yu. A. Zaharov, R. V. M. Pugachev, V. G. Dodonov, A. N. Popova // 2010 *«The 4-th conference* «Sviridov's *Proceeding*s» **6** (Minsk: BSU), 2010, 24-32

[9] E.A. Ilyinchik, V.V. Volkov, L.N. Mazalov // 2005 *J. of Structural Chemistry*, **46 (3),** 535-546

[10] Osamu Kitakami, Hisateru Sato, Yutaka Shimada et al. // 1997 *Phys. Rev. B.*, **56 (21)**, 13849-13854

[11] P.Fulmer, J. Kim et al. // 1999 *Nat. Res. Center for Plutonium: ANRCP*

[12] E.L. Dzidziguri, D.V. Kuznetsov, V.V. Levina, E.N. Sidorova // 200 *Perspective Materials,* **6,** 87-92

Acknowledgments
The work was supported by The Integration Grant SB RAS.

Journal of Physics: Conference Series **345** (2012) 012031 doi:10.1088/1742-6596/345/1/012031

Use of the hybrid recirculation scheme for lower power expenses

A Kudinov[1,2], D Timofeev[1,2] and N Laguntsov[1,2]

[1] Type National Research Nuclear University "MEPhI", Moscow, Russia
[2] JSC "Aquaservice", Moscow, Russia

E-mail: cruel_13@mail.ru

Abstract. In the present work generalised hybrid recyculation installation for manufacture highly enriched nitrogen (with concentration of nitrogen in a stream of a product above 99 %) is offered. This installation consists of a combination of membrane and absorbing gas separation blocks. Use of the hybrid scheme allows to raise the maintenance of a target component, extraction degree and also to lower power expenses. Functioning processes of membrane and sorption blocks hybrid gas separation systems are described. Lacks of single methods are described, and advantage of a hybrid combination of each of them resulted.

1. Results and discussion

At creation of the inert or drained atmospheres product improvement of quality (increase of concentration of nitrogen, humidity decrease) is essentially important at low power inputs, cost of installation and her service. These problems can be solved only on the basis of use of the hybrid technologies combining membrane and adsorption methods of separation. Thus optimum combination of advantages of a membrane method of separation (simplicity, high degree of extraction) and adsorption (high degree of enrichment by nitrogen) is probable.

Membrane gas adsorption is the new hybrid process combining advantages of membrane technology (compactness, modularity) and adsorptions (high selectivity in relation to target components). Thus the membrane acts not only as the selective device, but also as a partition dividing phases with the, independently varied hydraulic parameters and providing effective protection against dust particles of a sorbent.

As a result of the performed researches it is shown that the hybrid scheme of installation allows to increase degree of extraction of a product and to lower power expenses for 30-40 %. At hybrid schemes rather small mass-size characteristics, and also low level of power consumption of an order 0,5 kW/m3 nitrogen. Besides it these schemes possess high assembly readiness and simplicity in service and operation. For example, operation mode setup time is about 5-10 minutes. These hybrid schemes have spectrum of application, they can be used for domestic needs and in industrial [1].

Hybrid systems can be used for maintenance with dry nitrogen of cable highways, left-luggage offices of products, technological lines of food, chemical and other manufactures, creation of inert atmospheres over combustible fuels in gasoline storage tanks, cars and planes. since installation have second output stream, which enriched with oxygen, these installations also can be used for creation of comfortable conditions of ability to live of the person.

At present time researches in this field have research character, mainly at level of effects. Practically there are no hybrid gas separation technologies; there are no accurate technology requirements for the decision of those or other problems.

The hybrid technology developed within the limits of the project is supposed to be realized on a basis of nanostructured polymeric membranes that essentially expands area of applicability of a method.

2. Materials and methods

Hybrid combination of membrane and sorption methods with circulating contour allows to provide high concentration and high degree of extraction of target components in a product stream at rather small power expenses [2].

3. Conclusions

Thus in the work it is shown that application of the hybrid scheme allows to use to the full advantages of both separation processes, and to compensate their lacks. As a result the hybrid scheme of installation gives the chance to increase degree of extraction of a product and to lower power expenses for 30-40 %.

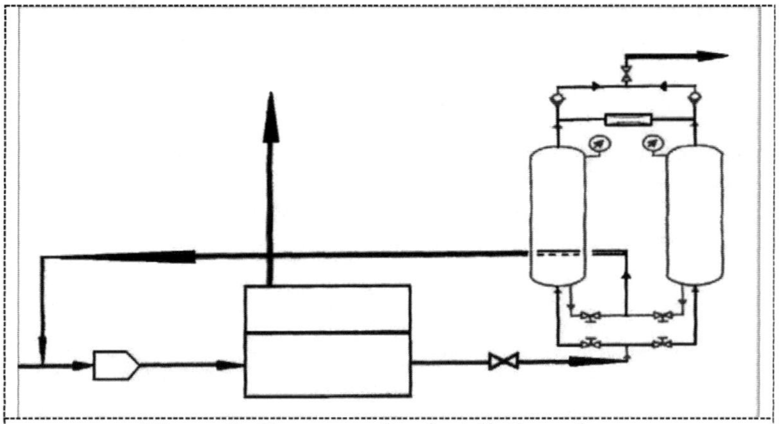

Figure 1. Gas separation installation scheme

4. Acknowledgements

Installation is completely certificated, and Sanitary-Epidemiological Conclusion for use in the food-processing industry and creations of artificial respiratory atmospheres including for use on Russian Defense Ministry objects received. Installation is protected by the patent for useful model «hybrid air separation installation» № 95547 under the demand № 2010107416 from 3/2/2010.

References

[1] Kudinov A N, Laguntsov N I and Kozhevnikov V Y 2010 *XI The Russian National Conference MEMBRANES-2010* **2** 180-181

[2] Laguntsov N I, Grusdev E B, Kosykh E V and Kozhevnikov V Y 1992 *Journal of Membrane Science* **67** 15-28

Journal of Physics: Conference Series **345** (2012) 012032 doi:10.1088/1742-6596/345/1/012032

Calculation method of direct and reverse effects anisotropy of permeability gas through the nanoporous membrane.

D.V. Timofeev A.V. Krukov

National Nuclear Research University MEPhI
JSC Aquaservice, 115409, Moskow, Kashirskoe h., 31,

E-mail: aquaserv@mail.ru

Abstract.

If implement the free-flow conditions in channels with a multilayer membrane inhomogeneous surface effects have been found to nonsymmetric gas transport. Made suggested [1-2] that the main reason for their occurrence is a specific interaction of gas molecules with the walls of the pores in the membrane.

1. Introduction

In [1-2] have been found to nonsymmetric gas transport effects (the effect of anisotropy of permeability of the membrane, the so-called "gas diode", changing the permeability of the membrane and the rate of catalytic reactions in the direction of motion of the gas). Such effects can be realized in the asymmetric nanoporous membranes - membranes, consisting of two or more layers with different values of porosity and pore sizes in the layers of the membrane [2]. Because these effects are observed in the free-molecular mode, has been suggested [1-4] that the main reason for their occurrence is the specific interaction of gas molecules with the rough walls of the pores in the membrane.

In this paper we consider a model that allows to calculate the anisotropic gas transport processes in the asymmetric nanoporous membranes. As the object of research was selected through a process of gas transport track membrane in the free flow regime. Membrane pores are cylindrical channel of diameter 50 nm for 11.5 micron and 0.5 micron layer of a length with a diameter of 30 nm. The results are compared with experiments conducted on the membranes, the data are given in Table (1).

Table 1. Characteristics of the investigated membranes.

Membrane	**Number of pores per**	**Thickness membrane**	**Thickness of the first layer**	**Pore diameter of the first layer**	**Pore diameter of the second layer**
MA-30/50-$2{\cdot}10^9$	$2{\cdot}10^9$ cm^2	12 microns	0,5 microns	30, nm	50, nm
MC-50/50-$9{\cdot}10^8$	$9{\cdot}10^8$ cm^2	12 microns	0,5 microns	50, nm	50, nm

Journal of Physics: Conference Series **345** (2012) 012032 doi:10.1088/1742-6596/345/1/012032

2. Materials and methods

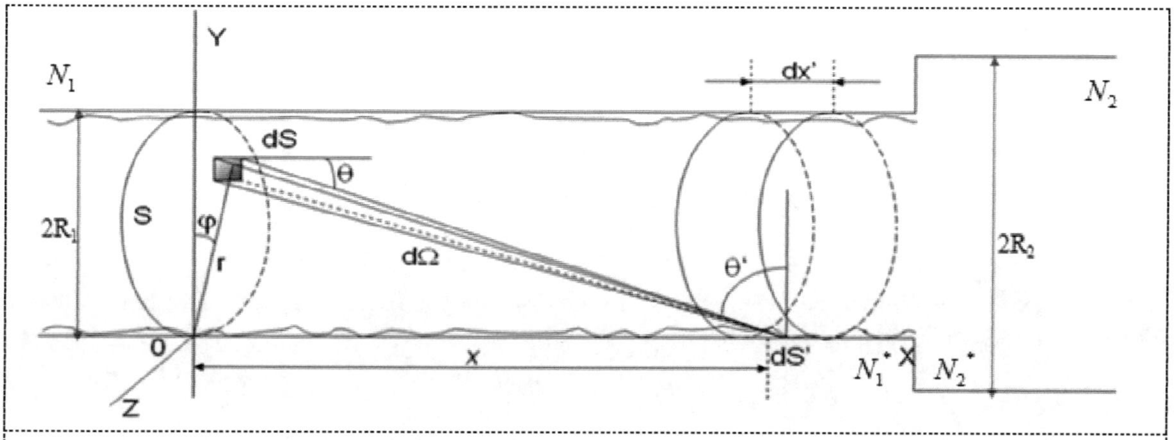

Figure 1. Calculation of the coefficients of conductivity in a channel with a rough surface.

We will consider the pore of the membrane in the form of two cylindrical channels of different widths corresponding to the pore radius of each layer of the membrane, as shown in Fig. 1. To describe the interaction of molecules with the surface of pores inside the channel model is used "white noise", according to which the distribution of molecules in the directions of departure from the surface of the pores is described by:

$$w(\theta) = \sqrt{\frac{A}{\pi}} \frac{\exp(-A\theta^2)}{erf(\frac{\pi}{2}\sqrt{A})}, \tag{1}$$

If the value of A <<1, the distribution is isotropic in directions that correspond to the diffuse model of the interaction of molecules with the surface (full accommodation). Once the molecule is sorbed by the surface, it departs from it at any angle (all directions are equally probable). The value of A>> 1 corresponds to a rough surface (the range of angles of departure from such a surface is limited, this means that the amplitude of the correlation function of white noise, the smaller the greater roughness).

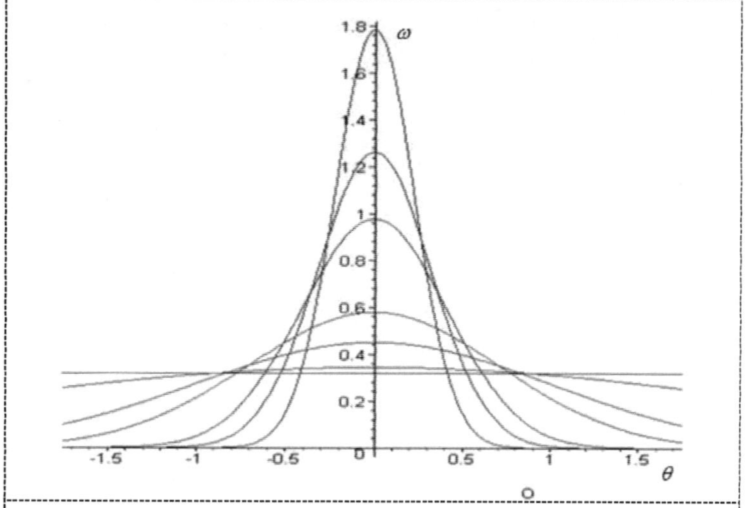

Figure 2. Distribution function at different interaction parameters.

Distribution function at different interaction parameters presented in Fig. 2. As seen in Figure 2, the greater the interaction parameter, the more preferable is the direction of reflection of molecules from the surface perpendicular to the surface of the channel.

3. Results

To calculate the flow in a channel used by an expression analogous to Fick's first law, written with respect to the gradient of the number of collisions with the surface [5]:

$$J = D\frac{dN_{cm}}{dx}, \tag{2}$$

Coefficient of conductivity (coefficient similar to the diffusion coefficient in Fick's first law, written down about the gradient of the collision with the surface):

$$D = 2R\int_{-\frac{\pi}{2}}^{+\frac{\pi}{2}}\omega_{\varphi}\cos\varphi d\varphi\int_{0}^{2R\cos\varphi} r^2 dr\int_{-Rtg\gamma}^{+Rtg\gamma}\frac{x^2}{(r^2+x^2)^2}dx, \tag{3}$$

The dependence of the number of collisions in the channel of the interaction parameter in the model of a white noise:

$$N_{cm} = n(x) < V_y >= n(x)V_h\int_0^{\pi/2} w(\theta)\cos(\theta)d\theta = n(x)V_h\int_0^{\pi/2}\sqrt{\frac{A}{\pi}}\frac{\exp(-A\theta^2)}{erf(\frac{\pi}{2}\sqrt{A})}\cos(\theta)d\theta, \tag{4}$$

It should be noted that due to the nonisotropic distribution there is a dependence of effective diffusion coefficient for the entire membrane of the pressure. This can be seen from Fig. 2[6-7]. Where the nonlinear dependence of the visible deviation of the distribution of molecules in the directions of motion upon application of pressure drop across the system, leading to the final formula for calculating the flux through the membrane bilayer rough:

$$J = \frac{\omega_{A1}(\theta_1)\omega_{A2}(\theta_2)(D_1D_2 - P_1D_1kV_{x2} - P_2D_2kV_{x1})}{D_2l_1\omega_{A2}(\theta_2) + D_1l_2\omega_{A1}(\theta_1) + k\omega_{A2}(\theta_2)V_{x2}l_1(P_2 - P_1) + k\omega_{A1}(\theta_1)V_{x1}l_2(P_1 - P_2)}(P_1 - P_2) \tag{5}$$

,where D_1, D_2 - conductivity coefficients calculated for each of the layers, $\omega_{A1}(\theta_1)$, $\omega_{A2}(\theta_2)$ - distribution of molecules in the directions of motion for each layer, coefficient k is determined from the same parameters of the two layers (the limit of a two-layer to single-layer system), average projection of the velocity of the molecules moving in the direction of flow$<V_{x1}>$:

$$< V_{x1} >= V_T\int_0^{\pi/2} w(\theta)\sin(\theta)d\theta \tag{6}$$

Conclusions

Thus, in a model that allows processes to calculate the forward and backward anisotropic gas transport in asymmetric nanoporous membranes (depending on the ratio of the interaction parameters A for each of the layers). The resulting model provided a diffuse model of interaction in the smooth channel ($A \to 0$) degenerates into a well-known model for resistance of the membrane bilayer[8].

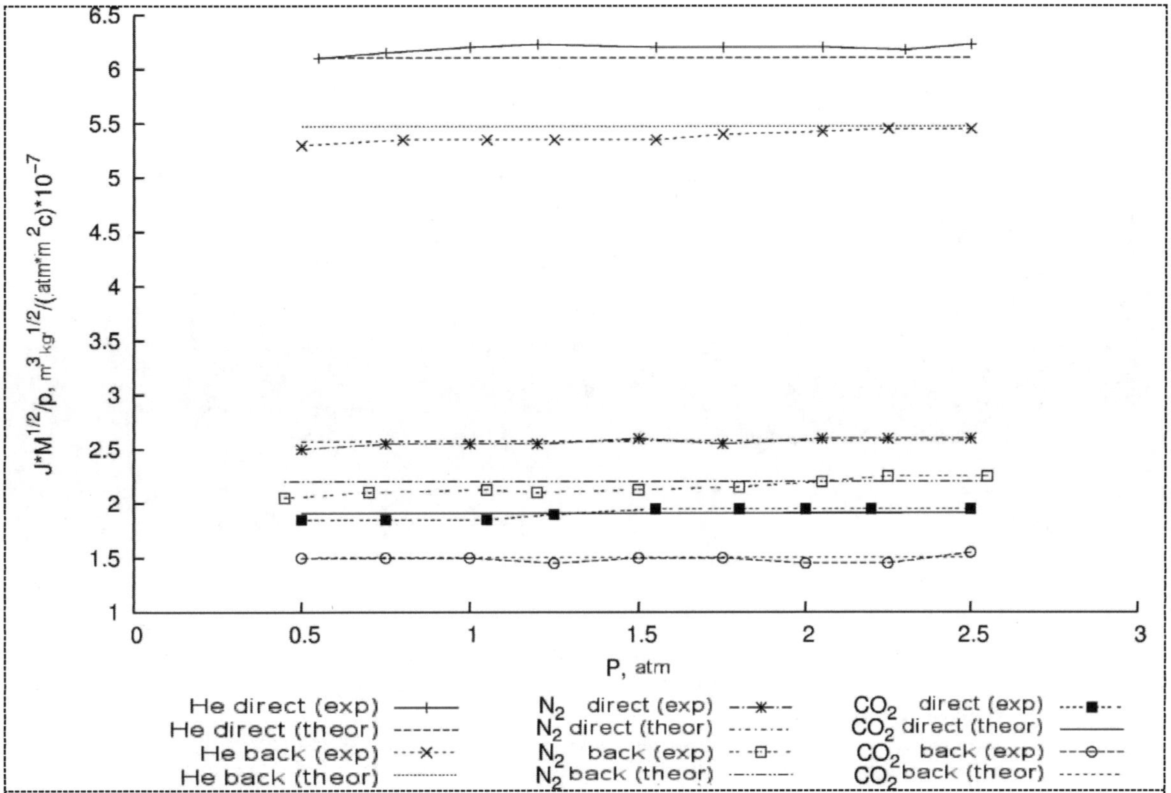

Figure 3. Experimental and theoretical dependence of the reduced flow from the pressure in the track membrane with the asymmetric shape of pores.

The good agreement between calculations and experimental data, which confirms the assumption that the asymmetry of the transport can be explained by specific interaction of molecules with the pore walls.

Acknowledgements
The work was supported by the grants of RFBR 10-08-01001-A.

References

[1] Kurchatov IM, Laguntsov NI, Tronin VN, Uvarov VI 2007 *Alternative Energy and Ecology*. № 5 (49). P.134-139.
[2] Kurchatov IM, Laguntsov N.I., Tsodikov MV, Fedotov AS, Moses II 2008 *Kinetics and Catalysis*,. T.49. Number 1. P.129-135.
[3] Kryukov AV, Kurchatov IM, Laguntsov NI, Tronin VN, Uvarov VI 2009 *Series. Critical technologies. The membrane.* № 1 (41). P.33-38
[4] Kurchatov IM, Laguntsov NI, Tronin VN, Uvarov VI, Borovinskaya IP 2008 *Reports of Academy of Sciences. Series: Physics*.. T.419. Number 1. P.38-40.
[5] LD Landau, EM Lifshitz *Theoretical physics: a tutorial*. T.10. *Physical Kinetics*. 2nd ed. M.: FML, 2002. P.87
[6] Kurchatov IM, Kryukov AV, Volkov VV, Laguntsov NI, Mchedlishvili BV 2009 *Series. Critical technologies. The membrane.*. № 3 (43). P.3-8.
[7] Kryukov AV, Kurchatov IM, Laguntsov NI 2010 *Alternative Energy and Ecology*. Vol.12. P.50-54.
[8] Pat. 4,230,463 US, 1980

HYBRID MEMBRANE CONTACTOR SYSTEM FOR CREATING SEMI-BREATHING AIR

Timofeev D.V.

National Nuclear Research University MEPhI

JSC Aquaservice, 115409, Moskow, Kashirskoe h., 31

dobtimofeev@gmail.com

Abstract. Typically, the equipment to create an artificial climate does not involve changing the composition of the respiratory air. In particular in medical institutions assumes the existence of plant of artificial climate and disinfection in operating rooms and intensive care wards.
The use of a hybrid membrane-absorption systems for the generation of artificial atmospheres are improving the respiratory system, blood is enriched or depleted of various gases, resulting in increased stamina, there is a better, faster or slower metabolism, improves concentration and memory. Application of the system contributes to easy and rapid recovery after the operation. By adding a special component, with drug activity, air ionization, and adjust its composition, you can create a special, more favorable for patients with the atmosphere. These factors allow for the treatment and rehabilitation of patients and reduce mortality of heavy patients.

1. Introduction

The system consists of a set of technologically connected membrane modules and membrane sorption-contactors (Fig. 1) [1]. The system has several of designs, can be located in the breathing zone and outside it possible that the part of the system is inside the zone, and part outside. In order to execute various therapeutic requirements the system provides the introduction of additional components in the air. The system can also be used for artificial ventilation of lungs. The composition of the locally-breathing zones can be adjusted within wide limits. For example, oxygen can be adjusted in the range of 15-30% carbon dioxide in the range from 0.03 to 1.0% [2], the humidity can be controlled in the range of 5 to 90% [3-4], and harmful organic impurities can be reduced by a factor of 2-4. Possible the addition of medicines needed by the patient. Performance is not less than 5 l / min. The same system can be created to regulate the temperature of the

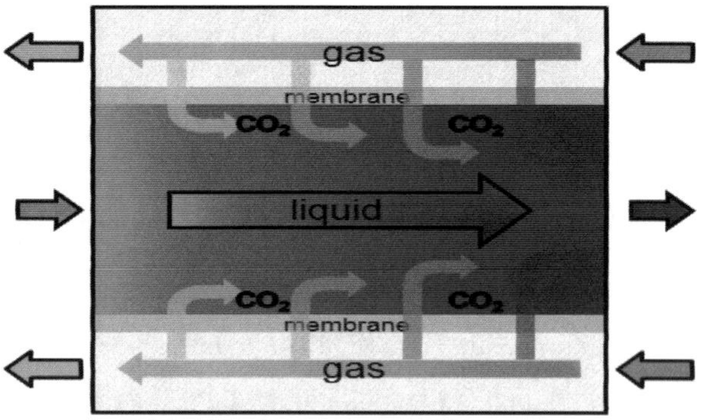

Figure 1. The scheme of flow in membrane sorption-contactor.

atmosphere. The intended use is modular, depending on the purpose of the plant.

The operation is based on a consistent treatment and enrichment of air with different gases, vapors, aerosols, medical components, as in the whole room and in separate areas. The system can be located, both indoors and outdoors, depending on the specific application. At the first stage the air fed into the membrane module, the outputs of which are obtained by two streams - the oxygen enriched to 40% and 10% deplete. There are also cleaning the air enriched in oxygen from organic impurities. In the second stage of membrane-sorption method controls the content of carbon dioxide in the range of 0,02-1,0%, thereby reducing or increasing the content of the gas in the blood. After that there is drainage or wetting the mixture[5-6], which allows for the desired mode of therapy. May then be added to a variety of aerosols, vapors, including medicines, can make air ionization.

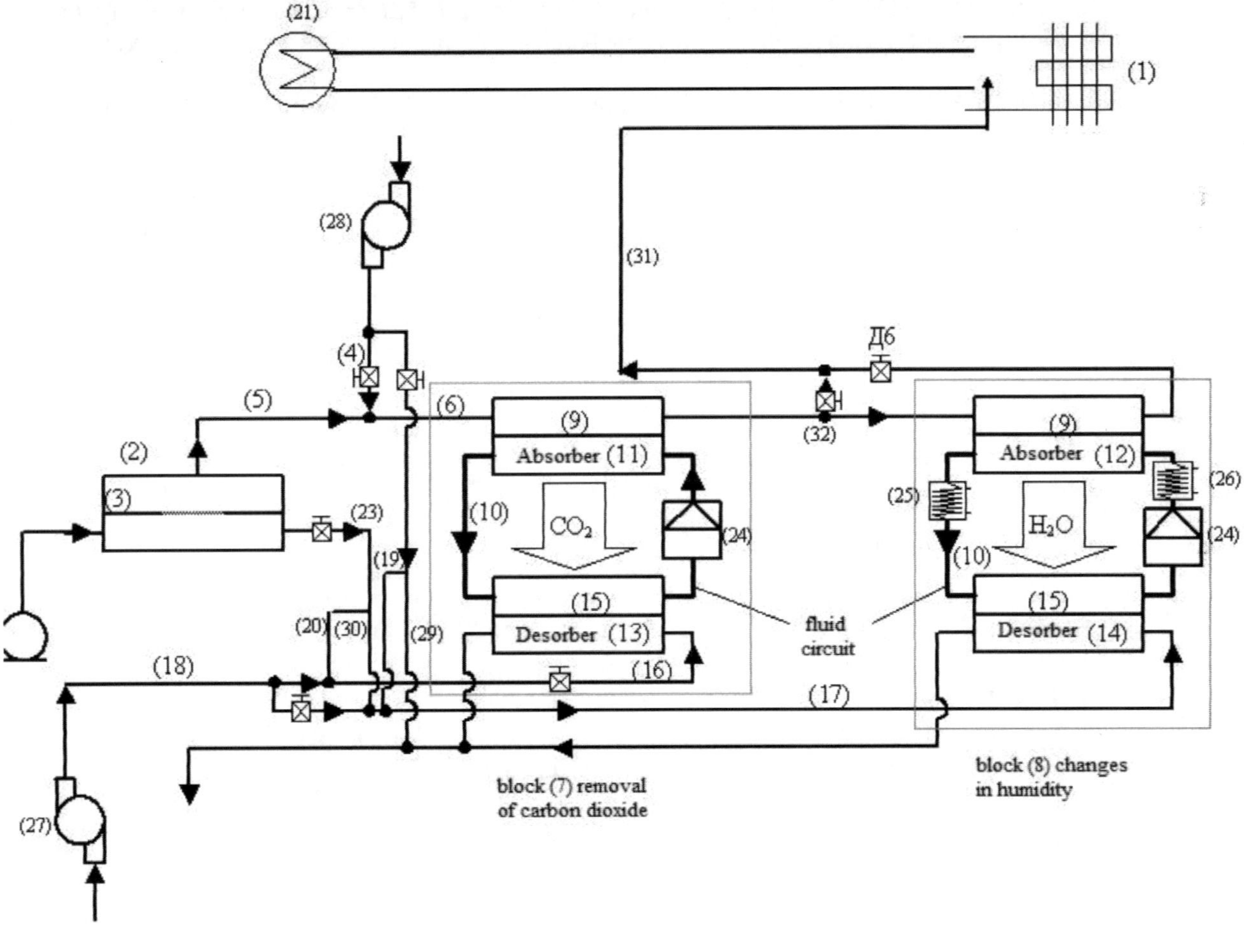

Figure 2. Scheme of membrane sorption plant for creation and management of semi-respiratory air.

2. Results

The use of a hybrid membrane-absorption systems for the generation of artificial atmospheres are improving the respiratory system, blood is enriched or depleted of various gases, resulting in increased stamina, there is a better, faster or slower metabolism, improves concentration and memory. Application of the system contributes to easy and rapid recovery after the operation. By adding a special component, with drug activity, air ionization, and adjust its composition, you can create a

Journal of Physics: Conference Series **345** (2012) 012033 doi:10.1088/1742-6596/345/1/012033

special, more favorable for patients with the atmosphere. These factors allow for the treatment and rehabilitation of patients and reduce mortality of heavy patients.

Adjustment of the content of all these components (oxygen, humidity, carbon dioxide) does not require any additional sources than the indoor and outdoor air. And both indoor and outdoor air before it enters the room is passed through a filter system, including non-porous and nanoporous membranes, which leads to the exclusion of the possibility of its dust and reduce the content of certain hazardous organic components. These measures help to maintain the composition of the atmosphere created is almost independent of the environment.

3. Conclusions

Based on this work it can be assumed that in the near future the preference for creating semi-breathing air can be given to modern energy efficient and environmentally friendly membrane technologies, among them a special place in membrane contactor technologies, including all the benefits of the absorption method (high performance cleaning and drying, the availability of reagents) and excluding their disadvantages (power consumption, low technological, environmental risk, the complexity of implementation)

Establishing an appropriate equipment will create the optimal composition of the artificial atmosphere, and the variation of the gas can make such a system is applicable both in office and administrative buildings (the creation of recreation rooms with a high oxygen content) as well as in medical and sporting purposes (the creation of both low and high oxygen content with simultaneous removal of excess moisture in the room). So far, this method of changing the composition of the atmosphere is practically not used, but in the coming years due to the high probability of various disasters and environmental degradation can find the widest possible dissemination.

The work was supported by RFBR, project 09-08-00972-A.

References

[1] V.Yu. Kojevnikov, N.I. Laguncov, E.V. Levin, A.Yu. Okunev, R.S. Hafizov Absorbcionno-desorbcionnoe ustroystvo. Patent Na Izobretenie №2304457 Po Zayavke №2005133111 Ot 28.10.2005, Byul. № 23 Ot 20.08.2007. (Russia)

[2] Timofeev D.V. Kurchatov I.M. Levin E.V. Laguntsov N.I. Membrane-Absorbption Methods for Biogas-Biomethane Reprocessing. Proceeding of the 5-th Membrane Science and Technology Conference of Visegrad Countries with Wider International Participation "PERMEA 2010", September 2 - 8, 2010, Tatranske Matliare, Slovakia, p. 82

[3] N.I. Laguntsov, A.Yu. Okunev, V.V. Teplyakov, V.V. Usachov. Membrane contactor air conditioning system: Experience And Prospects. Separation And Purification Technology, Volume 57, Issue 3, 15 November 2007, Pages 502-506)

[4] Timofeev D.V., Laguntsov N.I., Levin E.V., Conditioning of biogas by absorption/membrane method. Alternative energy and ecology, № 11, 2010, p. 95-100

[5] Laguntsov N., Levin E., Okunev A., Numerical Simulation of Mass- Transfer Process in Membrane Contactor for Gas/Vapor Separation//"Euromembrane 2004", 28 Sept. – 1 Oct., 2004, Hamburg, Germany, Book of Abstracts p.184

[6] Usachov V., Laguntsov N., Okunev A., Teplyakov V., Glukhov S. Experimental study of the membrane contactor systems for gas dehumidification // Ars Separatoria Acta. 2003 (2). P. 36-46.

Liquid flow in nanotubes

S A Chivilikhin[1], I Yu Popov[1], E M Aryslanova[1], D N Vavulin[1], V V Gusarov[2]

[1]St.-Petersburg National Research University of Information Technologies, Mechanics and Optics, Kronverksky, 49., St.-Petersburg, 197101, Russia

[2]St.-Petersburg State Technological Institute (Technical University), Moskovskiy, St.-Petersburg, 190013, Russia

E-mail: sergey.chivilikhin@gmail.com

Abstract. A model of fluid flow in the nanotube is proposed. The model based on the assumption about the existing of locally ordered nano-sized regions (crystallites) in the liquid. The flow in nanotube describes using the mass and energy balance. We take into account the slip boundary conditions on the wall of nanotube and the crystallite on the entrance and inside of nanotube. The effective viscosity of liquid in nanotube is calculated.

1. Introduction

Last years, fluid flows in nano-sized domains are intensively studied [1, 2]. At present, there are no general equations of nanofluidics. Usually, the molecular dynamics is used for computations [3]. As for analytical approaches, the simplest one involves introducing the slip condition at the boundary [4]. There is also a hybrid method, incorporating the continuous approach and the molecular dynamics (which is used to analyze the fluid structure and determine the rheological law) [5]. In [6] fluctuations are taken into account when deriving the hydrodynamic equations. Great increase of effective viscosity of water in a channel of nano-width is observed. For nanochannel flow the molecular structure of the medium plays a crucial role. Flows in nano-channels are influenced by local heterogeneity of molecular structure of the liquid if its size is compared with the channel width. A hypothesis about the existence of locally-ordered structures in liquid was put forward in [7]. During several decades scientists discuss whether there are local domains in liquid (crystallites) in which the molecular structure is similar to that for crystal [8, 9]. Investigations of fluid flows in nano-sized domains show that it is strongly influenced by local ordering of nano-sized scale. Experiments [10] show that the effective viscosity of water in nano-channel with hydrophilic walls is essentially greater than the corresponding macroscopic value. Computational experiments in [11] lead to appearance of ice-like states in nanotubes of small diameters under normal conditions. In this work we develop a model [12] based on the assumption about the influence of crystallites on the fluid flow in nanotube.

The character of the flow in the nanotube depends on the correlation between the locally equilibrium size of the crystallite and the nanotube diameter. We try to find the effective viscosity μ_{ef} of the fluid in the nanotube which is defined by the following way. Consider the correlation between our nanotube (possibly, containing crystallites) and classical tube (with the Poiseuille flow) of the same size and with the same pressure drop and flow rate. The viscosity of the Poiseuille flow having the same parameters is called the effective viscosity of the flow in the nanotube.

2. Theory

For small nanotube diameter ($D < h$) there is no crystallite inside the nanotube, but it is at the entrance, and to form the flow it should transform to liquid-like nonautonomous phase (with size h). Roughly speaking, the flow is formed by "melting" of the crystallite at the "entrance". Mass balance gives us $\pi H^2 V = \pi D^2 v$, where v is the average (for the cross-section) velocity of the flow in the nanotube, H and V are the characteristics size of crystallite and the crystallite velocity. Let us write down the energy balance equation. To do this we take into account that the power of the pressure forces $N_p = \pi D^2 v \Delta P / 4$ is equal to the sum of powers corresponding to viscous energy dissipation in the nanotube $N_\mu = 2\pi D^2 L \mu_0 v'^2$ and to crystallite transformation to the liquid-like non-autonomous phase (at the entrance of the nanotube) $N_q = \pi H^2 V q / 4$. Here ΔP is the pressure difference on the nanotube, μ_0 is the viscosity of the nonautonomous phase inside the nanotube, q is the specific heat of crystallite transformation to liquid-like non-autonomous phase (for unit volume), $v' = \dfrac{v}{0.5D + L_s}$ is the characteristic speed of liquid deformation inside the nanotube, L_s is the slip length of liquid on the wall of nanotube, L is the length of the nanotube.
The described algorithm leads to the following expression for the effective viscosity:

$$\mu_{ef} = \frac{qD^2}{32vL} + \mu_0 \left(\frac{D}{D + 2L_s} \right)^2 .$$

3. Conclusions

The model of fluid flow in nanotube was developed. This model based on the assumption about the influence of crystallites on the fluid flow in nanotube. Using this model the effective viscosity of liquid in the nanotube via the nanotube diameter was calculated.

Acknowledgements
The work was supported by Program "Development of Scientific Potential of Russian High School" (project 2.1.1/9425), Federal Targeted Program "Scientific and Educational Human Resources for Innovation-Driven Russia" (contracts P689 NK-526P, 14.740.11.0879, and 16.740.11.0030), grant 11-08-00267 of Russian Foundation for Basic Researches.

References

[1] Li D 2008 *Encyclopedia of microuidics and nanouidics* (New York: Springer)
[2] Rivera J L and Starr F W 2010 *J. Phys. Chem. C.* **114** 3737
[3] Kadau G at al 2004 *Proc.Nat. Acad. Sci. USA.* **101** 5851
[4] Joly L at al 2006 *Rev. Lett.* **2060** 96 046101
[5] Nie X B at al 2004 *J. Fluid. Mech.* 2004 **500** 55

[6] Kang W and Landman U. 2007 *Phys. Rev. Lett.* **98** 064504
[7] Frenkel Ya I 1975 *Kinetic Theory of Liquid* (Moscow: Nauka)
[8] Ubbelohde A R Melting and Crystal Structure 1965
[9] Huang C et al 2009 *Proc. Nat. Acad. Sci. USA* **106** 15214
[10] Li T D at al 2007 *Phys. Rev. B.* **75** 115415
[11] Mashl R J at al 2003 *Nano Letters* **3** 589
[12] Chivilikhin S A, Gusarov V V and Popov I Yu 2008 *Russian J. Math.* Phys. **15** 410

Journal of Physics: Conference Series **345** (2012) 012046 doi:10.1088/1742-6596/345/1/012046

Preferred oriented ZnO films growth on nonoriented substrates by CVD

A Abduev, A Akhmedov, A Asvarov, A Omaev

Institute of physics, Dagestan scientific centre of RAS, Yaragskogo str., 94, Makhachkala, 367003, Russia

cht-if-ran@mail.ru

Abstract. The processes of CVD synthesis of ZnO layers with different preferred orientation on non-crystalline substrates where studied. Synthesis was carried out in a hydrogen atmosphere on substrates with ZnO buffer layers with various preferred orientation, namely in the plane perpendicular as well as parallel to plane of the substrate. ZnO buffer layers were synthesized by magnetron sputtering of ZnO ceramic targets. Conditions for the ZnO layers formation by CVD with (002), (100), (110) preferred orientations and mixed (100) - (110) orientations were determined.

1. Introduction

The growing interest in zinc oxide oriented films is associated with the prospects of new LED, UV laser light sources, acousto-optic and other devices[1-3] This interest propels the demand for ZnO heteroepitaxial layers with low radiationless recombination channels. However, the synthesis of oriented layers implies the use of oriented substrates (such as sapphire, silicon carbide).

In this study we explored the synthesis of perfect ZnO layers by CVD with (002), (100) and (110) preferred orientation without the use of expensive crystalline substrates. Buffer layers are commonly used to manage the process of nucleation layers and contribute to improve the structure, characteristics and surface topography [4, 5]. This is due to the fact that the optimal conditions for the nucleation and subsequent growth of ZnO layers by CVD are significantly different: the ZnO nucleation occurs at relatively low temperatures, [6, 7], while the synthesis of perfect layers takes place in conditions close to equilibrium.

It is known that conditions (e.g. composition of the atmosphere Ar - O_2, substrate temperature and ion bombardment of the surface) for the formation of ZnO layers by ion-plasma methods can define layers preferred orientation [8-11]. In these studies the ZnO buffer layers (ZnO BLs) with preferred orientation were used to control the orientation of ZnO CVD layers. ZnO BLs were synthesized by magnetron sputtering at relatively low temperatures on non-oriented substrates. Subsequent CVD synthesis of ZnO layers was carried out in a flow of hydrogen by the method described in [6].

2. Experiment and materials

Initially ZnO BLs with the preferred orientation of perpendicular as well as parallel to the substrate plane were synthesized on the glass (Corning 1737) and oxidized Si substrates by altering mode of sputtering and bombardment conditions of the growing surface during the magnetron sputtering of ZnO ceramic target [10]. Thickness of the ZnO BLs was up to 300 nm. To improve the crystallinity of the ZnO BLs, they were further annealed in open atmosphere at about 600 °C for 1 hour.

Synthesis of ZnO layers was carried out via CVD in the reaction ZnO + $H_2 \leftrightarrow$ Zn + H_2O [6]. The temperature in the evaporation zone was 720 °C. The films were grown at different hydrogen pressures in the reactor and different temperatures in the deposition zone.

3. Results and discussions

3.1. ZnO buffer layers

The XRD and SEM data for two types of ZnO BLs are shown in Fig. 1. The ZnO BL with the *c* axis oriented perpendicular to the plane of the substrate, synthesized by dc-magnetron sputtering at moderate substrate temperatures (T = 200 °C), were clearly marked typical columnar structure with a column diameter of 100-200 nm (Fig. 1a).

The ZnO BL with a preferential orientation of the *c* axis of the crystallites in the plane of the substrate was prepared by rf-magnetron sputtering in the absence of substrate heating. These growth conditions favor the formation of crystallites with (100), (110) lying and (101) tilted orientations due to imbalances of the components (Zn and O) in the area of nucleation of the crystallites (Fig. 1b). cross-section view of ZnO film and the low intensity of the XRD peaks indicate a significant portion of the amorphous phase in the ZnO BL.

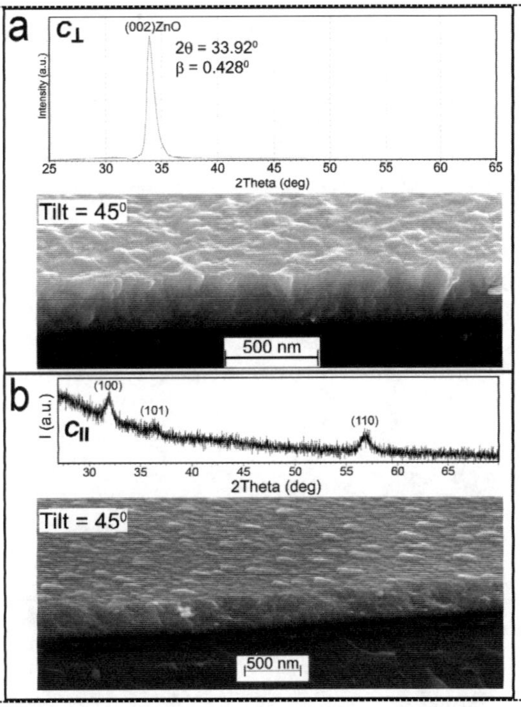

Figure 1. XRD patterns and SEM images of ZnO BLs with the preferred orientation of perpendicular (a) and parallel (b) to plane of the substrate.

3.2. Synthesis of ZnO layers with preferred (002) orientation by CVD

Fig. 2 shows the XRD spectrum of the ZnO layers, synthesized by CVD on the substrate with basic (002) preferred orientation of ZnO BL at a hydrogen pressure of 100 Torr and various substrate temperatures. For the ZnO layer deposited at 300 °C the reflections from the basic (002) plane and tilted (10*l*) planes are dominated. With increasing synthesis temperature to 600 °C in the XRD pattern begins to dominate the tilted (103) orientation of the plane which is the least densely packed opposite atoms. Thus, in the case of the CVD synthesis of ZnO layers at high pressures of H_2, we found no orienting action of the ZnO BL.

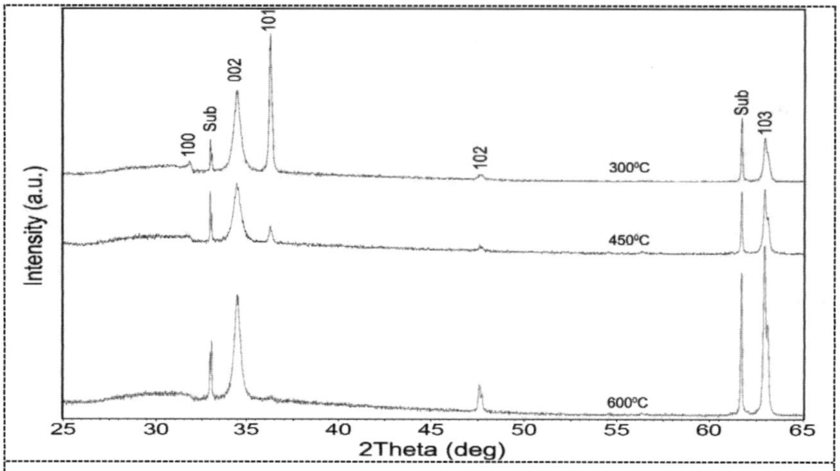

Figure 2. XRD patterns of ZnO layers grown by CVD at H_2 pressure of 100 Torr.

The decrease of the hydrogen pressure of 30 Torr in CVD reactor leads to a significant ordering of the structure of ZnO layer (Fig. 3). Even at the synthesis temperature of 300 °C in the XRD pattern of the ZnO layer, synthesized on (002) ZnO BLs, there are intense (002) peak and minor reflections from lying (100) and tilted (101), (103), (112) planes. Increasing the synthesis temperature results in increased intensity of the (002)ZnO peak, followed by a decrease in its FWHM, as well as minimize any reflections corresponding to the tilted planes.

In addition, with increasing temperature synthesis has been a shift in (002) and (004) peaks positions toward smaller angles to the tabulated values for bulk ZnO and decrease the ratio of their FWHMs, which indicates an improvement of crystalline perfection of the ZnO layer.

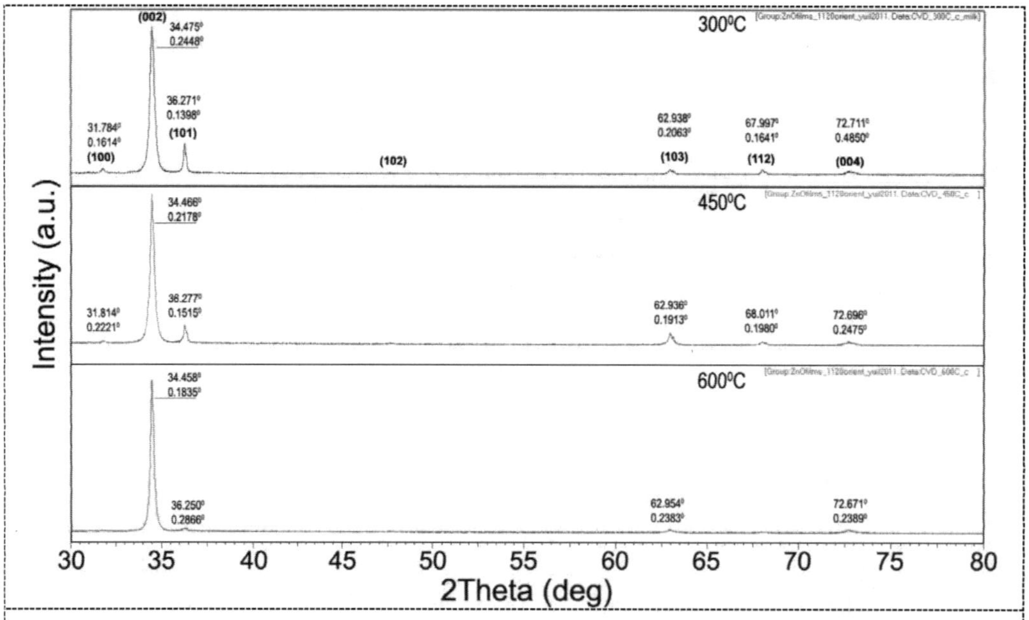

Figure 3. XRD patterns of ZnO layers grown by CVD at H_2 pressure of 30 Torr.

Journal of Physics: Conference Series **345** (2012) 012046 doi:10.1088/1742-6596/345/1/012046

3.3. Synthesis of ZnO layers with preferred lying orientation by CVD

Fig. 4 shows the XRD patterns of the ZnO layers, synthesized by CVD at a hydrogen pressure of about 30 Torr on the ZnO BLs with preferred lying orientations. It is shown that layers undergo significant structural changes with increased substrate temperature from 310 to 610 °C.

At a substrate temperature of 310 °C, (100) and (110) textured ZnO layers are observed, i.e., the c axis is oriented parallel to the substrate plane. Increasing the temperature to 450 °C leads to an appearance of the additional tilted (101) orientation. And finally, at 600 °C, a polycrystalline structure with a dominant tilted (101) orientation is observed. The increase in the synthesis temperature is accompanied by increased size and structural perfection of the ZnO crystallites with tilted (10*l*) orientations and deterioration of the structural characteristics for the ZnO crystallite with lying (100), (110) orientations.

We suggest that the transformation of the spectra with increasing temperature is associated with (i) a change in the ratio of Zn and ZnO in the gas phase as well as (ii) the differences in the densities of surface energy (001), (100), (110) and (101) planes [7, 8]. A detailed elucidation of the mechanisms requires further investigation.

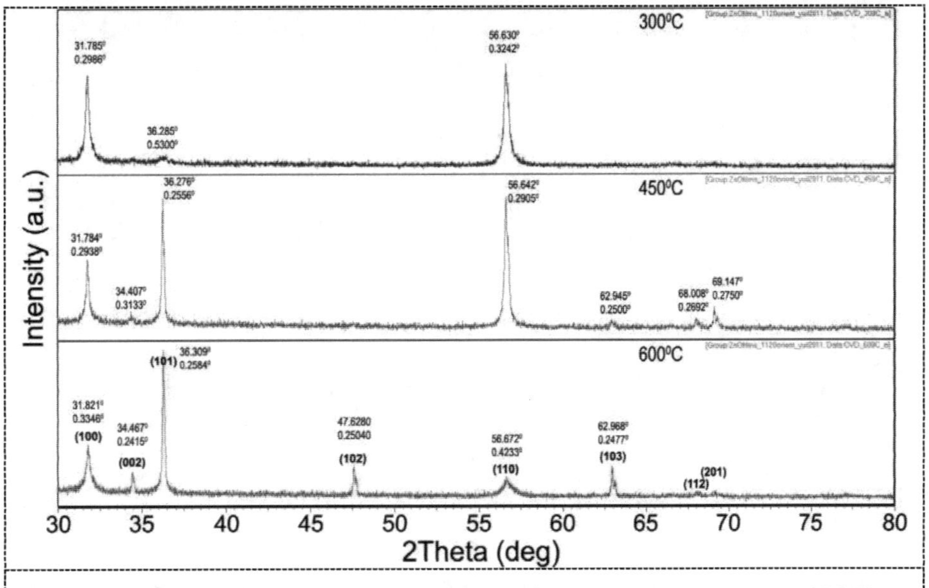

Figure 4. XRD patterns of ZnO layers grown by CVD at H_2 pressure of 30 Torr on the ZnO BLs with preferred lying orientations.

In Fig. 5 shows the plan of SEM micrographs of ZnO layers synthesized by CVD on ZnO BLs with different preferred orientations at 300, 450 and 600 ° C and a hydrogen pressure of 30 Torr (basic and lying preferred orientations of the ZnO BLs are marked with icons $c_\perp$ and $c_{||}$ respectively). The SEM micrographs clearly show that for ZnO layers deposited on a lying ZnO BL the grain size increases significantly already at the synthesis temperature of 450 ° C, whereas for the ZnO layers on the basic (002)ZnO BL the crystallite size increases noticeably only at 600 ° C. This is due to the fact that equilibrium growth conditions are reached earlier for the ZnO layers with nonbasic orientation. It is known that the formation of ZnO layers with the lying and tilting orientations occurs during the oxidation of zinc atoms on the growth surface, whereas the formation of ZnO layers with (002) preferred orientation – from the ZnO molecule flow [7]. In addition, the migratory ability for Zn atoms on the ZnO growing surface increases significantly at about 400 ° C [12]. Thus, the equilibrium conditions for the formation of ZnO layers from the flow of oxygen and zinc atoms (i.e. ZnO layers with non-basic orientation) is achieved at lower temperatures than the formation of (002)ZnO layers.

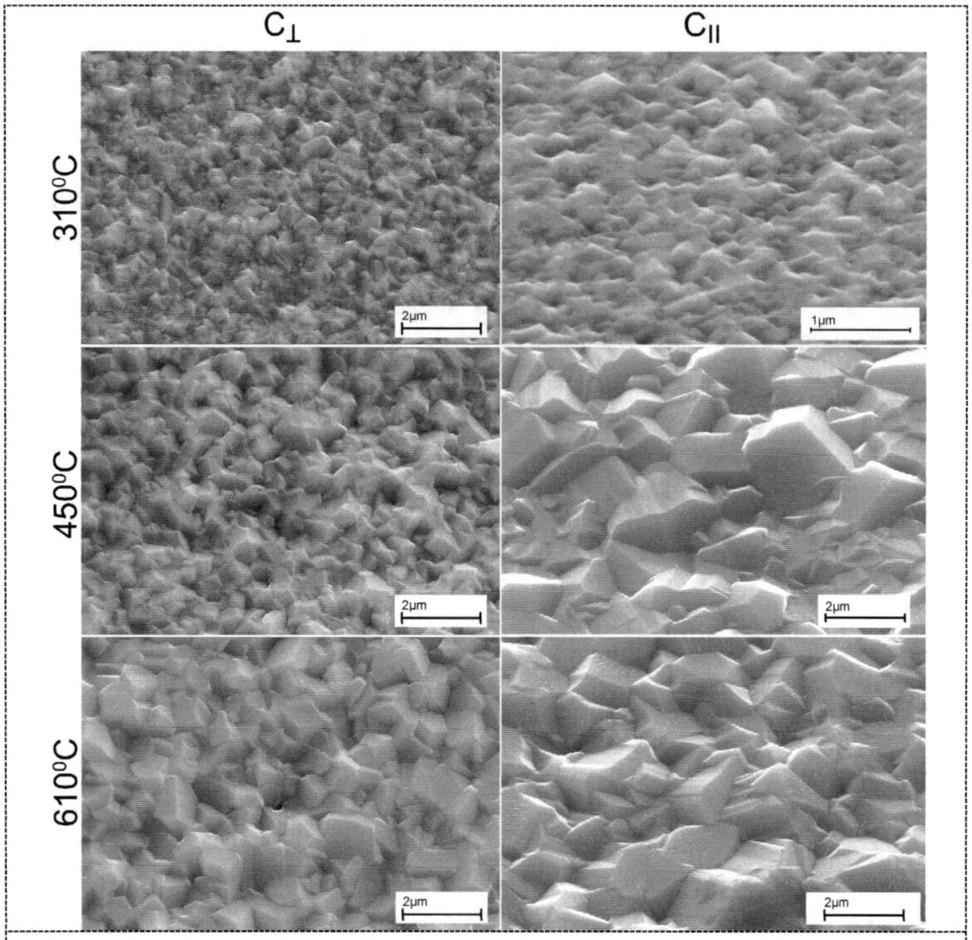

Figure 5. SEM images of ZnO layers grown by CVD at H_2 pressure of 30 Torr on the ZnO BL with preferred basic ($c_\perp$) and lying ($c_{||}$) orientations.

4. Conclusions

Results of the study show the possibility of formation of advanced CVD ZnO films with preferred orientation of the c-axis parallel or perpendicular to the plane of the nonoriented substrate.

Optimization of magnetron synthesis ZnO buffer layers with given preferential orientation leads to the formation of perfect crystal layers of ZnO with a different orientations on amorphous substrates.

There is reason to believe that by the optimization of nucleation and synthesis regimes will be possible to achieve the crystalline perfection of the ZnO layers on non-oriented substrates like one of epitaxial layers.

Acknowledgements

The work was supported by Grant of Presidium of RAS (Program "Innovation and Development"). The authors acknowledge the use of equipments of the Analytical Shared Center of Dagestan Scientific Center of RAS.

References

[1] Ohtomo A, Kawasaki M, Sakurai Y, Yoshida Y, Koinuma H, Yu P, Tang Z K, Wong G K L and Segawa Y 1998 *Materials Science and Engineering: B* **54**(1-2) 24
[2] Gorla C R, Emanetoglu N W, Liang S, Mayo W E, Lu Y, Wraback M and Shen H 1999 *J. Appl.*

Phys. **85** 2595
[3] Özgür Ü, Alivov Y I, Liu C, Teke A, Reshchikov M A, Doğan S, Avrutin V, Cho S-J and Morkoçd H 2005 *J. Appl. Phys.* **98** 041301
[4] Zeng Z, Liu Y, Yuan H, Mei Z, Du X, Jia J, Xue Q, Zhang Z and Salamo G 2010 *J. Phys. D: Appl. Phys.* **43** 1
[5] Park S, Choi C, Kim K and Kim S 2003 *J. of Electron. Mater.* **32**(11) 1148
[6] Abduev A Kh, Ataev B M and Bagamadova A M 1987 *Bulletin of the Soviet Academy of Sciences, Inorganic materials* **11** 1928
[7] Pant P, Budai J D, Aggarwal R, Narayan R J and Narayan J 2009 *J. Phys. D: Appl. Phys.* **42** 105409
[8] C.R. Aita, A.J. Purdes, R.J. Lad, P.D. Funkenbusch. The effect of O_2 on reactively sputtered zinc oxide. J. Appl. Phys. V. 51, N 10, 1980,
[9] Grankin I M, Kalnaya G I, Pogrebnyak V P and Lopushenko V K 1981 *Ukrainski Physicheski Journal* **26**(6) 1040
[10] Fujimura N, Nishihara T, Goto S, Xu J and Ito T 1993 *J. of Cryst Growth* **130** 269
[11] Kawamoto T, Yanagitani T, Matsukawa M, Watanabe Y, Mori Y, Sasaki S and Oba M 2009 *Proceedings of Symposium on Ultrasonic Electronics* **30** 479
[12] Abduev A Kh, Akhmedov A K, Asvarov A Sh, Kamilov I K and Sulyanov S N 2005 *NATO Science Series II: Mathematics, Physics and Chemistry* **194** 15

Journal of Physics: Conference Series **345** (2012) 012020 doi:10.1088/1742-6596/345/1/012020

High-speed nanocrystallization in a laser plasma

Bogonosov K.A.[1], Maximovsky S.N.[1, 2]

[1]Moscow State University Of Technology and Management by K.G. Razumovsky
[2]P.N.Lebedev Physical Institute RAS

kbogonosov@gmail.com

Abstract. The article briefly describes high-speed low-temperature nanocrystallization in a laser plasma. It also made an evaluation of some parameters of calculation of the crystallization process. Describes one possible course of crystallization process on a liquid substrate.

The In the Physical Institute of the Academy of Sciences a new phenomenon - high-speed nanocrystallization of amorphous metal substrate in a gaseous medium was discovered, which occurs in ~ 10^{-8} sec in the zone of existence of low-temperature plasma produced by a pulse of laser light. On the basis of this phenomenon carried out new technology. It can be used to obtain in the natural environment of ordered micro- and nanostructures for the transfer, conversion, storage, or generation of information signals, special types of printing to protect against fraud, as well as to create on its basis of flexible solar cells for solar panels.

According to the above technology to the surface of the sheet material (substrate) is applied to a solution containing a material for forming single crystals, and then act on the given points of the substrate laser pulses.
Low temperature plasma, arising inside substrate is creating oxide-reduced atmosphere in which we can controlling pure metal and atomic hydrogen.
The process of crystallization can take place either on the surface substrate (sheet),and within the channels formed in the body of the substrate.
The advantages of this technology in comparison with analogues are:
- No expensive special equipment installed in clean rooms and has a working chamber with reduced pressure;
- High performance, the order of 100 m/sec, the process of growing single crystals of metal, metal alloys, semiconductors;
- Not need to have crystallographic correlation between crystallized material and substrate.
At present this technology is implemented as an example of high-speed growth of single crystals of copper in the body of the paper sheet (notes). The results are shown in Figs.

The uniqueness of the photographs submitted as single crystals due to the fact that they are formed in the body of the paper sheet 10 ns, and the fact that the list may still be in the natural environment, rather than being placed in a cell with a special environment

This technology is protected by a patent of the Russian Federation. Currently patented in the EU countries, USA, Canada, Japan, Korea and China.

References

[1] K. A. Bogonosov and S. N. Maksimovskii. Doklady Physics, Vol. 56, No. 8, pp. 419–422, 2011
[2] S. N. Maximovsky and G. A. Radutsky, RF Patent 000264, 2004.
[3] S. N. Maximovsky and G. A. Radutsky, Jpn. Patent 4575393, 2004.

Journal of Physics: Conference Series **345** (2012) 012021 doi:10.1088/1742-6596/345/1/012021

Determination of nanoscale inhomogeneities in scattering and absorbing medium

D Vavulin[1], A Alfimov, E Aryslanova, A Panteleev, S Chivilikhin

Department of Photonics and Optoinformatics, Saint-Petersburg National Research University of Information Technologies, Mechanics and Optics, Kronverksky Avenue, 49., Saint-Petersburg, Russian Federation

E-mail: dima-vavulin@mail.ru

Abstract. In this paper, we propose an original method for determining the size of nanoscale inhomogeneities in scattering and absorbing medium by comparing the theoretical and experimental dependences of the transmission medium on the wavelength of light. The method was tested on a sample of nanoporous glasses and gave satisfactory results. Were also carried out numerical evaluation of this phenomenon by using the software package CST Microwave Studio.

1. Introduction

Determination of the characteristic size of inhomogeneities of the medium is an important scientific and technical challenge that attracts the attention of many researchers [1,3,4,11]. Currently known methods of solving this problem are often unsuitable because of its complexity or high cost (for example, small-angle x-ray scattering). We have proposed a simple and cheap method for determining the size of the inhomogeneities in scattering and absorbing media. The method was tested on a sample plate of absorbing nanoporous, in which nanoscale inhomogeneities (pores with an average diameter of 17 nm, filled with water) served as the scatter-particles.

To implement this method was obtained by the theoretical dependence of the transmittance of the scattering and absorption coefficients and the experimental dependence of transmittance on the wavelength of light.

The main idea of this method is as follows: at long wavelengths the absorption of the medium becomes negligible and the attenuation of light occurs due to scattering by inhomogeneities. Thus it is possible to determine experimentally the scattering coefficient by examining the transmission of the plate in the range of longer wavelengths. Since the characteristic size of inhomogeneities of the medium is much smaller than the wavelength, the scattering is Rayleigh in nature. Consequently, the scattering coefficient is inversely proportional to the fourth power of wavelength. Further, the proportionality coefficient and the absorption coefficient chosen in such a way as to ensure the best agreement between theoretical and experimental dependences of the scattering coefficient on wavelength. Since Rayleigh scattering is character, then the resulting proportionality factor depends on the sixth degree of pore diameter. From the results obtained is determined by the average pore size.

[1] To whom any correspondence should be addressed.

Journal of Physics: Conference Series **345** (2012) 012021 doi:10.1088/1742-6596/345/1/012021

2. Theoretical derivation of the transmission coefficient of the system parameters

To receive our dependence transmittance from scattering and absorption coefficients, one must understand the process of light propagation in scattering and absorbing medium. We first derive for this equation of radiation diffusion.

Assuming that the size of scattering particles is much smaller than the wavelength of light [4], the radiative transfer equation has the form:

$$\mathbf{\Omega}\nabla I = -(\gamma_a + \gamma_s)I + \frac{\gamma_s}{4\pi}\int I d\omega , \tag{1}$$

Where I - the intensity of radiation, $\mathbf{\Omega}$ - the unit vector in the direction of the beam, γ_a and γ_s - absorption and scattering coefficients of the medium (which are assumed to be constant), $d\omega$ - an element of solid angle. The value of a constant damping system $(\gamma_a + \gamma_s)^{-1}$, that is a characteristic spatial scale of intensity. We assume that this quantity is small compared with typical dimensions of the system. We introduce the total intensity of radiation incident on this element of the medium:

$$Q = \int I d\omega \tag{2}$$

and apply to (1) Fourier transform. Then we have:

$$i\mathbf{k\Omega}I_{\mathbf{k}} = -(\gamma_a + \gamma_s)I_{\mathbf{k}} + \frac{\gamma_s}{4\pi}Q_{\mathbf{k}} ,$$

where $I_{\mathbf{k}} = \int I e^{-i\mathbf{kx}} d^3x$. Thus:

$$I_{\mathbf{k}} = \frac{\gamma_s}{4\pi}\frac{Q_{\mathbf{k}}}{\gamma_a + \gamma_s + i\mathbf{k\Omega}} . \tag{3}$$

Integrating (3) of the total solid angle and using (2), we obtain:

$$Q_{\mathbf{k}} = \frac{\gamma_s}{4\pi}Q_{\mathbf{k}}\int\frac{d\omega}{\gamma_a + \gamma_s + i\mathbf{k\Omega}} ,$$

or:

$$\left[1 - \frac{\gamma_s}{k} arctg\left(\frac{k}{\gamma_a + \gamma_s}\right)\right]Q_{\mathbf{k}} = 0 , \tag{4}$$

where $k = |\mathbf{k}|$. Equation (4) represents the Fourier-representation of pseudodifferential equation [7] for the value Q. In the short-wave approximation $\frac{k}{\gamma_a + \gamma_s} << 1$, this equation becomes:

$$\left[1 + \frac{k^2\gamma_s}{3\gamma_a(\gamma_a + \gamma_s)^2}\right]Q_{\mathbf{k}} = 0 . \tag{5}$$

Confine ourselves to the media in which the scattering coefficient γ_s is much higher than the absorption coefficient γ_a. Then, moving (5) in coordinate representation, ie performing the inverse Fourier transform, we obtain the desired equation of radiation diffusion [4]:

$$\left(\Delta - 3\gamma_a\gamma_s\right)Q = 0. \tag{6}$$

Under the same assumptions can be obtained by the boundary condition to equation (6):

$$Q - \frac{1}{2\gamma_s}\frac{\partial Q}{\partial n} = 2Q_e, \tag{7}$$

where $\frac{\partial}{\partial n}$ - derivative along the outward normal to the area, Q_e - the total intensity of radiation incident outside the border area.

2.1. Transmittance equation

Now we get directly the transmittance of the medium in the case of plane-parallel plate nanoporous glass of thickness h. Then we can consider the change in light intensity in only one spatial direction (along the axis x).

According to (6), (7), the diffusion equation and radiation boundary conditions for him are as follows:

$$\frac{d^2Q}{dx^2} - 3\gamma_a\gamma_s Q = 0, \tag{8}$$

$$\left[Q - \frac{1}{2\gamma_s}\frac{dQ}{dx}\right]_{x=0} = 2Q_e, \qquad \left[Q + \frac{1}{2\gamma_s}\frac{dQ}{dx}\right]_{x=h} = 2Q_e. \tag{9}$$

Integrating (8) using (9), we obtain

$$Q = A\,ch(\alpha x) + B\,sh(\alpha x) \tag{10}$$

where

$$A = \frac{sh(\alpha) + \beta\,ch(\alpha)}{(1+\beta^2)\,sh(\alpha) + 2\beta\,ch(\alpha)}2Q_e, \quad B = \frac{ch(\alpha) + \beta\,sh(\alpha)}{(1+\beta^2)\,sh(\alpha) + 2\beta\,ch(\alpha)}2Q_e,$$

$$\alpha = h\sqrt{3\gamma_a\gamma_s}, \quad \beta = 0.5\alpha/\gamma_s$$

Knowing Q, we can determine the intensity of the radiation propagating at an angle ϑ:

$$I = \frac{1}{4\pi}\left(Q - \frac{\cos(\vartheta)}{\gamma_s}\frac{dQ}{dx}\right) \tag{11}$$

and the total intensity of radiation propagating in the positive direction along the axis x:

$$Q_+ = 2\pi \int_0^{\pi/2} I \sin \vartheta \, d\vartheta . \qquad (12)$$

Using (10), (11), (12), we calculate the desired transmittance $T = \frac{Q_+(h)}{Q_+(0)}$ plate nanoporous glass:

$$T = \frac{1}{sh\left(h\sqrt{3\gamma_a \gamma_s}\right) + \sqrt{\frac{\gamma_s}{3\gamma_a}}\, ch\left(h\sqrt{3\gamma_a \gamma_s}\right)} . \qquad (13)$$

From this relation together with the experimental results of our proposed method to determine the characteristic pore size.

3. Experimental details

For the experimental measurement of the transmission coefficient of the sample scattering and absorbing medium, as a model object was taken polished disc nanoporous glass NPS-17 15 mm in diameter and 1 mm thick, with an average pore size of 17 nm and a relative pore volume of 58%.

Before the measurements the pores were filled with water. For this sample has been submerged for days in distilled water. The measurements were performed on a standard spectrophotometer Evolution-300. Further measured the transmittance of the sample at various wavelengths in the range 350 - 1100 nm. The measurement results are shown in Fig. 1.

Nanoporous matrices based on silica glass are a unique tool for studying physical and chemical processes to a limited extent commensurate with the scale of the processes and the size of the objects being studied. Space constraints, and effective contact with the pore walls cause significant features of the condition and properties of the filling material compared with the case of its location in the void volume.

The use of nanoporous glasses at the moment, mainly due to the transparency of these glasses in the visible spectrum and the possibility of obtaining samples of optical quality [8,9]. Optical density of samples in the near UV spectral region due to scattering and absorption in the structure of the material.

When considering the optical properties of porous glasses in the visible and near-infrared absorption in the samples can be neglected because of the smallness of this quantity, and the effective optical constants of samples are determined only by attenuation due to scattering by the porous structure and its irregularities. This property of nanoporous glasses and formed the basis of this method.

4. Results and discussion

The experimental and calculated dependences of the transmission and absorption are shown in Fig. 1.

Scattering coefficient is inversely proportional to the fourth power of the wavelength of incident light $\gamma_s = \frac{C}{\lambda^4}$. Coefficient C and the spectral absorption coefficient γ_a chosen so as to ensure the best agreement between theoretical and experimental dependences of transmittance on the wavelength of light. Obtained by this coefficient C can approximately estimate the characteristic size of inhomogeneity $D \sim C^{1/6}$.

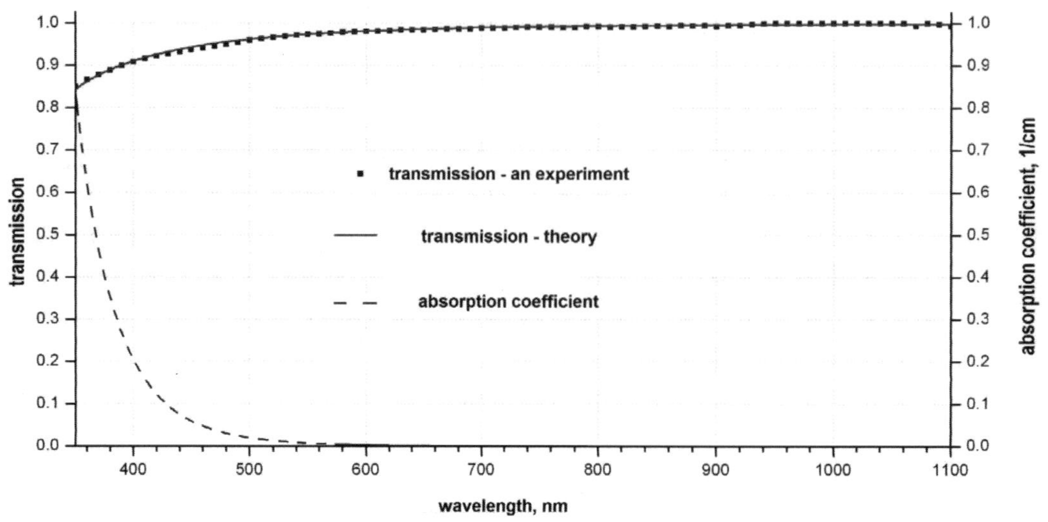

Fig. 1. Experimental and calculated dependence of the transmission and absorption of nanoporous glass on the wavelength of light.

Estimated value of the average pore diameter $D = 13.5$ nm close to the average pore size (17 nm) nanoporous glass. This suggests that the proposed method has a relatively high degree of accuracy in its simplicity and clarity.

5. Numerical Simulation

Numerical simulation using the software package CST Microwave Studio has qualitatively confirmed these results in the approximation of exponential dependence of the medium attenuation coefficient of the wavelength. A small element of the medium (glass), filled the pores with water by 58% in volume terms, was simulated in this package. After this item was held incident electromagnetic wave (port 1 in Fig. 2). Part of the electromagnetic wave, which passing through this element, is recorded on the other side in port 2. Further, the results were processed and compared with theoretical and experimental dependences.

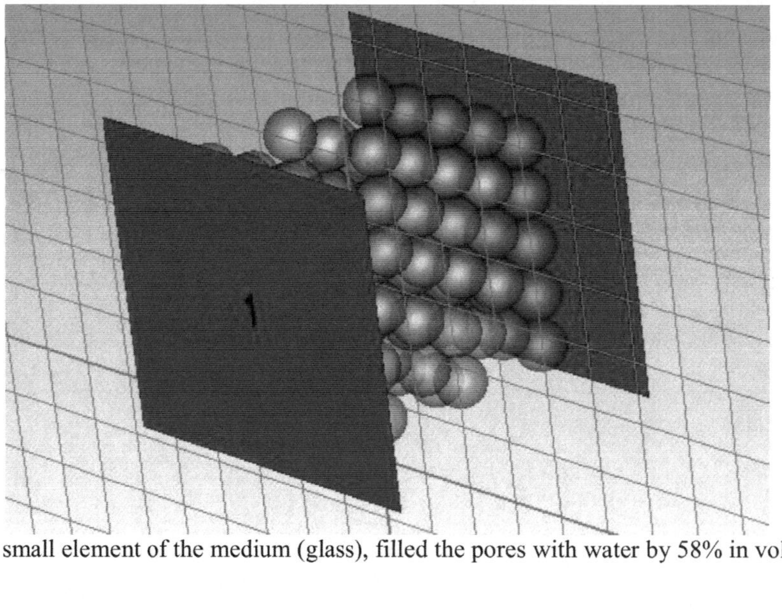

Fig. 2. A small element of the medium (glass), filled the pores with water by 58% in volume terms.

In Fig.3 is shown numerical dependence of the transmission coefficient, except the wavelength 432 nm (T = 0.80), 1017 nm (T = 0.86) and 1064 nm (T = 0.67), since at these wavelengths are contained plasmon modes, associated, as expected, from the Fano antiresonances. These antiresonancess were observed in the numerical simulation due to the not randomize structure (Fig.2), which in a real experiment can't be observed, and therefore not of interest for our case.

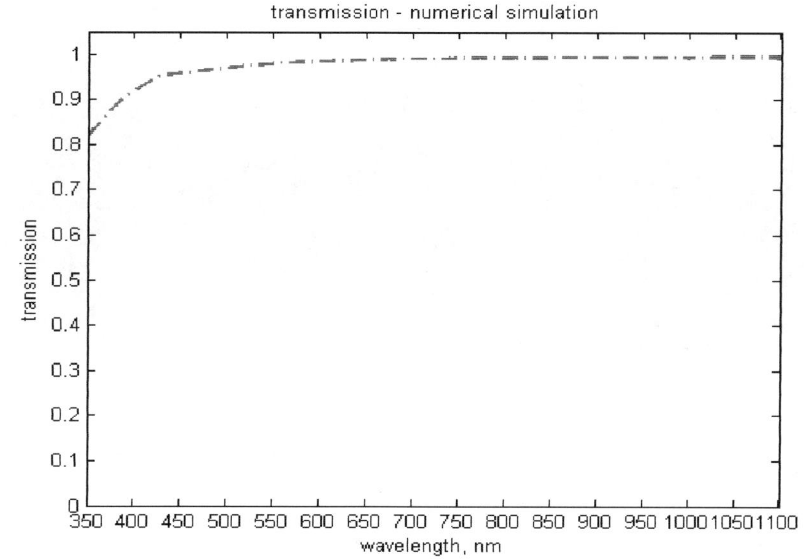

Fig. 2. Qualitatively confirmation of results in the approximation of exponential dependence of the medium attenuation coefficient of the wavelength.

6. Conclusion

Proposed a method for determining the average size of characteristic inhomogeneities of the medium. The method was tested on a sample of nanoporous glasses and gave satisfactory results. Carried out numerical simulation of the process, which also turned out to be in good agreement with experiment. Thus, the method has relatively good accuracy at low cost and complexity, but also allows us to understand the process of light propagation in scattering and absorbing medium.

References

[1] Alfimov A V, Aryslanova E M, Vavulin D N, Andreeva O V, Temnova D D, Lesnichiy V V, Almyashev V I, Kirillova S A, Chivilihin S A, Popov I Yu and Gusarov V V Diffusion of light in a medium with nanoscale inhomogeneities // *Scientific and Technical Bulletin of St. Petersburg State University of Information Technologies, Mechanics and Optics*. 2011. **N72.** p.105-109.

[2] Van de Hulst. Light *Scattering by Small Particles*// Moscow: IL, 1961.

[3] S.. Chandrasekhar, *Radiative Transfer*// M: IL, 1953

[4] Shifrin K.S. *Scattering of light in a turbid medium* - M-L: Gos. Ed. Tehn.-Thm. lit., 1951.

[5] Ishimaru A. *Wave Propagation and Scattering in Random Media*/ /**V.1.** - M.: Mir, 1981.

[6] Zel'dovich Ya.B., Raizer Yu.P. *Physics of Shock Waves and High-hydrodynamic phenomena*// M.: Nauka, 1966.

[7] Schartl W. *Light scattering frompolymer solutions and nanoparticle dispersions*//Springer Laboratory. **V. XIV**, 2007.

[8] Maslov V.P. *Operator methods*//M.: Nauka, 1973.

[9] Andreeva O.V., Obyknovennaya E.I. *Nanoporous matrices - the basis of composites of different purposes for optical experiments*// The writings of Research Center of Photonics and Optoinformatics: a collection of articles / edited by I.P. Gurov and S.A. Kozlov - Spb: ITMO, 2009, p. 303-322.

[10] O.V. Andreeva, I.E. Obyknovennaya *Nanoporous matrix NPC-7 and NPC-17 - can be used in optical experiments*/ / Nanosystems: physics, chemistry, mathematics. 2010, **V.1, № 1**, p.37-53.

[11] J.M. Pitarke, V.M. Silkin, E.V. Chulkov, P.M. Echenique. *Theory of surface plasmons and surface-plasmon polaritons*// INSTITUTE OF PHYSICS PUBLISHING REPORTS ON PROGRESS IN PHYSICS Rep. Prog. Phys. **70** (2007) 1-87.

Journal of Physics: Conference Series **345** (2012) 012022 doi:10.1088/1742-6596/345/1/012022

Nanoscale ceria for new functional materials

I V Zagaynov, E A Trusova and V V Belousov

Laboratory of Functional Ceramics, A.A. Baikov Institute of Metallurgy and Materials Science of RAS, Leninskii pr. 49, Moscow, Russia

E-mail: igorscience@gmail.com, trusova@imet.ac.ru

Abstract. Ultra dispersed ceria powders with crystallite size from 10 to 20 nm were synthesized by modified sol-gel and hydrothermal methods using cerium (III) nitrate as source of metal. N,N-dimethyloctylamine, monoethanolamine or tetraethylammonium hydroxide were used as a surfactant in syntheses. It was shown that increase of surfactant/Ce molar ratio value resulted in decreasing dispersion of obtained cerium oxides. It was shown that surfactant nature in the hydrothermal method had no significant effect on the dispersion of ceria obtained. Ceria had mesoporous structure with pore size less than 15 nm.

1. Introduction

Nanostructured inorganic particles are promising systems as oxidation catalysts due to the high surface to volume ratio. Especially, nanoscale ceria containing composites are very attractive three-way catalyst systems for soot combustion reactions [1], SOFC [2], biological applications, solar cells, electrochromic devices, UV radiation filters [3] because of the high mobility of lattice oxygen. The reason for the effect is oxygen donation of ceria caused by partial Ce^{4+}/Ce^{3+} reduction on the particle surface [4].

There are many different methods to prepare nanoscale ceria in attempt to optimize its properties for various applications. New techniques have become available due to nanotechnology and have resulted in some progress toward achieving this goal. Several methods were developed for the preparation of CeO_2 nanoparticles, especially such as sol-gel, hydrothermal and microemulsion because they are more promising [5]. We have used sol-gel and hydrothermal methods to establish a correlation between synthesis and thermo treatment conditions from one hand and physical-chemical properties of obtained ceria powders from other hand.

2. Experimental procedure

2.1. Synthesis

- a) sol-gel: cerium (III) nitrate was dissolve in deionized water (0.05M solution) and then after stirring ethanol and a surfactant such as N,N-dimethyloctylamine (DMOA), monoethanolamine (MEA) or tetraethylammonium hydroxide (TEAH) were added, all procedures were carried out at 60-80°C;
- b) hydrothermal: as-prepared sol (as described above) was treatment at 130°C during 3h.

Initial molar ratio value surfactant/Ce in the reaction mixture was changed in the range 1-20. At the final stage of preparation obtained substances were stepwise calcined at 500°C.

2.2. Characterization

The powders prepared by these methods were characterized by X-ray diffraction (XRD) using CuK_α radiation (DRON-3M, Russia). Specific surface areas of the catalysts were measured by a conventional (BET) nitrogen sorption method (Tri Star 3000 Micromeritics).

Testing of catalytic activity of obtained ceria powders was carried out in modeling CO oxidation reaction with use the fixed-bed quartz tube reactor at 50-400°C. The inlet gas composition was 3.8 vol.% CO, 7.3 vol.% O_2 and N_2 as balance, and the flow rate was adjusted to 30 mL/min (volume rate was 1800 h^{-1}). Analysis of gas reaction products was carried out by gas chromatography (Chrom-5).

3. Results and discussion

Figure 1 shows the typical XRD pattern of the synthesized ceria nanopowder. All reflexes in the XRD pattern were perfectly indexed as the pure cubic phase (Fm3m, JCPDS-34-0394) of CeO_2. According to XRD data calculation by Scherer formula the size of ceria crystallites was ≤20 nm for all samples (table 1).

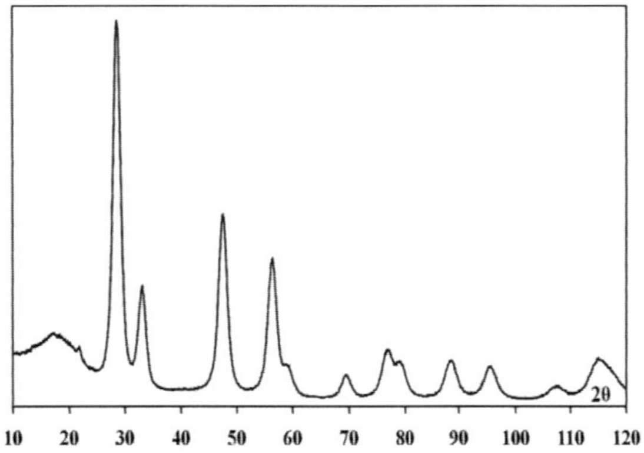

Figure 1. XRD pattern of CeO_2 (table 1, № 1).

Table 1. Synthesis conditions and average crystallites size of obtained CeO_2 samples (surfactant/Ce=1).

№	Method	Surfactant	D_{CeO2}, nm
1	sol-gel	DMOA	12
2		TEAH	14.5
3		MEA	21
4	hydrothermal	DMOA	12
5		TEAH	11
6		MEA	12

It was found that with increasing molar ratio of DMOA/Ce from 1 to 20 the crystallite size dependence of this ratio is nonlinear (figure 2*a*). The minimum crystallite size of 12 nm was observed at DMOA/Ce=1.

The comparison of ceria dispersibility was carried out for powders obtained by use of different surfactant in sol-gel synthesis at the molar ratio value surfactant/Ce = 1. It is possible that smaller crystallite size (12 nm) and highest BET surface (figure 2*b*) were achieved due to the presence of relatively long hydrocarbon chain in the DMOA molecule. The analysis of figure 2*b* also shows that the use of TEAH and MEA as surfactant results in larger CeO_2 particles. It is interesting that using of TEAH instead of DMOA ceria particle size increased slightly (~20%), however, the BET surface value reduced by more than 30%. Replacing of DMOA by MEA leads to doubling of ceria crystallite size and at the same time to reduce BET surface value almost 4 times. Apparently, the steric hindrance caused by DMOA and TEAH presence in the reaction mixture prevented the formation of larger crystallites. In addition the availability of long hydrocarbon chain in DMOA molecule promotes the formation of a more developed surface than the availability of TEAH or MEA, but the use of MEA resulted in a more narrow distribution of pores than the use of TEAH or DMOA (figure 3).

In the case of hydrothermal method when molar ratio surfactant/Ce value was 1 the crystallite size was almost the same (about 11-12 nm) for all surfactants (table 1). The all obtained ceria samples had mesoporous structure.

Ceria powders were tested in model CO oxidation reaction. It was shown that the onset temperatures (CO conversion was 10-15%) for all catalysts were in the interval 100-180°C, at temperatures 350-370°C CO conversion reached 100% (table 2).

Journal of Physics: Conference Series **345** (2012) 012022 doi:10.1088/1742-6596/345/1/012022

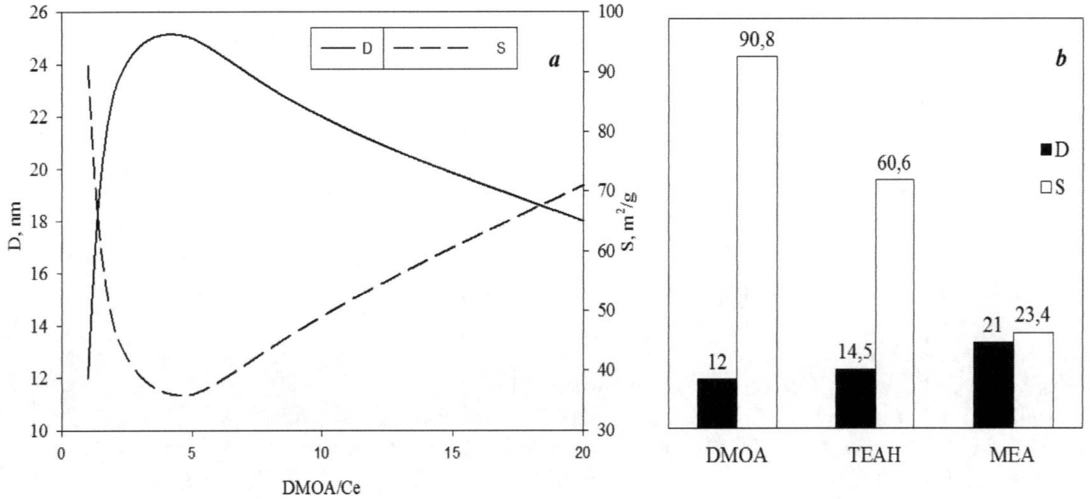

Figure 2. Impact the amount (a) and type (surfactant/Ce=1) (b) of surfactants on the average crystallite size (D) and surface area (S) of ceria powders obtained by sol-gel technique.

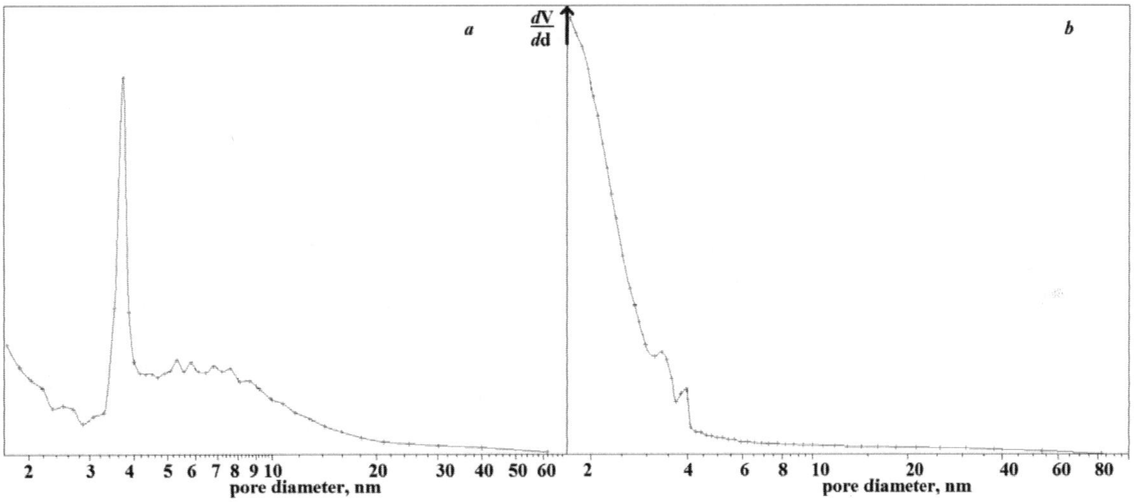

Figure 3. Mesoporous (a) and micro-mesoporous (b) structures of obtained ceria powders (number 3 and 2 according to table 1).

Table 2. Reaction conditions and catalytic activity of CeO_2 obtained by sol-gel technique.

№	Surfactant	Surfactant/Ce	$T_{25\%}$, °C	$T_{50\%}$, °C	$T_{99\%}$, °C
1	DMOA	1	220	260	348
2	DMOA	5	238	267	363
3	DMOA	10	260	285	358
4	DMOA	20	203	240	347
5	TEAH	1	195	250	333
6*	MEA	1	253	290	397

volume rate was 1800 h^{-1} and *5000 h^{-1}

4. Conclusion

The family of nanoscale ceria powders with average crystallite size less 20 nm was obtained by sol-gel and hydrothermal methods. All synthesized powders have mesoporous or micro-mesoporous structure with pore size not exceeding 15 nm. It was found that with increasing molar ratio DMOA/Ce value from 1 to 20 the crystallite size and specific surface area dependences of this ratio are nonlinear and have the upper and the lower extremes, accordingly, on their curves.

It was shown that the hydrocarbon chain length and three-dimensional configuration of surfactants using the sol-gel synthesis affects the dispersion and specific surface of ceria obtained. Contrary to using the hydrothermal method, nature of N-containing surfactants has no significant effect on the dispersion of ceria obtained, and its particle size reached low values.

References

[1] Blanco G, Calvino J J et al. 1999 *Chem. Mater.* **11** 3610
[2] Milliken Ch, Guruswamy S, Khandkar A 2002 *J. Am. Ceram. Soc.* **85** 2479
[3] Eliseev A A , Lukashin A V 2010 *Functional Nanomaterials* (Moscow: Physmatlit)
[4] Kockrick E, Schrage Ch, Grigas A, Geiger D, Kaskel St 2008 *J. Solid State Chem.* **181** 1614
[5] Ivanov V K, Polezhaeva O S, Tret'yakov Yu D 2010 *Russ. J. Gen. Chem.* **80** 604
[6] Zagainov I V, Trusova E A 2010 Composites based on mesoporous aluminium-silicates: synthesis by sol-gel method and physical-chemical properties *III International Competition of Scientific Papers in Nanotechnology for Young Researchers (Moscow, Russia, 1-3 November 2010)* s.5 № 22

Acknowledgments

This work was supported by grant RFBR № 09-08-00917-a, Program of Basic Research № 22 of the Presidium of RAS, and Program of Department of Chemistry and Materials Sciences № 7 of RAS.

Journal of Physics: Conference Series **345** (2012) 012023 doi:10.1088/1742-6596/345/1/012023

Nanocomposite materials based on ZnO and Fe_3O_4 nanoparticles in a polymer martix

M A Zaporozhets[1] , C G Rustamova [2], N S Timoshenko[2], V I Nikolaichik[3], K A Dembo[1], S V Savilov[4], I I Khodos[3], A S Avilov[1], S P Gubin[2]

[1] Shubnikov Institute of Crystallography of the Russian Academy of Sciences, 59 Leninsky prospect, Moscow 119333, Russia

[2] Kurnakov Institute of General and Inorganic Chemistry of the Russian Academy of Sciences, 31 Leninsky prospect, Moscow 119071, Russia

[3] Institute of Microelectronics Technology of RAS, Chernogolovka, Moscow District 142432, Russia

[4] Lomonosov Moscow State University, GSP-1, Leninskie Gory, Moscow 119991, Russia

E-mail: zaporozhets_m.a@mail.ru

Abstract. The techniques of sample synthesis of four types (the original liquid-phase dispersions of ZnO nanoparticles (NPs) and Fe_3O_4 NPs, solid nanocomposites ZnO-high-density polyethylene (HDPE) and Fe_3O_4-HDPE) have been developed. The dispersion of ZnO NPs was obtained by alkaline hydrolysis of zinc acetate potassium hydroxide in isopropyl alcohol. The dispersion of Fe_3O_4 NPs was prepared via the reaction of aqueous solutions of the iron salts ($FeCl_2$ and $FeCl_3$). After removal of impurity ions the dispersions were introduced into HDPE. A comparative analysis of the characteristics and properties of ZnO and Fe_3O_4 NPs in the original dispersions and in the polyethylene matrices was carried out by a complex of structural and physical methods (X-ray and electron diffraction, transmission and scanning electron microscopy, X-ray small-angle scattering, photoluminescence, electron magnetic resonance).

1. Introduction

One of the main directions of actively developed nanotechnology is to obtain stable, including to external influences, nanomaterials with reproducible properties that can be directly used in various fields of science and technology (electronics, catalysis, medicine, etc.). A promising way to create materials based on nanoparticles (NPs) is to introduce them into a polymer matrix, especially in a stable and widely available polymer such as polyethylene. However, due to the fact that the formation of a nanocomposite is carried out at relatively harsh chemical and technological conditions, the question remains open whether it is possible to preserve the composition, structure and specific physical characteristics of the original NPs in a nanocomposite during the introduction stage.

Nanocomposite materials based on ZnO and Fe_3O_4 NPs, which attract much attention from scientific and practical points of view, have been studied in the present paper. ZnO NPs are optically active materials as well as having high brightness, narrow emission spectrum, high photostability [1]. They have the prospect of application in optoelectronics and photocatalysis as biological labels and sensors. The interest in Fe_3O_4 NPs is due to the fact that they are widely used in systems for recording and storing information, in new permanent magnets, in magnetic cooling systems, and as magnetic sensors [2].

2. Materials and methods

We obtained samples of four types: the original liquid-phase dispersions of ZnO and Fe_3O_4 NPs, and solid nanocomposites ZnO-high-density polyethylene (HDPE) and Fe_3O_4-HDPE. The dispersion of ZnO NPs was obtained by alkaline hydrolysis of zinc acetate potassium hydroxide in isopropyl alcohol. The dispersion of Fe_3O_4 NPs was prepared via the reaction of aqueous solutions of the iron salts ($FeCl_2$ and $FeCl_3$). After removal of impurity ions the dispersions were introduced into HDPE. The comparative analysis of the characteristics and properties of ZnO and Fe_3O_4 NPs in the original dispersions and in the polyethylene matrices was carried out by a complex of structural and physical methods (X-ray and electron diffraction, transmission and scanning electron microscopy, X-ray small-angle scattering, photoluminescence, electron magnetic resonance).

3. Results

3.1. ZnO samples

The analysis of X-ray (figure 1) and electron diffraction patterns revealed that ZnO NPs isolated from the dispersion in isopropanol and NPs in the polyethylene matrix are of single-phase state. NPs have the wurtzite structure (the $P6_3mc$ space group).

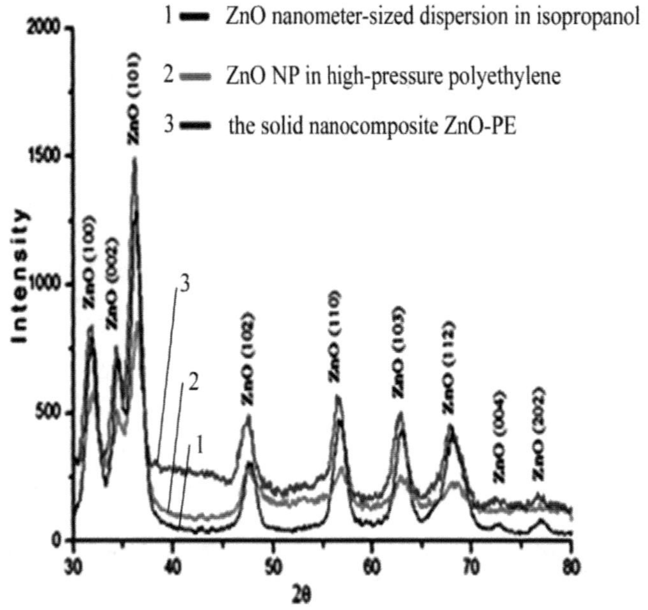

Figure 1. X-ray patterns of ZnO NPs.

Transmission electron microscopy (figure 2) and X-ray small-angle scattering revealed that the ZnO samples contained NPs of spherical shape with narrow size distribution. The average size of ZnO

NPs was 5±1.5 nm in the original dispersion and 6±1.5 nm in a polyethylene matrix. Some broadening of the size distribution in the matrix is probably due to high temperature needed for the introduction of ZnO NPs into the matrix that leads to a partial agglomeration of the NPs.

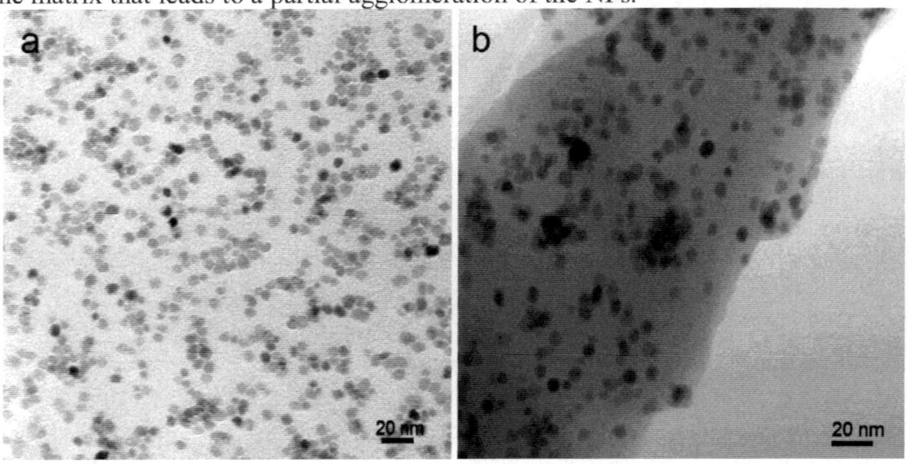

Figure 2. Electron microscope image of ZnO NPs in the original dispersion (a) and in the polyethylene matrix (b).

To determine the effect of the HDPE matrix on the optical properties of embedded ZnO NPs, photoluminescence spectra for samples of the original dispersion of NPs and the NPs in the matrix were measured (figure 3).

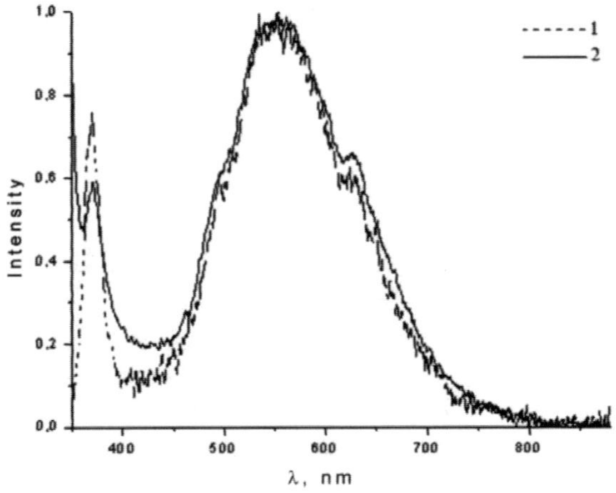

Figure 3. Photoluminescence spectra of ZnO NPs s in the original dispersion (1) and in the polyethylene matrix (2).

The obtained spectra of two samples are largely repetitive. The positions and half-widths of both peaks (375 nm and 560 nm) look nearly identical. No significant changes in the spectra of NPs after the introduction into the composite indicate that the matrix has no appreciable effect on the exciton recombination mechanism. The matrix acts as a chemically stable preservative of NPs, retaining their spectral features.

3.2. Fe_3O_4 samples

The average size of Fe_3O_4 NPs was 8.5±2.0 nm in the original dispersion and 9±2.5 nm in the polyethylene matrix. Analysis of X-ray and electron diffraction patterns revealed that samples containing Fe_3O_4 NPs are single-phase. The crystal structure of Fe_3O_4 NPs is described in terms of a cubic lattice (the Fd3m space group, a=8.374 Å).

Figure 4 shows the dependence of room temperature magnetization of Fe_3O_4 NPs, that were isolated from the dispersion (figure 4a) and introduced into the polyethylene matrix (figure 4b), on external magnetic field. The measurement results indicate that these two samples are of superparamagnetic behavior, since there is no residual magnetization and, consequently, a hysteresis loop.

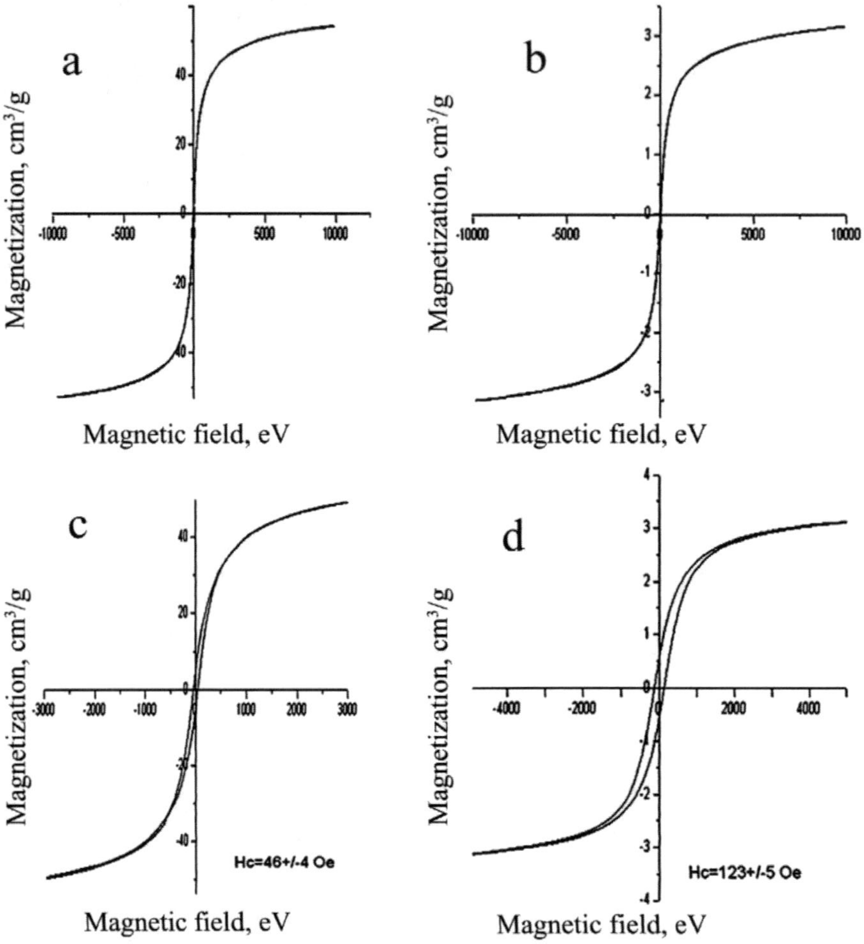

Figure 4. Magnetic field dependence of magnetization of Fe_3O_4 NPs isolated from the dispersion (a,c) and NPs introduced into the polyethylene matrix (b,d). T=298 K (a,b), T=80 K (c,d).

The maximum magnetization of non-covered magnetite Fe_3O_4 (a) is much higher than that of magnetite in the polymer matrix (b). This is due to the fact that magnetization is directly proportional to the weight fraction of magnetic component in a sample. In converting the magnetization per atom, the results for the two samples are roughly equal, that demonstrates a conservation of the magnetic state after the transition from the dispersion into the solid matrix. The same dependence of the magnetization on the magnetic field was recorded for the two samples at liquid nitrogen temperature. The results are shown in figure 4 (c,d). There is a residual magnetization and hysteresis in this case, which suggests that the samples lose their superparamagnetic properties and become ferrimagnetics at low temperature.

Journal of Physics: Conference Series **345** (2012) 012023 doi:10.1088/1742-6596/345/1/012023

The sample with magnetite NPs introduced into the polyethylene matrix was cooled in zero and an applied external magnetic field (figure 5). The blocking temperature was obtained at the intersection of two curves, it amounted to approximately 300 K for the sample. The sample is in a stable superparamagnetic state above the blocking temperature. Below this temperature, thermal energy is insufficient to overcome the magnetic anisotropy and intermolecular interactions, so paramagnetism disappears, and the sample is ferrimagnetic.

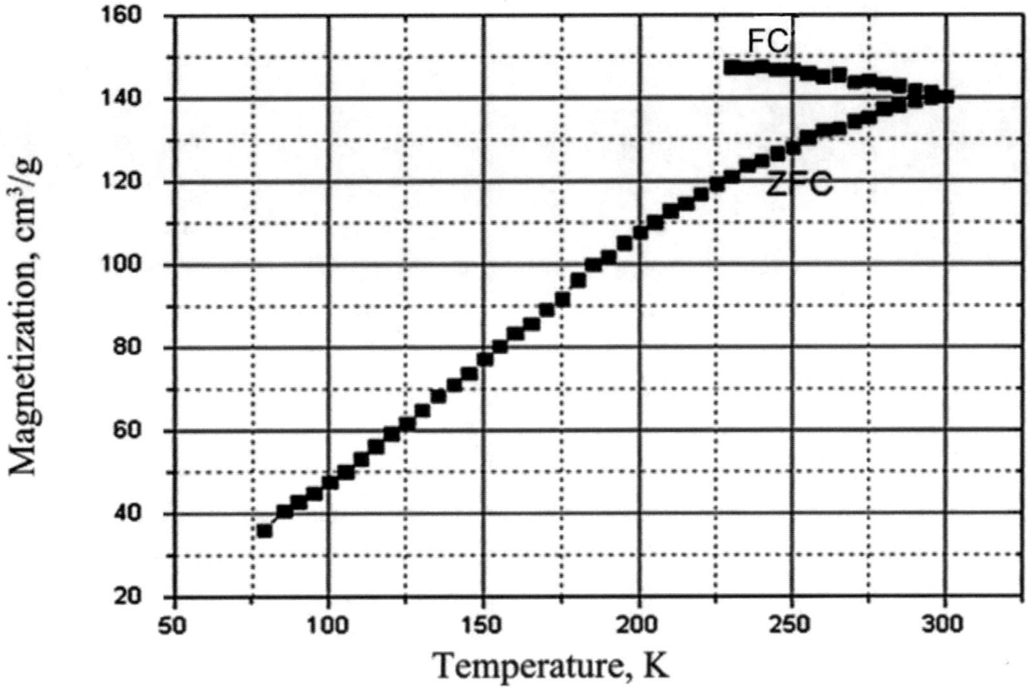

Figure 5. The temperature dependence of magnetization for Fe_3O_4 HDPE in the presence (FC) and absence (ZFC) of an external magnetic field.

4. Conclusions

The technology for production of nanomaterials based on ZnO and Fe_3O_4 NPs in a polymer matrix has been developed. The use of a complex of structural and physical methods has shown that the structure and properties of ZnO and Fe_3O_4 NPs are retained when they are introduced into a polyethylene matrix. The use of the synthesis technology of ZnO-HDPE nanocomposite is proposed for the development of standards of fluorescent emission in UV and visible ranges.

Acknowledgements

The work has been made with the support of the Russian Federation President's Grant MK-3695.2011.3.

References

[1] Singh S, Thiyagarajan P, Mohan K K, Anita D, Thirupathiah S, Rama N, Tiwari B, Kottaisamy M and Ramachandra R M S 2007 *J. Phys. D Appl. Phys.* **40** 6312

[2] Baraton M I 2002 *Synthesis, Functionalization, and Surface Treatment of Nanoparticles*. (Am. Sci., Los-Angeles)

Journal of Physics: Conference Series **345** (2012) 012025 doi:10.1088/1742-6596/345/1/012025

Design and computer modeling of the supracrystalls

A A Karenin
Department of Pphysics, Ulyanovsk State Technical Universaty, 32 Severny Venetz, Ulyanovsc 432027, Russia

E-mail: aspralx86@mail.ru

Abstract. The possibility of solid state crystalline structures, which we called supracrystals, is shown. Unlike ordinary crystals, there are no separate atoms or ions in the nodes of crystalline lattice. They are replaced by symmetric atomic associates. Symmetry classes of 2D- and 3D-supracrystals are determined. The bond length, energy per one atom, bond energy, and gap energy of 2D- and 3D-supracrystals consist of C, Si, B, N,S are calculated by the program ABINIT-5.8.4 in Hartree–Fock approximation. Unusual properties of supracrystals are of interest for their production and practical use.
Energy stability and electrical properties of the nanotubes produced from 2D-sypracrystals suggested by the authors before are researched by computers modeling technic. Supracrystaline nanotubes from the atoms of carbon, silicon, boron and nitrogen, suphfur are considered. It is shown that the electrical properties of supracrystals nanotubes may be controlled in large range (from metallic to dielectric) changing their chemical composition, structure, diameter and chirality .

1. Introduction

The idea on the possibility of supracrystals existing has been advanced by us in paper [1]. In nodes of ordinary crystals the atoms or ions are located, in supracrystals they are replaced by symmetrically organized atomic associates. For all this one can always to pick out a supracell and to construct the whole crystalline lattice by its translation along epy ribs. There are two-dimensional (2D) and three-dimensional (3D) supracrystals. 2D-supracrystals represent the generalization of two-dimensional square or sixangular, in particular graphenic, nets. 3D- supracrystals may be taken into consideration as the generalization of cubic crystals. The impotent peculiarity of supracrystals is their periodicity both in large cells (supracells) and in small cells (cells of enclosing) as the nodes elements. It permits to combine the different, at times, mutually exceptive properties in one and the same crystal, for example, high conductivity and small heat conduction. Moreover, unlike the ordinary crystals, supracrystals are more porous and less dense materials. In supracells of “host’s” crystalline lattices “quest” atoms or ions may be located, what approaches such supracrystals to clathrates [2]. However the atoms (ions) in supracrystals are banded by valent chemical bonds, but not hydrogen, electrostatic or coordination bonds as in the clathrates. Therefore, supracrystals are more thermally stable then clathrates. In this paper, we for first time attempted theoretically to calculate the main physical properties of supracrystals: cell parameters, band energy, gap energy et al. All calculations have been provided in Hartree-Fock approximation by the program ABINIT-5.8.4. [3]. For the k-points generation in Brilloun band Monchorst-Packs algorithm [4] has been used. The Broiden iteration procedure has been used as a mathematical approach. The cut energy has been automatically taken for k=6.

2. 2D-supracrystals

Two-dimensional mosaics formed only regular geometric figures (so called Kepler's nets) are well-known [5]. There are 11 varieties of such nets (see figure1) but only 5 of them may be realized in supracrystals structure. This conclusion follows from the necessity to secure the building of a whole two-dimensional crystalline lattice by the supracell's translation along two basic directions on a plane. Besides that, all bonds between the atoms must be equivalent.

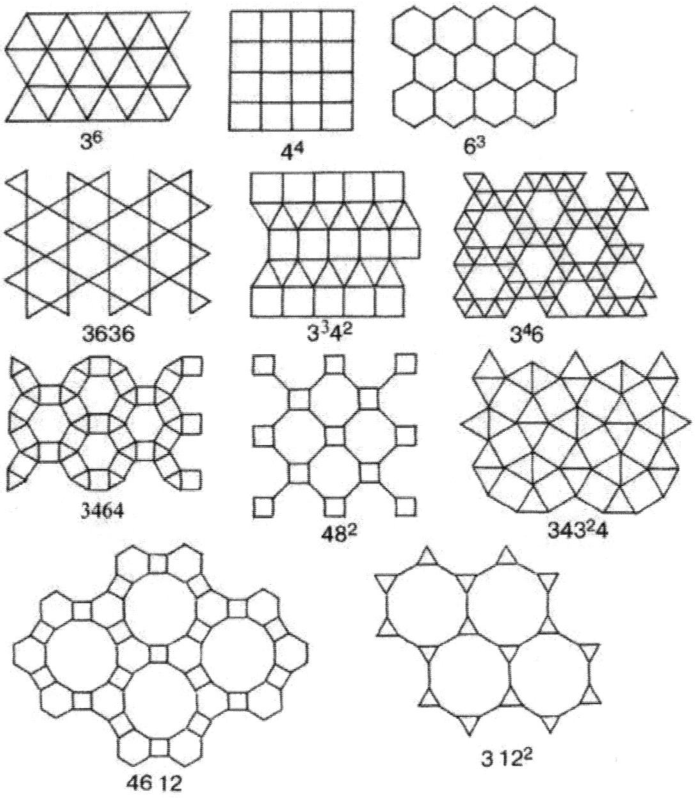

Figure 1. Kepler's nets [5]. Integers at the bottom at the figure (Schlafli symbols) denotes the number and the type of regular polygons meeting in each net's node. The powers index shows a number of identical polygons meeting in the one point and contacting with each other by their ribs.

Figure 2. shows the permissible types of supracrystalline cells forming 2D-supracrystals. Because all the bonds are equivalent, they must be formed by either identical atoms or atoms of two different kinds. In second case, homogeneous bonds as energetically less profitable, hence heterogeneous ones [6] must be absent. It means that supracrystals which consist of the different atoms don't include triangular cells of enclosing. Here we also suggested a new, more convenient for the supracrystals description, notations. Letters X or Y in parentheses determine the symbol of chemical element. Indexes behind the parentheses are located in following order: the first index (4 or 6) determines the type of supracell, the next indexes describe the cell's of enclosing type. At first the number of sides of node's cell are shown, then the same are shown for surrounding cells (if there are).

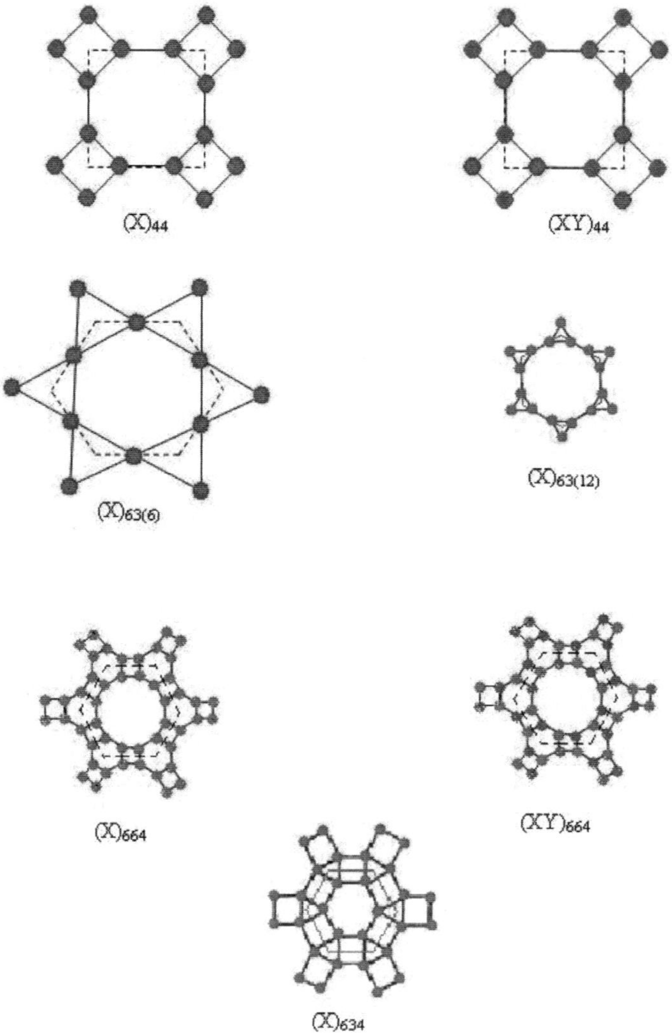

Figure 2. Supracells of 2D-supracrystals and their notations. In $(X)_{63(6)}$ and $(X)_{63(12)}$ supracells the integer in parentheses indicate the type of polygon in centre of the cell.

The possibility of carbon planar and nanotubular structures consist of five- and sevenangles, or five-, six-, and sevenangles, was theoretically demonstrated in Ref. [7]. These structures have been called

Journal of Physics: Conference Series **345** (2012) 012025 doi:10.1088/1742-6596/345/1/012025

"Haekelites". Later this name were spread to other two-dimensional structures containing non-hexagonal elements, including non-carbon ones [8,9]. Note, that 2D-supracrystals with supracell of $(XY)_{44}$ type are one of the forms of haekelitic planar structures studed in Ref. [8,9]. This structure consists of alternative atoms of boron and nitrogen forming squares and regular eightangles. At the same time, 2D-supracrystals and haekelitic structures are not equivalent each other, thought either can be rolled up to nanotubes. Firstly, not all haekelitic structures are 2D-supracrystals. Secondly, the particular properties of supracrystals best of all are observed on the level of 3D-supracrystals. In Figure 1 you can see that all 2D-sypracrystals belong to tetragonal and hexagonal singonies, moreover $(XY)_{44}$ – to class 4,$(X)_{44}$ – to class 4mm, $(XY)_{664}$ – to class 6, and the others – to class 6mm.

3. 3D-supracrystals

3D-supracrystals are the cubic lattices, in which nodes symmetric atomic associates forming regular or semi regular polyhedral (Platonic and Archimedean bodies accordingly [10]) are located. Supracell of 3D-supracrystals always are cubic, and permissible types of their node's elements are given in figure 3.

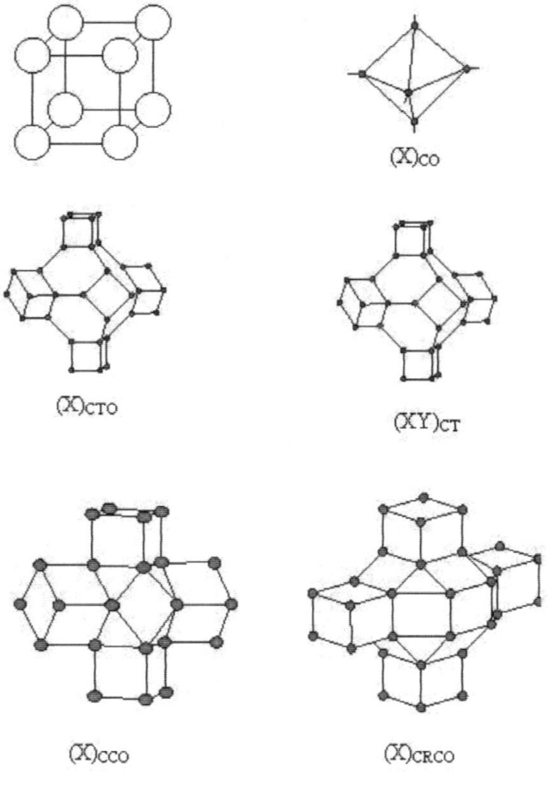

Figure3. Supracell of 3D-supracrystals (above left), and permissible types of node's elements. Cubes of the binding atoms oriented to the observer for non-octahedric node's elements are not shown.

Here we used the same notations as for 2D-supracrystals. Letters X and Y denote the symbols of chemical elements. The first index "C" behind parenthes meters determines the type of supracell (cubic). The next indexes denote regular or semi-regular polyhedron forming node's element: O – octahedron, TO –

truncated octahedron, CO – cuboctahedron, RCO – rhombicubooctahedron. As for 2D-supracrystalls, the node's elements consist of two different atoms are also possible here, if there are no triangles. Note, that 2D-supracrystals may be obtained by the crystallization of 4-, 5-, or 6- valent atoms, but it is necessary to use 3- or 4- valent atoms for 2D-supracrystals construction. All 3D-supracrystals belong to the cubic singony, moreover, $(XY)_{CTO}$ – to class 432, and others – to class m3m.

4. Physical properties of 2D-supracrystals

In table 1 the results of supracell's parameters optimization, total energy per one atom calculations, and bond energy for 2D-supracrystals are represented. The possibility of boron, nitrogen, carbon, and sulfur showing among the others valency equal 3 and 4 for 2D-supracrystals construction are studies.

Table1. Calculated values of the main parameters of 2D-supracrystalls (bond length l, energy per one atom E, bond energy E_b).

Structure	$l, \overset{0}{A}$	E, eV/atom	E_b, eV
$(C)_{44}$	1,43	- 12,9	- 4,9
$(Si)_{44}$	1,82	- 6,3	- 1,7
$(S)_{44}$	1,51	- 15,6	- 3,4
$(BN)_{44}$	1,38	- 10,2	- 2,3
$(C)_{63(6)}$	1,86	- 14,2	- 2,9
$(Si)_{63(6)}$	2,07	- 7,3	- 2,2
$(S)_{63(6)}$	1,11	- 17,2	- 5,0
$(C)_{63(12)}$	1,62	- 12,3	- 4,3
$(Si)_{63(12)}$	2,18	- 9,3	- 3,7
$(S)_{63(12)}$	1,08	- 17,8	- 5,6
$(C)_{664}$	1,56	- 11,3	- 1,2
$(Si)_{664}$	2,27	- 8,7	- 3,5
$(S)_{664}$	1,14	- 14,4	- 5,2
$(BN)_{664}$	1,44	- 9,1	- 2,3
$(C)_{634}$	2,17	- 13,3	- 2,6
$(Si)_{634}$	2,03	- 9,8	- 4,2
$(S)_{634}$	1,01	- 16,6	- 4,1

For comparison we'll give the values of energy per one atom for graphenic nets calculated by the same method made from: carbon's atoms – -15,3eV/atom, silicon's atoms – -8.7eV/atom, boron's and nitrogen's atoms – -11,2eV/atom. Hence, all types of considered 2D-supracrystals are stable, but energetically less profitable with comparison of corresponding graphenic layers. The comparison between different 2D-supracrystlls shows that more stable are the crystals consist of sulfur's atoms and less stable are silicon structures. As a rule, more stable are $(X)_{63(12)}$ and $(X)_{63(6)}$ 2D-supracrystals containing triangular cells of enclosing. It may be explained by their greater stability to deformations. The rules of crystallographic directions choice (in terms of Miller's indexes) in supracell are shown in figure 4. The results of gap energy calculations for the same 2D-supracrystals as in table 1 are represented in table 2.

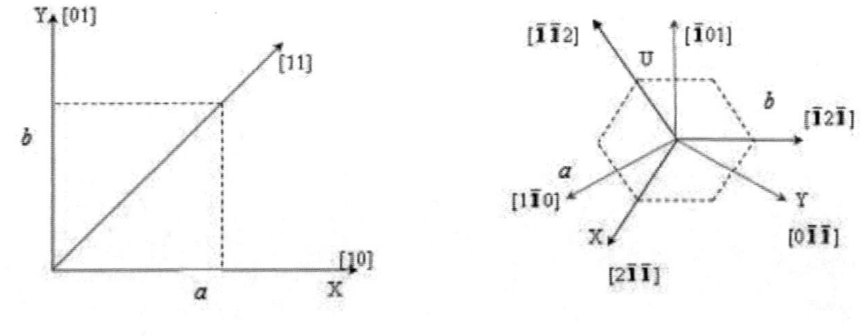

Figure 4. The rules of crystallographic directions choice in 2D-supracrystals with square and sixangular supracells. The direction [10] and [01] are equivalent and can be denoted as <10>. By analogy the directions [211], [121], and [112] can be denoted by a common symbol <211>, and [011], [101], [110] as <011>. Symbols X and Y denote the axes of crystallophysical coordinates.

Table 2. Calculated values of gap energy (E_g) for basic directions in 2D-supracrystals.

Structure	$E_g^{<10>}$, eV	$E_g^{<11>}$, eV	$E_g^{<211>}$, eV	$E_g^{<011>}$, eV
$(C)_{44}$	0,06	0,09		
$(Si)_{44}$	0,8	1,3		
$(S)_{44}$	2,1	2,7		
$(BN)_{44}$	0,7	1,2		
$(C)_{63(6)}$			1,37	1,45
$(Si)_{63(6)}$			1,5	1,9
$(S)_{63(6)}$			2,2	3,3
$(C)_{63(12)}$			0,69	0,91
$(Si)_{62(12)}$			1,0	1,3
$(S)_{63(12)}$			2,4	3,6
$(C)_{664}$			1,72	2,21
$(Si)_{664}$			2,2	2,4
$(S)_{664}$			6,1	8,2
$(BN)_{664}$			2,0	2,5
$(C)_{634}$			1,12	1,41
$(Si)_{634}$			1,3	1,6
$(S)_{634}$			2,9	3,7

In table 2 you can see that with respect to electric current all considered structures, except dielectrical ones $(X)_{664}$ and $(XY)_{664}$, show semiconductors properties. Only sulfur already forms dielectric 2D-supracrystals.

5. Physical properties of 3D-supracrystals

In table 2 the results of supracell's parameters, energy per one atom, and bond energy for 3D-supracrystals calculations are represented. The calculations have been carried out for the carbon's atoms possessing the valency equal 4, sulfur's atoms forming, among others, 4-, 5-, and 6- valent bonds, and 5-valent phosphorous.

Table 3. Calculated values of bond length l, energy per one atom E and Bond energy E_b for 3D-supracrystals.

Structure	l, $\overset{0}{A}$	E, eV/atom	E_b, eV
$(S)_{CO}$	2,1	- 7,2	- 6,6
$(P)_{CO}$	2,7	- 5,8	- 5,2
$(C)_{CTO}$	1,3	- 5,8	- 4,2
$(S)_{CTO}$	1,7	- 6,3	- 5,4
$(S)_{CCO}$	0,9	- 5,8	- 4,9
$(Si)_{CRCO}$	1,5	- 6,1	- 5,2
$(P)_{CRCO}$	2,7	- 4,6	- 4,2

Compare the results given in table 1 and table 3 we can see that energy per one atom in 3D-supracrystals are essentially higher then in 2D-supracrystals, i.e. the potential pits where there atoms are located are less deep, and 3D-supracrystals are less stable then 2D-supracrystals.

However, the bond energy is commensurable with the bond energy for 3D-supracrystals made up from corresponding atoms. In figure 5 you can see the crystallographic directions for which in table 4 the calculated values of gap energy for the same 3D-supracrystals as in table 3 are represented.

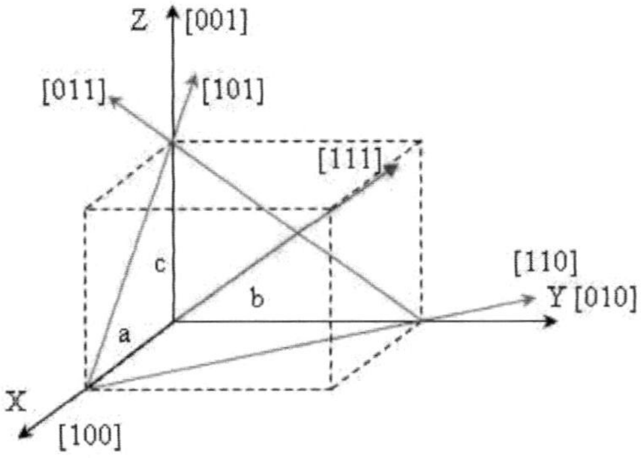

Figure 5. The basic crystallographic directions in 3D-supracrystals. Axial directions [100], [010], [001] are equivalent and further are denoted as <100>. Diagonal directions of the type [110], [011], [101] also are equivalent and are denoted as <010>. The directions of a cube's space diagonals of the type [111] are denoted by symbol <111>.

Table 4. Calculated values of gap energy (E_g) for basic direction in 3D-supracrystals

Structure	$E_g^{<100>}$, eV	$E_g^{<010>}$, eV	$E_g^{<111>}$, eV
$(S)_{CO}$	6,6	6,8	7,3
$(P)_{CO}$	2,2	4,0	4,4
$(C)_{CTO}$	1,1	2,2	2,6
$(S)_{CTO}$	6,2	6,6	6,8
$(S)_{CCO}$	4,4	6,1	6,8
$(Si)_{CRCO}$	4,3	6,2	6,7
$(P)_{CRCO}$	1,9	2,6	3,8

According to the results given in table 4 for the considered 3D-supracrystals only carbon and, by a some stretch, phosphorous can demonstrate semiconductors properties in the structures $(C)_{CTO,}$ $(P)_{CRCO}$, and even in the case of axial directions of electric current propagation. In all other cases the considered above 3D-supracrystals are dielectrics.

6. Supracrystals nanotubes

The nanoribbons are defined as two-dimensional nanostructures in the form of flat strips of 10–100 nm width and of 100 μm and more length.

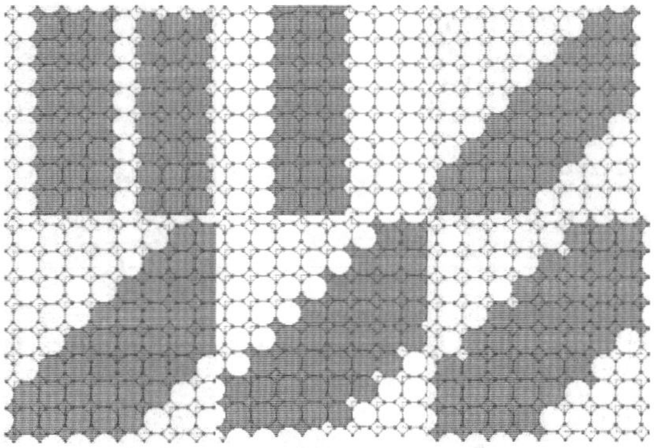

Figure 6. Types of nanoribbons $(X)_{44}$

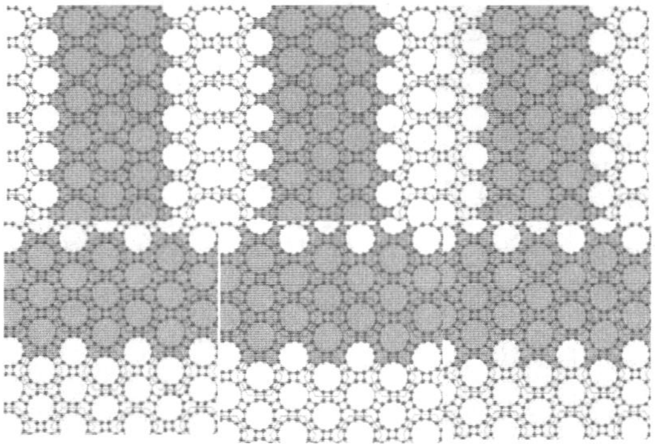

Figure 7. Types of nanoribbons $(X)_{664}$

The asymmetrical nanoribbons may be rolled up to the nanotubes. If one roll up the nanoribbon around the axis not coinciding with the ribbon's axis, the nanotubes will appear with chirality angel differ from 0° and 45° (in the case of square supracells). Except of represented on these figures nanoribbons one can cut nanoribbons with more complicated boundaries.

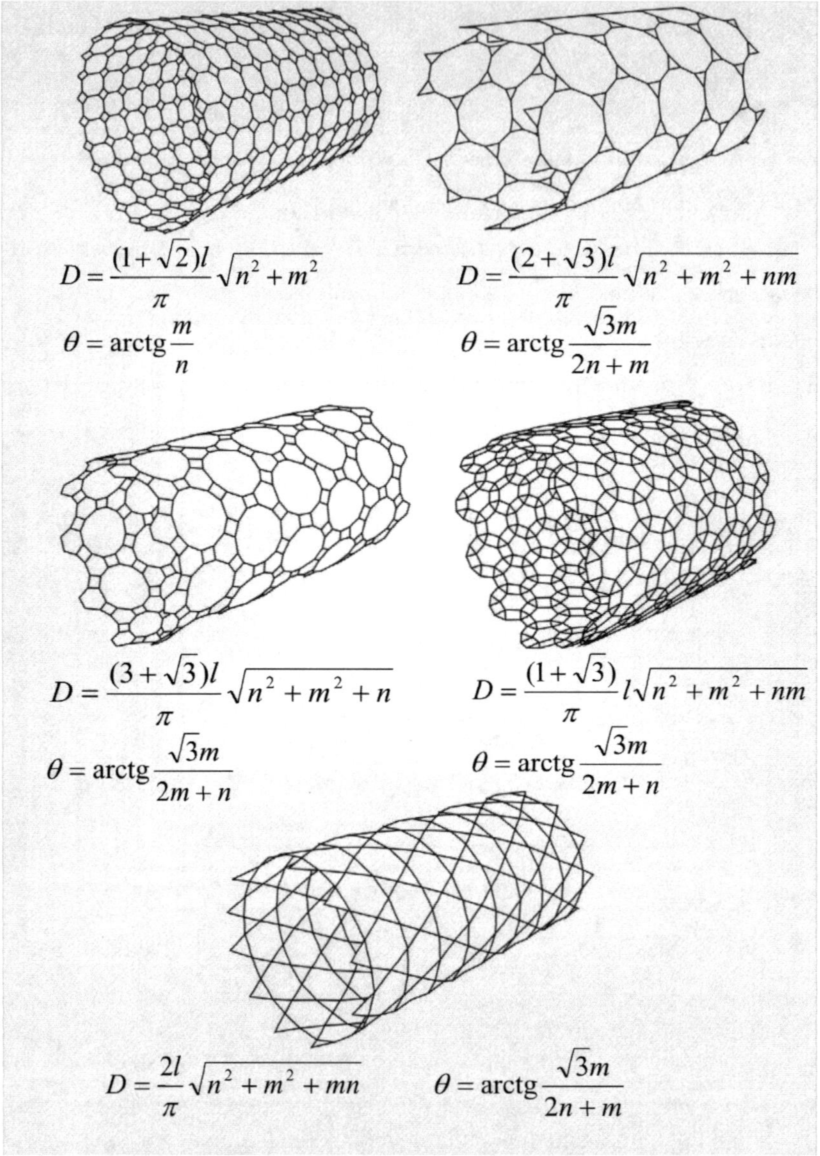

$$D = \frac{(1+\sqrt{2})l}{\pi}\sqrt{n^2+m^2}, \quad \theta = \operatorname{arctg}\frac{m}{n}$$

$$D = \frac{(2+\sqrt{3})l}{\pi}\sqrt{n^2+m^2+nm}, \quad \theta = \operatorname{arctg}\frac{\sqrt{3}m}{2n+m}$$

$$D = \frac{(3+\sqrt{3})l}{\pi}\sqrt{n^2+m^2+n}, \quad \theta = \operatorname{arctg}\frac{\sqrt{3}m}{2m+n}$$

$$D = \frac{(1+\sqrt{3})}{\pi}l\sqrt{n^2+m^2+nm}, \quad \theta = \operatorname{arctg}\frac{\sqrt{3}m}{2m+n}$$

$$D = \frac{2l}{\pi}\sqrt{n^2+m^2+mn} \qquad \theta = \operatorname{arctg}\frac{\sqrt{3}m}{2n+m}$$

Figure 7. Geometric parameters of supracrystalline nanotubes

Table 5. Calculated values of supracrystal naniotubes parametres

structure	Type of atoms	$\lvert E_{HT}-E_{2D}\rvert$ eV/at	Number of supercell on diametr	E_{gap} eV
1	2	3	4	5
$(X)_{44}$ (n.0)	Si	0.068 0.059 0.051 0.031	(6) (8) (9) (11)	0 0,11 0,29 0,59 {0,8}
	C	0,076 0.065 0,062 0,041	(6) (8) (9) (11)	0 0 0,27 0,36 {0,06}
	B-N	0,063 0.051 0,045 0,028	(6) (8) (9) (11)	0,11 0,31 0,42 0,63 {0,7}
	S	0,045 0.037 0,032 0,011	(6) (8) (9) (11)	1,51 1,79 2,02 2,16 {2,1}
$(X)_{44}$ (n.n)	Si	0.101 0.091 0.089 0.073	(6) (8) (9) (11)	0 0.49 0.62 0.85 {0.93}
	C	0.112 0.098 0.086 0.077	(6) (8) (9) (11)	0 0 0 0.06 {0.09}
	B-N	0.097 0.087 0.076 0.069	(6) (8) (9) (11)	0.50 0.70 0.82 0.97 {1.2}

Table 5. Continues

1	2	*3*	4	*5*
		0.067	(6)	2.10
$(X)_{44}$		0.056	(8)	2.50
(n.n)	S	0.047	(9)	2.60
		0.032	(11)	2.70
				{2.7}
$(X)_{63(12)}$		0,043	(6)	0
	Si	0.032	(8)	0,27
	(n,0)	0,027	(9)	0,38
		0,021	(11)	0,59
				{1,0}
		0,029	(6)	0.12
		0.024	(8)	0.32
	(n,n)	0,025	(9)	0.75
		0,019	(11)	1.15
				{1.3}
	C	0,057	(6)	0
	(n,0)	0.042	(8)	0,27
		0,035	(9)	0,38
		0,027	(11)	0,59
				{0,69}
	(n,n)	0,037	(6)	0.11
		0.033	(8)	0.19
		0,031	(9)	0.44
		0,022	(11)	0.67
				{0.91}
	S	0,059	(6)	2,16
	(n,0)	0.036	(8)	2,42
		0,031	(9)	2,71
		0,023	(11)	3,17
$(X)_{63(6)}$	Si	0,027	(6)	0.67
	(n,0)	0.025	(8)	0.83
		0,024	(9)	1,17
		0,017	(11)	1,33
				{1,5}
		0,028	(6)	1.12
	(n,n)	0.026	(8)	1.18
		0,026	(9)	1.60
		0,015	(11)	1.76
				{1.9}

Journal of Physics: Conference Series **345** (2012) 012025 doi:10.1088/1742-6596/345/1/012025

Table 5. Continues

1	2	3	4	5
	C	0,035	(6)	0.50
$(X)_{63(6)}$	(n,0)	0.031	(8)	0.70
		0,030	(9)	1.00
		0,020	(11)	1,23
				{1,37}
		0,034	(6)	0.97
		0.024	(8)	1.00
	(n,n)	0,021	(9)	1.20
		0,018	(11)	1.37
				{1.45}
$(X)_{664}$	Si	0,050	(6)	1,25
	(n,0)	0.032	(8)	1,57
		0,028	(9)	1,89
		0,025	(11)	2,07
				{2,2}
	(n,n)	0,050	(6)	1.70
		0.037	(8)	1.90
		0,027	(9)	2.14
		0,024	(11)	2.17
				{2.4}
	C	0,062	(6)	0,08
	(n,0)	0.047	(8)	0,86
		0,034	(9)	1,28
		0,032	(11)	1,57
				{1,72}
	(n,n)	0,061	(6)	0.20
		0.044	(8)	0.90
		0,034	(9)	1.40
		0,031	(11)	1.92
				{2.21}
	B-N	0,044	(6)	0,45
	(n,0)	0.037	(8)	0,62
		0,025	(9)	1,05
		0,021	(11)	1,42
				{2,0}

In figure parentheses gives E_{gap} value for planar structure.

So,

The energy per one atom for 2D supracrystals is of the same order as for graphene and haekelites one. Hence, such crystals may exist at the room temperature.

The electric properties of 2D supracrystals depend of its chemical composition and symmetry type, and may vary from metallic up to dielectric.

2D supracrystalline sheets can be cut into the nanoribbons and rolled up to the nanotubes. The less the diameter of nanotube the greater it energy stability and the less gap energy. Changing the composition, structure, diameter and chirality one can obtain the nanotubes with necessary value of conductivity: from dielectric to metallic.

3D-supracrystals also may exist at the room temperature, vary their electric properties from semiconducting up to dielectric and have more loose structure then usual crystals.

7. Conclusions

First of all we'll note that some of 2D-supracrystals suggested, in particular, $(BN)_{44}$ already were described before witling the framework of haekelitic structures research. It's thermal stability has been proved, what indirectly confirm the validity of our results. As for 3D-supracrystals, our investigation in this area seems to be pioneer one. Thermal stability, semiconductoral properties and friability, what is the reason of a high electrous velocity, first of all, carbonic and boron-nitrogen supracrystals, permits to hope that they will come in place of traditionally used silicon in the micro- and nanoelectronic devices. Supracrystalline nanotubes can found the application as for semiconductor devices as for dielectric, for example, from sulfur, covers of nanowires. Friable, with large cameras 3D-supracrystals may be used as containers for composite materials and hydrogen's accumulators in fuel elements of hydrogen ebergetics.

Acknowledgments

We acknowledge the Russian Foundation for Basic Research for providing financial support (Project № 10-02_97002-*r*_povolzhye_a) and the authors of program ABINIT for useful consultations.

References

[1] Brazhe R A and Karenin A A 2009 *Proc. 7 th Int. Conf. on mathematical modeling of Physical, Economic, Technical, Social System and Processes* (Ulyanovsk) p 40 [in Russian]

[2] Kvenvolden, K. (1995). A review of the geochemistry of methane in natural gas hydrate. *Organic Geochemistry* **23** (11-12): 997–1008

[3] URL http://www.abinit.org

[4] Monkhorst H B and Pack J D 1976 *Phys.Rev*. B **13** 12 5188

[5] URL http://geo.web.ru/db/msg.html?mid=1163834&uri=06.htm

[6] Browser J R, Jelski D A and George T F 1992 *Inorg. Chem*. **31** 154

[7] Terrones H, Terrones M, Hernandez E et al. 2000 *Phys. Rev. Lett.* **84** 1716

[8] Lisenkov S B, Vinogradov G A, Astakhova T Yu and Lebedev N G 2005 *JETP letters* **81** *7* 431 [in Russian]

[9]Lisenkov S B, Vinogradov G A, Astakhova T YU and Lebedev N G 2006 *Solid State Physics* **48** *1* 179 [in Russian]

[10] Wenninger M *Polyhedron models*, Cambridge, Cambridge University Press 1971

BIOTRANSFORMATION AND METABOLISM OF MAGNETIC NANOPARTICLES IN AN ORGANISM FROM MÖSSBAUER SPECTROSCOPY

I. Mischenko [1], M. Chuev [2]

[1] National Research Centre "Kurchatov Institute", 123182 Moscow, Russia

[2] Institute of Physics and Technology, Russian Academy of Sciences, 117218 Moscow, Russia

IlyaMischenko@rambler.ru

Abstract. Mössbauer spectroscopy is a powerful method for investigation structural, magnetic and thermodynamic properties of magnetic nanoparticles, in particular those delivered in a body. Multiform temperature- and field-dependent Mössbauer absorption spectra of fine particles provides the researcher with rich information about physical characteristics inherent to such particles staying in different environment. With that the ^{57}Fe gamma-resonant spectroscopy is efficiently used to study spin states, electronic and dynamical properties of iron-containing proteins in a living organism. For quantitative estimates of characteristics of the magnetic nanoparticles it is required to define a model of the magnetic dynamics. We have developed such a model and performed consistent least-square fitting procedure for the set of temperature- and magnetic field-dependent spectra as well as magnetization curves of nanoparticles injected into mice. This allowed us to reliably evaluate changes in the nanoparticles characteristics and the chemical transformation of the iron to paramagnetic ferritin-like forms in mouse's organs as a function of time after injection of nanoparticles. In fact, the approach allows the researcher to quantitatively characterize biotransformation and metabolism of magnetic nanoparticles injected into a body.

1. Introduction

A multitude of techniques are used to characterize structural, magnetic and thermodynamic properties of magnetic nanomaterials, among which one of the most informative is the Mössbauer spectroscopy. Today one of the most promising applications of this method seems to be an analysis of absorption curves of iron-containing magnetic nanoparticles injected into a living organism. The main puzzle in such investigation is to decompose the spectra into partial contributions of exogenous iron atoms in nanoparticles and endogenous iron atoms included, for example, in ferritin of the organism [1]. The temperature evolution of the spectral shape for initial nanoparticles is similar with the conventional behavior of the spectra of an ensemble of single-domain particles with magnetic anisotropy. With that a significant lines broadening in the spectra of the mouse liver and spleen after injection of nanoparticles as well as an additional doublet of lines are observed. The doublet parameters are typical for Mössbauer spectra of iron in ferritin-like proteins [2]. The formal analysis generally used for such spectra and based on the consideration of continuous distributions of the hyperfine field $\mathbf{H}_{hf}$ at iron nuclei [3], results in only qualitative treatment. Meantime, the most accurate model for analysis the temperature-dependent Mössbauer spectra of single-domain particles is the multi-level relaxation model that takes into account their magnetic anisotropy and diffusion of uniform magnetization [4]. This model was successfully used to fit self-consistently the experimental spectra of an ensemble of magnetic nanoparticles at various temperatures [5, 6], including nanoparticles injected into a living organism [1]. Moreover, there is another highly informative, long ago known and very easy method, i.e., measurements of the gamma-resonant spectra of nanoparticles in a weak static magnetic field. But an adequate theory for describing the field-dependent shape of the spectra has been developed only in the recent past [7]. This formalism can be extended within the multi-level relaxation model [4] on the basis of a model of magnetic dynamics developed in [7].

The main purpose of this paper is to demonstrate the advantages of the theoretical approach to the self-consistent treatment of the temperature- and field-dependent Mössbauer spectra and magnetization curves for characterizing the biotransformation and metabolism of magnetic nanoparticles in an organism by the example of behavior of nanoparticles injected into mice.

2. Samples and experiment

We have used industrial ferrofluid “fluidMAG-ARA-250” (Chemicell Gmbh, Germany). The suspension of the magnetic nanoparticles was injected into mice. On the expiry of fixed intervals of time the mice were destroyed, organs pulled out and lyophilized. The dried initial ferrofluid and lyophilized mouse’s organs were then triturated and the powder samples were prepared for Mössbauer exploration. The ^{57}Fe gamma-resonant absorption spectra of the samples were measured at liquid nitrogen and room temperatures on a conventional spectrometer in the linear-speed regime and transmission configuration with a ^{57}Co/(Rh) radioactive source. The spectra of the samples were also taken in the cross (with respect to the gamma-beam direction) magnetic field of the intensity H = 3.4 kOe using a permanent magnet. Magnetization curves for the samples were measured on the magnetometer LakeShore in a field up to 5 kOe. The ^{57}Fe Mössbauer spectra and magnetization curve of initial nanoparticles as well as the spectra of mouse liver two days after injection are shown in figures. 1 and 2.

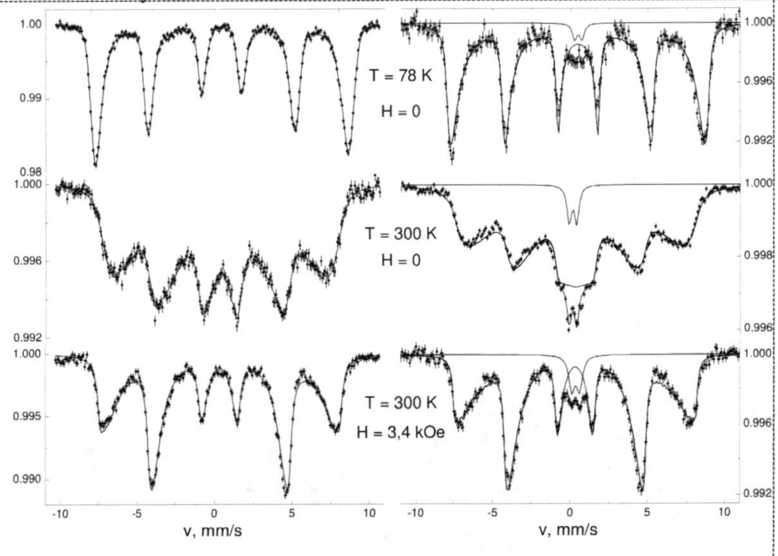

Figure 1. ^{57}Fe Mössbauer spectra of initial nanoparticles (left panel) and a mouse liver two days after nanoparticles injection (right panel). Solid lines represent the partial spectra of nanoparticles, calculated within the multi-level relaxation model, and the ferritin-like contribution as well as the resulting spectra.

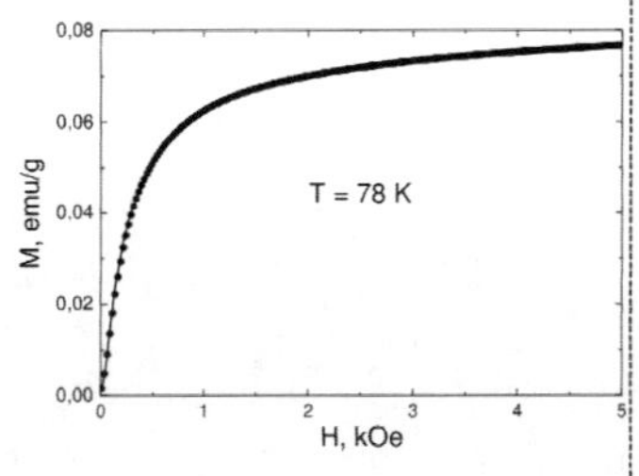

Figure 2. Magnetization curve for initial nanoparticles .

3. Formalism for calculating the relaxation Mössbauer spectra

Nowadays the most accurate model for accounting relaxation processes in the ensemble of single-domain particles is founded on the quantum-mechanical description of nanoparticle with the volume V and the uniform magnetization M_0 (or the total spin S) which vector $\mathbf{M}$ changes its orientation randomly in the anisotropy field [4], so that we have the following expression for the particle energy (figure 3):

$$E = -KV\cos^2\theta = -KV\,m^2/S^2 = E_m\,. \qquad (1)$$

Here, K is the axial magnetic anisotropy constant, θ is the angle between the magnetization vector and the easy magnetic axis of the particle and m is one of $2S$+1 possible spin projections onto this axis. The populations of the energy levels E_m are determined by Gibbs' distribution:

$$W_m = e^{-E_m/k_BT} \Big/ \sum_k e^{-E_k/k_BT}\,, \qquad (2)$$

where T is the absolute temperature and k_B is the Boltzmann constant. The transitions between these states can be specified under the assumption that relaxation is associated with the transverse components of the random magnetic field [4].

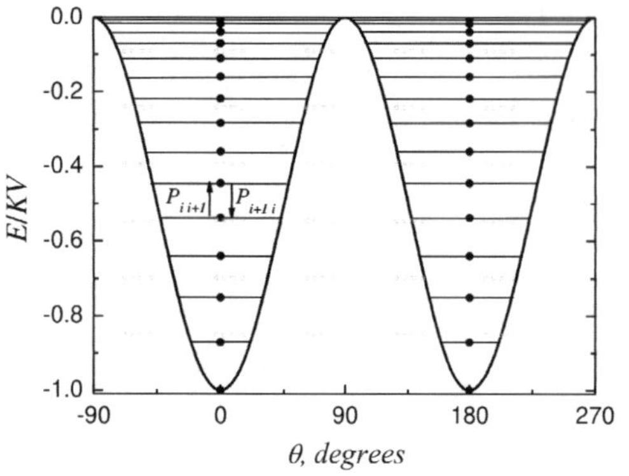

Figure 3. Schematic of the multi-level relaxation model for a single-domain particle.

Calculations of the Mössbauer spectra within the multi-level relaxation model can be performed in terms of the stochastic approach which results in the following formula for the absorption spectrum for gamma-quantum with energy $E_\gamma = \hbar\omega$ [1,4]:

$$\sigma(\omega) = -\frac{\sigma_a \Gamma_0^2}{4} \operatorname{Im} \sum_\alpha |C_\alpha|^2 \langle W | \hat{A}_\alpha^{-1}(\omega) | 1 \rangle \ . \tag{3}$$

Here, $\omega \equiv \omega - E_0/\hbar$, E_0 is the energy of the resonance transition, $\Gamma_0 \equiv \Gamma_0/\hbar$ is the natural line width, $\alpha = (m_e, m_g)$ specifies hyperfine transition with the nuclear spin projections m_e and m_g onto the $\mathbf{H}_{hf}$ direction for the excited (e) and ground (g) states, C_α determines the intensity of α-th hyperfine transition, $\langle W|$ is the row vector of the equilibrium populations of the stochastic states and $|1\rangle$ is the column vector with all components equal to unity. The operator

$$\hat{A}_\alpha(\omega) = \omega + i\Gamma_0/2 - \hat{\omega}_\alpha + i\hat{P} \tag{4}$$

is defined by the diagonal matrix of hyperfine interaction

$$(\hat{\omega}_\alpha)_{mm'} = \delta_{mm'}\omega_\alpha m/S \, , \tag{5}$$

where

$$\omega_\alpha = (g_e m_e - g_g m_g)\mu_N \mathrm{H}_{hf}/\hbar \tag{6}$$

specify the resonance transitions between the ground and exited states ($g_{g,e}$ is their g-factors, μ_N is the nuclear magneton), and the tridiagonal relaxation matrix

$$P_{m\,m\pm1} = -p_{m\,m\pm1} f_{m\,m\pm1}, \quad P_{mm} = -P_{m\,m-1} - P_{m\,m+1}, \tag{7}$$

$$p_{m\,m+1} = p_{m+1\,m} = D[S(S+1) - m(m+1)], \tag{8}$$

$$f_{ij} = \begin{cases} \exp[-(E_j - E_i)/k_B T], & E_j > E_i \\ 1, & E_j < E_i \end{cases}, \tag{9}$$

specified by the diffusion constant D [4].

Journal of Physics: Conference Series **345** (2012) 012026 doi:10.1088/1742-6596/345/1/012026

The quantitative analysis of experimental spectra requires averaging over particles sizes in the sample studied, e.g., over the Gaussian distribution of diameters:

$$P(d>0) \propto \exp\left(-(d/\bar{d}-1)^2 / 2\gamma_d^2\right). \tag{10}$$

In this case, the resulting spectra is expressed as

$$\bar{\sigma}(\omega)=\int \sigma(\omega)x^3P(x)dx \Big/ \int x^3P(x)dx\,. \tag{11}$$

Here, we have taken into account different number of atoms in the particles with different sizes.

4. Calculating the Mössbauer spectra in a magnetic field

Recently we have developed a model for describing nanoparticle's magnetic dynamics in the external magnetic field **H** [7], when the particle's energy is given by the expression:

$$E(\theta,\varphi,\Theta) = -KV\cos^2\theta - \mathbf{HM}V\,, \tag{12}$$

where θ and φ are the polar and asimuth angles in respect with the particle's easy axes, Θ is the angle between the directions of this axes and the external magnetic field. For the stationary states of the particle we have choosen the precesion orbits C_E of the magnetization vector's end in the effective field $\mathbf{H_{eff}}(\theta,\varphi,\Theta) = -\nabla E(\theta,\varphi,\Theta)/M_0$ (figure 4). These orbits correspond with the fixed values $E \equiv E_i$ of the particles energy and can be caracterized by the mean values of magnetization $\overline{\boldsymbol{M}}_i(\Theta) \equiv \overline{\boldsymbol{M}}(\Theta, E)$, which can be calculated as the curve integrals along the trajectories C_E :

$$\overline{M}_{z,x}(\Theta,E)=\int_{C_E}\frac{M_{z,x}dm_z}{\sqrt{1-m_x^2-m_z^2}}\Big/\int_{C_E}\frac{dm_z}{\sqrt{1-m_x^2-m_z^2}}\,,\qquad \overline{M}_y(\Theta,E)=0\,, \tag{13}$$

$$m_{x,z} \equiv M_{x,z}(E,\Theta)/M_0\,. \tag{14}$$

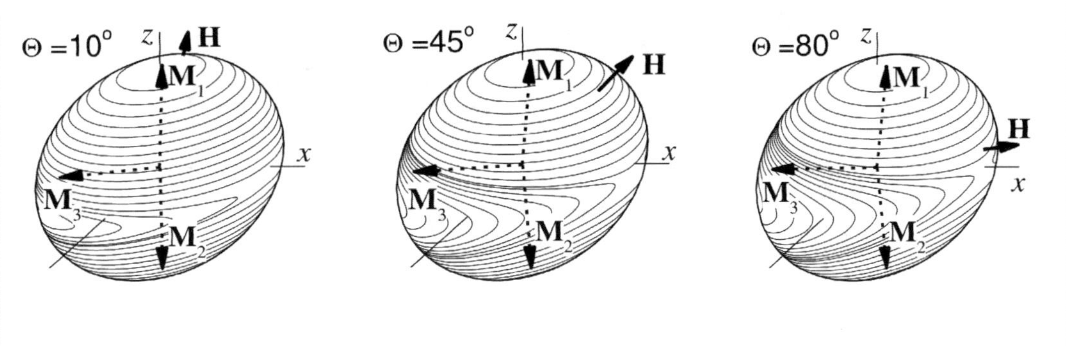

Figure 4. Precession orbits of the magnetization vector's end (constant energy levels) for a single-domain particle with the axial magnetic anisotropy in a magnetic field **H** with different orientations.

The Mössbauer spectra of nanoparticles in this case can be described within a similar stochastic approach with operators of more general type [1,7]:

$$\sigma(\omega,\Theta) = -\frac{\sigma_a\Gamma_0^2}{4}\,\mathrm{Im}\sum_{\eta}\mathrm{Sp}\left(\hat{V}_\eta\langle\mathbf{W}|\,\hat{\mathbf{A}}^{-1}(\omega,\Theta)|\mathbf{1}\rangle\hat{V}_\eta^+\right). \tag{15}$$

Here, $\hat{V}_\eta$ is the operator for the interaction of the gamma-quantum with a given polarization η and the nucleus, $\langle\mathbf{W}| = \langle W|\otimes\hat{\mathbf{1}}_\eta$, $|\mathbf{1}\rangle = |1\rangle\otimes\hat{\mathbf{1}}_\eta$, where $\hat{\mathbf{1}}_\eta$ is the identity operator in the space of $(2I_g+1)(2I_e+1)$ nuclear variables (m_e, m_g), where $I_{g,e}$ are the nuclear spins for the ground and excited states. The superoperator

$$\hat{\mathbf{A}}(\omega,\Theta) = \omega + i\Gamma_0/2 - \hat{\mathbf{L}}_{hf}(\Theta) + i\hat{\mathbf{P}}(\Theta) \tag{16}$$

is defined by the Liouville operator of hyperfine interaction that is diagonal over the stochastic states:

$$\langle i|\hat{\mathbf{L}}_{hf}(\Theta)|\,j\rangle = \hat{\mathbf{L}}_{\hat{H}}(\overline{\boldsymbol{M}}_i(\Theta))\delta_{ij} \tag{17}$$

and the relaxation matrix

$$\hat{\mathbf{P}} = \hat{P}\otimes\hat{\mathbf{1}}_\eta\ , \tag{18}$$

$$\hat{P}_{ii} = -\sum_{j\neq i}\hat{P}_{ij}\,. \tag{19}$$

[1] Here, the superoperator $\hat{\mathbf{L}}_{\hat{H}}(\overline{\boldsymbol{M}}_i(\Theta))$ is determined by the matrix elements of the Hamiltonians of hyperfine interaction for the ground and excited states:

$$(\hat{\mathbf{L}}_{\hat{H}})_{m_e m_g m'_e m'_g} = H^{(e)}_{m_e m'_e}\delta_{m_g m'_g} - H^{(g)}_{m_g m'_g}\delta_{m_e m'_e}\ , \tag{20}$$

$$\hat{H}_i^{(g,e)}(\Theta) = -g_{g,e}\mu_N H_{hf}^{(0)}\,\hat{\boldsymbol{I}}^{(g,e)}\overline{\boldsymbol{M}}_i(\Theta)/M_0\,, \tag{21}$$

where $H_{hf}^{(0)}$ is the hyperfine field at extremely low temperature.

The resulting spectrum is defined by averaging the partial spectra over the chaotic orientations of the anisotropy axes:

$$\sigma(\omega) = \int\sigma(\omega,\Theta)\sin\Theta d\Theta\,. \tag{22}$$

5. Slow diffusion limit

When the external field increases the particle's energy levels move apart and the stochastic transitions between them become less. So, in a field strong enough one can neglect the stochastic averaging in the basis expression (15). As a result, the absorption spectrum is determined by the Gibbs distribution of effective hyperfine field over the equilibrium population of stochastic states [6]:

$$\sigma(\omega,\Theta) = \sum_i L_i(\omega,\Theta)W_i(\Theta) \tag{23}$$

where

$$L_i(\omega,\Theta) = \frac{\sigma_a \Gamma_0^2}{4} \sum_\alpha \frac{\left|C_\alpha\right|^2 F_\alpha^{(i)}(\Theta)}{\left(\omega - \omega_\alpha^{(i)}(\Theta)\right)^2 + \Gamma_0^2/4}, \tag{24}$$

$$\omega_\alpha^{(i)}(\Theta) = \left(m_e \omega_e - m_g \omega_g\right) \overline{M}_i(\Theta)/M_0 . \tag{25}$$

Here, $F_\alpha^{(i)}(\Theta)$ are the polar-angular distributions of the radiation intensity, averaged by the azimuth angle and the polarization. This intensities for unpolarized radiation have the form:

$$F^{(i)}_{m_g, m_g}(\Theta) = \frac{3}{2}\left(1 - \beta_i(\Theta)\right), \quad F^{(i)}_{m_g \pm 1, m_g}(\Theta) = \frac{3}{4}\left(1 + \beta_i(\Theta)\right), \tag{26}$$

$$\beta_i(\Theta) = \cos^2\theta_\gamma \cos^2\tilde{\Theta}_i(\Theta) + \sin^2\theta_\gamma \sin^2\tilde{\Theta}_i(\Theta)/2 , \tag{27}$$

where θ_γ is the angle between the directions of gamma-ray beam and applied magnetic field, $\tilde{\Theta}_i(\Theta)$ is the angle between the field direction and the averaged magnetization $\overline{\boldsymbol{M}}_i(\Theta)$ for the i-th stationary state.

Note, that the consideration described above can be easily used to calculate the average magnetization of the ensemble of magnetic nanoparticles in a field [7]:

$$M(\mathrm{H}) = \int \sin\Theta d\Theta \sum_i W_i(\Theta) \mathbf{H} \overline{\boldsymbol{M}}_i(\Theta) / \mathrm{H} . \tag{28}$$

6. Results of data treating

Using the worded formalism we have managed a least-squares fitting procedure to analyze simultaneously the whole set of experimental data for initial nanoparticles including the Mössbauer spectra measured at different temperatures and in a magnetic field as well as the magnetization curve on the basis of the same multi-level relaxation model. The results of the self-consistent fitting are shown in figure 1 (left panel) and figure 2 by solid lines. This allows us to reliably characterize the sample with initial nanoparticles with the following estimates (in brackets we point errors in the least significant digit of the parameters values, by 'and' we divide values for T = 78 K and 300 K): the magnetic characteristic, (i) the mean magnetic anisotropy energy $K\overline{V}/k_B$ = 350(40) K, the average particle diameter $\overline{d}$ = 10.6(1) nm, the relative Gaussian width of particle's size distribution $\sigma_d/\overline{d}$ = 0.21(3), the critical field of magnetization reversal H_C = 1.62(8) kOe, D (mm/s) = 0.47(7) and 0.95(5), and the Mössbauer characteristics, (ii) the ^{57}Fe concentration $n_{NP} \cdot 10^{18}$ = 6.3(4) sm^{-3}, the isomer shift δ (mm/s) = 0.475(3) and 0.335(5), H_{hf} (kOe) = 524(1) and 491.5(8).

A similar fitting procedure has been applied to the gamma-resonant spectra of the samples of mouse liver at different stages of biodegradation (figure 1, right panel). The only difference is a necessity to include additional variable parameters of a quadrupolar doublet, indicating the contribution of ferritin-like specimens: the partial area, the isomer shift δ_f, and the quadrupolar splitting $2q$. The following principal parameters of the sample of mouse liver two days after nanoparticles injection have been evaluated: nanoparticles characteristics, (i)

$K\bar{V}/k_B = 50(20)$ K, $\sigma_d/\bar{d} = 0.7(1)$, $H_C = 1.17(3)$ kOe, and the 'ferritin' characteristics (ii) δ_f (mm/s) = 0.37(2) and 0.47(4), $2q$ (mm/s) = 0.54(3) and 0.46(7).

In particular, such an analysis allowed us to quantitatively characterize biodegradation and biotransformation of the initial particles in ferritin-like forms in a mouse organism, what is represented in figure 5 by the graph of time-evolution of ^{57}Fe concentrations in nanoparticles and iron-containing proteins.

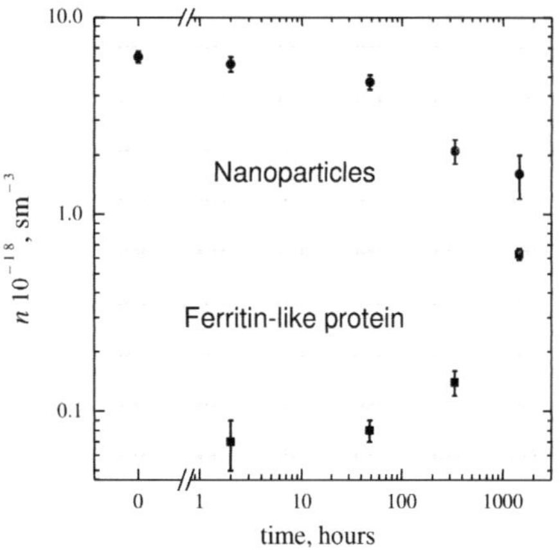

Figure 5. Time-evolution of the ^{57}Fe-consentrations in nanoparticles and ferritin-like proteins.

7. Summary

Thus, a powerful technique for describing biotransformation and metabolism of magnetic fine particles in a living organism is developed and realized on the base of self-consistent treatment of a minimal set of experimental data, which include three Mössbauer spectra measured at different temperatures and in a magnetic field as well as the magnetization curve.

8. Acknowledgments

This work is supported by the Russian Foundation for Basic Research.

References

[1] M.A. Chuev, V.M. Cherepanov, S.M. Deyev, I.N. Mischenko, M.P. Nikitin, M.A. Polikarpov and V.Y. Panchenko, AIP Conf. Proc. 1311 (2010) 322-328.

[2] G.C. Papaefthymiou, Biochimica et Biophysica Acta 1800 (2010) 886-897.

[3] M.A. Chuev and J. Hesse, in: K. B. Tamayo (Ed.), Magnetic Properties of Solids, Nova Science Publishers, New York, 2009, pp. 1-104.

[4] D.H. Jones and K.K.P. Srivastava, Phys. Rev. B34 (1986) 7542-7548; M.A. Chuev, JETP Lett. 83 (2006) 572-577.

[5] J. van Lierop and D.H. Ryan, Phys. Rev. B63 (2001) 064406 (8pp); R.D. Desautels, J.M. Cadogan and J. van Lierop, J. Appl. Phys. 105 (2009) 07B506 (9pp).

[6] M.A. Chuev, M.A. Polikarpov, V.M. Cherepanov, Doklady Phys. 55 (2010) 6-12; M.A. Chuev, V.M. Cherepanov, M.A. PolikarpovJETP Lett. 92 (2010) 21-27.

[7] M.A. ChuevJ. Phys.: Condens. Matter 20 (2008) 505201 (10pp); M.A. Chuev, JETP 108 (2009) 249-259.

Journal of Physics: Conference Series **345** (2012) 012027 doi:10.1088/1742-6596/345/1/012027

Hydrogenation catalyst based on modified carbon nanofibers

Osipov N.N., Klyuev M.V.

Ivanovo State University, Ivanovo, ul. Ermak 39

E-mail: nicolay-00@yandex.ru

Abstract. The aim of this work was to study the palladium-carboxylated carbon nanofibers (CNF) as a catalyst for the hydrogenation of nitrobenzene model reaction. It is shown that the efficiency of the catalyst obtained more than 6 times higher than that of the industrial counterpart (Pd/C).

Development and reduce the cost of production of carbon nanomaterials (CNM), expands the range of applications. Surface modification of functional groups of CNM offers new opportunities to use them, for example, as a carrier for catalysts of different chemical processes [4,5,7,8].

Used in the carbon nanofibers (CNF) were obtained in the Vladimir State University, the pyrolysis of propane-butane mixture in a copper-nickel catalyst at a temperature of 500 °C. Specific surface area of the material used was 124 m^2/g, and the CNF content in the material - about 90%. CNF have a coaxial-conical shape (Fig. 1).

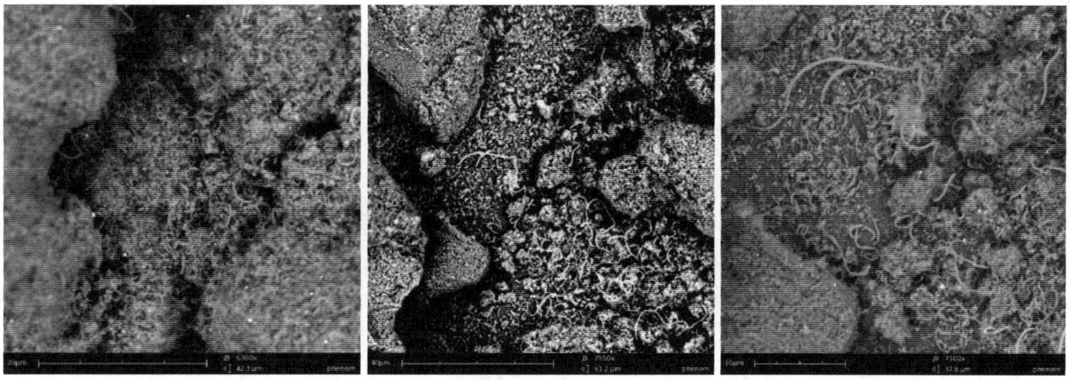

Fig. 1 Photos palladium carboxylated SAM CNF Phenom™ G2.

Initially, the CNF do not exhibit catalytic activity and are not binding transition metals. In order to be secure metal to CNF is necessary to modify their surface functional groups [1,2]. Functionalization of CNF was carried out by treatment with concentrated sulfuric acid at 70 °C with vigorous stirring for 24 hours. Then the material was filtered, washed successively with water, ethanol, and dried in air to air-dry state.

The resulting material was characterized by IR- and X-ray photoelectron spectroscopy. The IR spectra of CNF signal appears in the 1713 °cm^{-1}, indicating the presence of carboxyl groups in the samples [3]. In the XPS spectra of C 1s peak can be seen in the area of 285 eV, indicating that carbon is sp3 hybridization, which is typical for two-dimensional grids of graphite, graphene and nanotubes [6]. In the 288 eV peak is recorded indicating the presence of carboxyl groups on the surface of the CNF (Fig. 2).

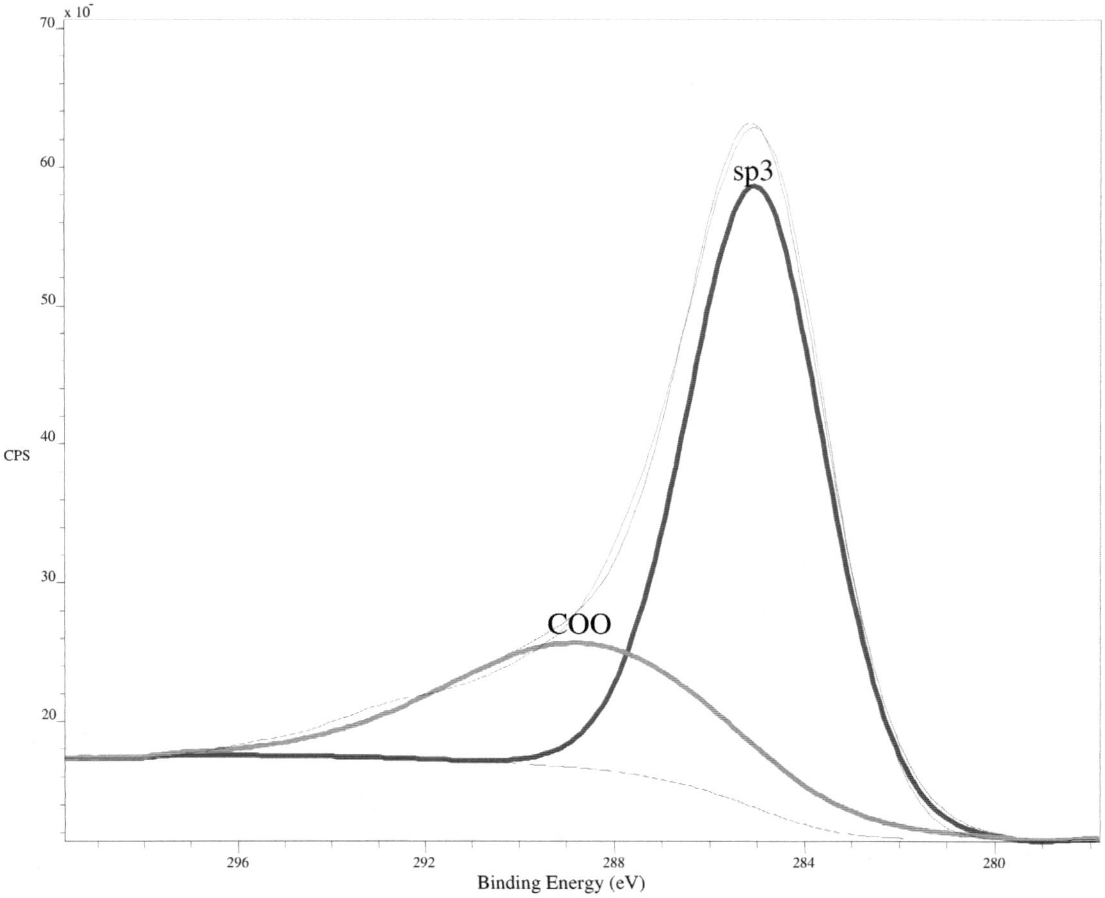

Fig. 2 XPS spectrum C 1s

To fix the palladium-modified CNF placed in a 2% solution of $PdCl_2$ in 1n HCl and stirred with a magnetic stirrer for 60 min at 20 °C. The material was then filtered, washed successively with water, ethanol, and dried in air to air-dry state. The number of entrenched palladium was determined by XPS (Fig. 3, Table 1.). As can be seen from the data, initially assigned as a palladium Pd (II) and Pd (0), and their ratio close to 1:1. Simultaneously, the sample contains chlorine.

Journal of Physics: Conference Series **345** (2012) 012027 doi:10.1088/1742-6596/345/1/012027

Table 1

The content of Pd and its valence state (wt.%) (according to XPS)

	The content of Pd wt.%	Pd^{2+}	Pd^{0}	Pd^{2+}/Pd^{0}
Prior to the activation	1,55	0,83	0,72	1:1
after activating	1,48	0,95	0,53	1:1,8
after the reaction	1,38	0,85	0,53	1:1,6

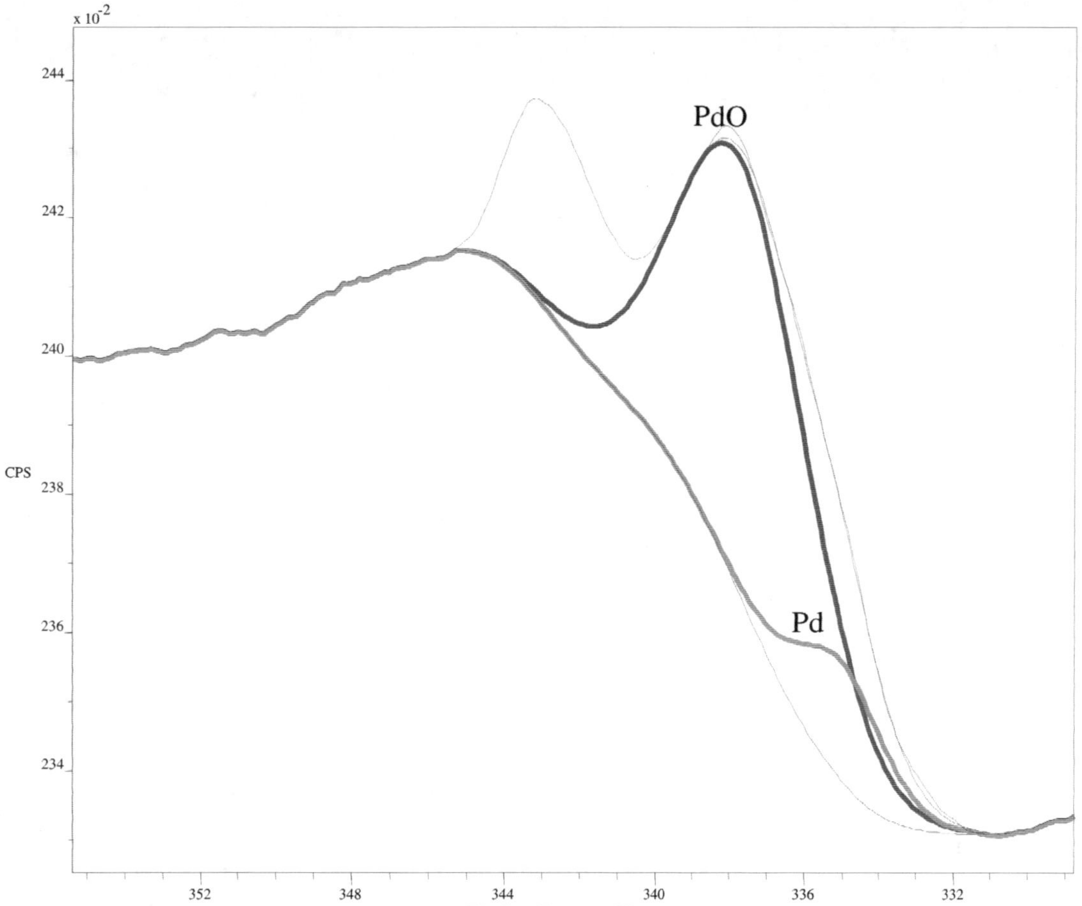

Fig. 3 XPS spectrum of Pd 3d

Further activation was carried out, for which the material was placed in ethanol (25 mL) and treated with 10 mg of $NaBH_4$ in hydrogen at a temperature of 45 °C for 60 min. After that, it was filtered, washed successively with water, ethanol, and dried in air to air-dry state. Activation of catalyst is necessary in order that would remove the chlorine from the sample. Analysis of the activated material by XPS shows that the relation Pd-Cl, most likely, not, however, the total number of Pd decreased and the ratio of Pd (II) and Pd (0) was 1.8:1. Apparently, this is due to the fact that the sample is oxidized in air. This is evidenced by the peak of 529 eV at the O 1s line in XPS spectrum of

CNF after activation (Fig. 4). The same peaks in the region 534 eV and 532 eV confirm the presence of carboxyl groups in the sample.

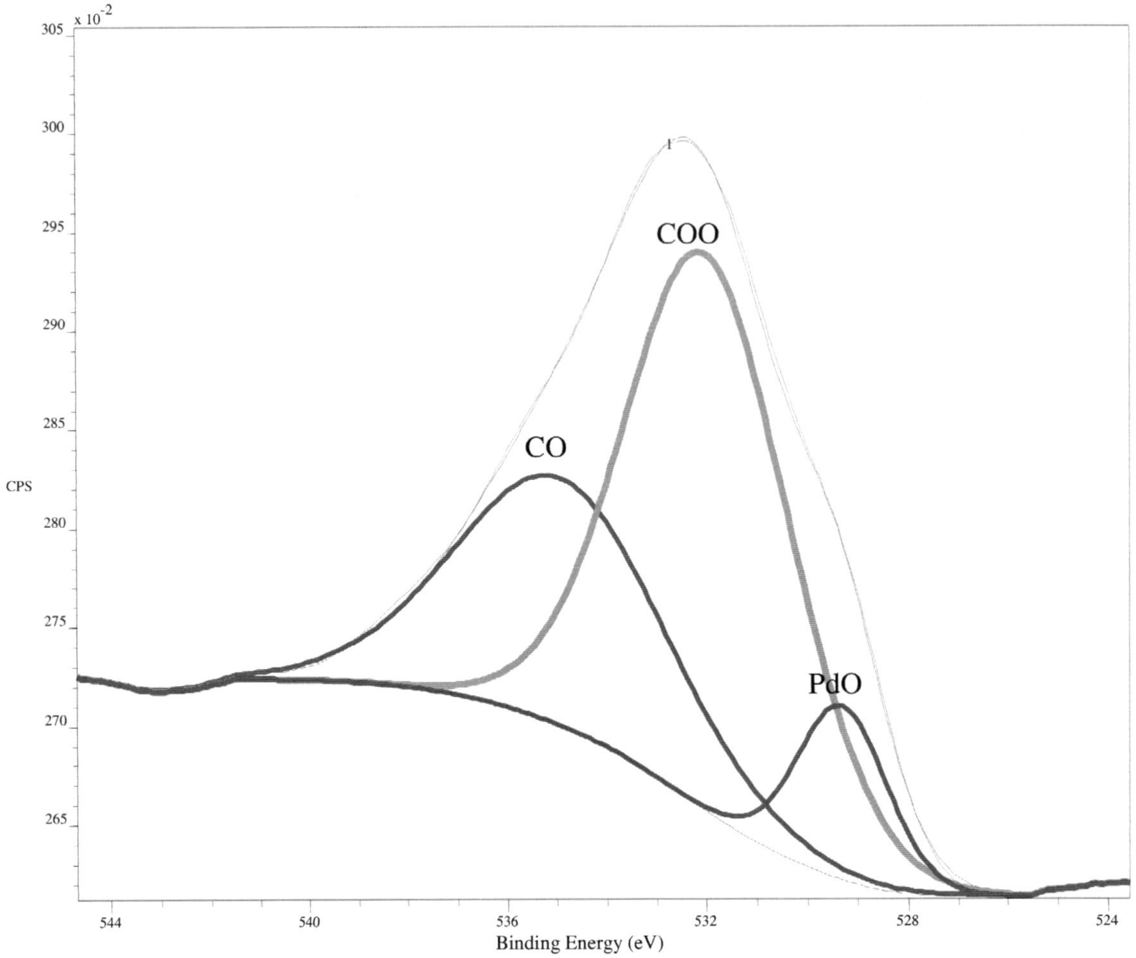

Fig. 4 XPS spectrum of O 1s

Hydrogenation reaction was carried out as follows: in the reactor under a layer of solvent (ethanol, 25 ml) was placed portion of the catalyst (Pd/CNF, 40 mg) and $NaBH_4$ (10 mg) and stirred under hydrogen at 45 °C for 10 min. Then, in a hydrogen contributed 1 mmol of nitrobenzene hydrogenation and reacted with molecular hydrogen at T=45°C, P_{H_2}=0,1 МПа. In the triple repetition of the reaction we received palladium catalyst fixed on the surface of carboxylated carbon nanofibers showed showed stability by giving the same results.

According to the analysis of reaction products, made with a gas-liquid chromatography, nitrobenzene reacted completely. By-products were found. Hydrogenation products of the aromatic ring and incomplete recovery of the nitro group were found.

In Table. 2 shows the efficiency of palladium-carboxylated carbon nanofibers as catalyst hydrogenation of nitrobenzene in comparison with the commercial (industrial) Pd/C.

Table 2

The effectiveness of Pd / CNF and Pd / C in the hydrogenation of nitrobenzene

catalyst	The content of Pd wt.%	The reaction rate 10-4 mol/(l*s)	TN mol/(g*atom Pd*min)
Pd/C	1	1.02	4.0
Pd/УНВ	1.5	8.60	23.4

Thus, the palladium catalyst attached to the carboxylated carbon nanofibers is almost 6 times more effective than Pd/C.

The authors are grateful to Vaganova V.E. Director of Research and Education and Innovation Centre "Nanotechnologies and Nanomaterials" Vladimir State University (VlSU) for providing the samples of carbon nanomaterials.

References

[1] Glebov, NV, AA Nechitailo Functionalization of the surface of multiwalled carbon nanotubes / / Technical Physics Letters. 2010. T. 36. No. 19.

[2] Yeletsky A. Carbon nanotubes / / Successes of physical sciences. 1997. T. 167. Number 9. S. 945-971.

[3] Kazitsyna LA, NB Kupletskaya Application of UV, IR, NMR and mass spectroscopy in organic chemistry. MA, Univ. Moscow State University, 1979, 240. silt.

[4] Magdalinova NA, Klyuev MV Volkova, TG Platinum-carbon nanomaterials as catalysts for hydrogenation and hydrogenation amination / / Journal "Alternative Energy and Ecology". 2009. № 10 (78). S. 89-93

[5] Ukrainians VB, KA Khokhryakov Hydrogenation catalysts based on nanosized palladium deposited on the nano-carbon materials / / Journal. Grew up. chem. of the Society to them. DI Mendeleev. 2006. T. 50. Number 4. S. 154-156

[6] Okpalugo TIT, Papakonstantinou P., Murphy H., McLaughlin J., Brown NMD High resolution XPS characterization of chemical functionalized MWCNTs and SWCNTs / / Carbon 43 (2005) 153-161

[7] Hongkun He, Chao Gao. Synthesis of Fe3O4/Pt Nanoparticles Decorated Carbon Nanotubes and Their Use as Magnetically Recyclable Catalysts / / Journal of Nanomaterials Volume 2011, Article ID 193510, 10 pages

[8] Steven Crossley, Jimmy Faria, Min Shen, Daniel E. Resasco. Solid Nanoparticles that Catalyze Biofuel Upgrade Reactions at the Water / Oil Interface / / science vol 327, 2010, 68-72

Diagnostic and analysis of aggregation stability of magnetic fluids for biomedical applications by small-angle neutron scattering

V I Petrenko[1,2]*, M V Avdeev[1,3], L A Bulavin[2], L Vekas[4], L Rosta[5], V M Garamus[6], R Willumeit[6], V L Aksenov[3,1]

[1] Joint Institute for Nuclear Research, Dubna, Russia

[2] Kyiv Taras Shevchenko National University, Ukraine

[3] National Research Centre Kurchatov Institute, Moscow, Russia

[4] Center for Fundamental and Advanced Technical Research, Timisoara, Romania

[5] Research Institute for Solid State Physics and Optics, Budapest, Hungary

[6] Helmholtz Zentrum Geesthacht, Germany

*E-mail: vip@nf.jinr.ru

Abstract. Diagnostics of aggregation and determination of the aggregation regimes and their control in biocompatible magnetic fluids are necessary for their development in biomedical applications. Small-angle neutron scattering (SANS) method was applied in the structure analysis of various types of magnetic fluids for biomedical applications. Additionally the interaction characteristics between surfactant/polymer molecules used in stabilization of magnetic fluids were investigated, which is very important for understanding the synthesis procedure of highly stable magnetic fluids with controllable properties.

1. Introduction

Cancer remains one of the most widespread diseases and leading cause of death worldwide, despite a certain progress in diagnosis and treatment methods in recent years. During last decade researchers have been investigating applications of colloidal particles (or nanoparticles) that could act as delivery systems for targeted cancer drugs [1,2]. The nanoparticles with radii between 1 – 10 nm provides large specific surface for functionalization (e.g. chemically active substances interacting with cancer cells). Along with it, if the magnetic nanoparticles are used, then, their transport and concentration in a given place can be additionally controlled by an external magnetic field, thus increasing the efficiency of the treatment, and, avoiding the spread of the nanoparticles in healthy tissues in the organism. Moreover, new possibilities for employing magnetic properties of the particles in medical diagnostics (magnetic resonance tomography) and therapy (magnetic hyperthermia) of cancer tumors appear [1,2].

Biomedical applications require that magnetic nanoparticles are to be placed in liquid conditions. The corresponding systems are known as magnetic fluids (or ferrofluids) [3]. For colloidal stabilization of magnetic fluids, the particles are coated with special shells of surfactants or polymers. Despite the fact that magnetic fluids are actively used in technical applications for rather long time, still it is not an easy task to place magnetic nanoparticles in highly polar water without aggregation [4]. The main problem is a strong interaction of surfactants with water, which is competitive with their adsorption on magnetic nanoparticles. At the moment, there are several ways to stabilize water-based ferrofluids, but what concerns the biological media it is not possible to prevent aggregation completely. In biomedical context the formation of aggregates has side effects relating to the difficult elimination of nanoparticles from organisms, the possible appearance of blood clots, as well as the reduce in the therapeutic efficiency. Thus, knowledge of aggregation regimes in magnetic fluids is a key point for their development in biomedical applications. In this connection, important goal is a reliable diagnostics of aggregation and determination of the aggregation regimes and their control in biocompatible magnetic fluids.

For this purpose we develop the method of small-angle neutron scattering (SANS), which is quite sensitive to the aggregation processes in nanosystems. SANS is actively applied in structure research of magnetic fluids [5]. Wide possibilities of the contrast variation (hydrogen/deuterium isotopic substitution) in neutron experiments allow us to 'look' inside the aggregates. Also, the additional magnetic scattering of neutrons can be used for studying magnetic correlations in nanosystems with magnetic inclusions. Taking into account the complexity of magnetic fluids (which are mostly polydisperse), reasonable tasks for SANS in this case are to find out the type of aggregates formed in them under different conditions, conclude about their inner nuclear, as well as magnetic structures, and obtain information about interaction between their different components. The present work shows how the SANS method is applied in the structure analysis of various types of magnetic fluids for biomedical applications. Additionally the interaction characteristics between surfactant/polymer molecules used in stabilization of magnetic fluids are investigated, which is very important for understanding the synthesis procedure of highly stable magnetic fluids with controllable properties.

2. Experiments

SANS experiments were performed at the Yellow Submarine small-angle instrument at the steady-state reactor of the Budapest Neutron Centre, Hungary and the SANS-1 instrument located at the Neutron Facility at the Helmholtz Zentrum Geesthacht, Germany. The differential cross-section per sample volume (hereafter referred to as scattered intensity) was obtained as a function of the scattering vector module, $q = (4\pi/\lambda)\sin(\theta/2)$, where λ is the incident neutron wavelength and θ is the scattering angle. At the Yellow Submarine instrument the fixed wavelength of λ=0.386 nm and 1.2 nm (monochromatization $\Delta\lambda/\lambda$ = 20%) and sample-detector distances of 1.3 and 5.6 m (detector size 64 × 64 cm^2; resolution 1 × 1 cm^2) were used to cover the q-interval of 0.15-4.5 nm^{-1}. In the measurements performed on the SANS-1 instrument, we used the wavelength λ = 0.81 nm (monochromaticity $\Delta\lambda/\lambda$ = 10%) and the sample–detector distances ranging from 0.7 to 9.0 m (detector size 55 × 55 cm^2; resolution 0.7 × 0.7 cm^2) and, therefore, the measurements were performed in the range of scattering vectors from 0.04 to 2.00 nm^{-1}. All SANS measurements were performed at 25 °C. The calibration on 1-mm water sample was made after the background, buffer and empty cuvette corrections in a standard way [6].

3. Results and discussion

Different aggregate classes depending on the stabilization mechanism (steric, electrostatic and steric/electrostatic stabilization) of magnetic particles (magnetite) in physiological conditions are studied. In figure 1 example of the SANS contrast variation application [7,8] is given for water-based magnetic fluids used as a source of magnetic nanoparticles in the therapy of the brain cancer glioblastoma. These fluids are characterized by record achievable concentrations of magnetic material (up to 10 vol. %) with keeping high stability [9]. The nanomagnetite was stabilized by double layer of fatty mono-carboxylic acids with short alkene chains, namely lauric (C12) and myristic (C14) acids. Despite high concentration and long time stability, SANS shows the presence of aggregates in the systems. The found effective match points (see insets to figure 1) are shifted and show a difference in the composition of the aggregates. Schematic views of the fluid structures in figure 1 illustrate the main important conclusion of the analysis: there is a different rate in the surfactant coating of the particles in the aggregates, which is consistent with the longer chain of myrstic acid. Among the studied systems in this way are water-based magnetic fluids with substitution of sodium oleate as surfactant by biocompatible polymer polyethyleneglycol at the magnetite surface, which aims at the increase in the life time of magnetic nanoparticles in living organisms by reducing the response of their immune systems. This fluid was used as an initial component in the synthesis of the magnetic carrier containing anticancer drug Taxol. It is revealed that quite large amounts of polymer in the fluid structure results in a decrease in the aggregation stability, thus requiring that an optimal polymer content to be chosen.

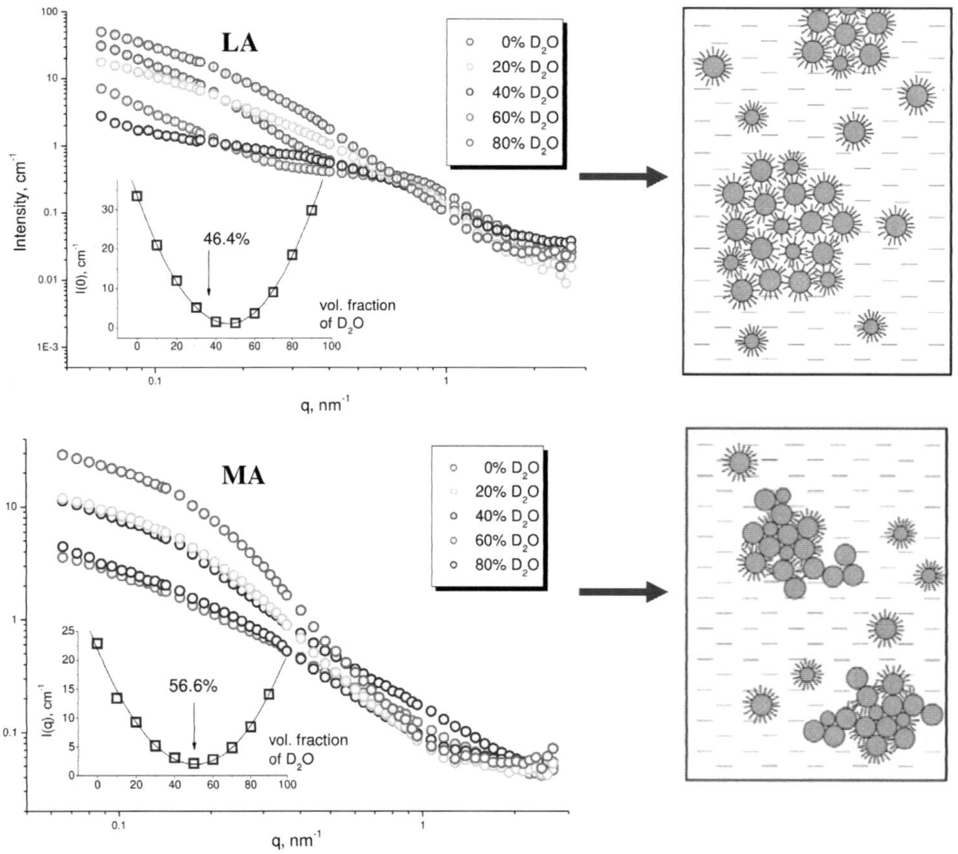

Figure 1. SANS contrast variation for 1% biocompatible aqueous magnetic fluids with magnetite coated by double layer of lauric (LA) or myristic (MA) acids. Changes in the scattering curves are followed when varying the content of heavy water (D_2O) in the solvent. In insets the effective match points are found from the change in the forward scattering intensity. The revealed aggregation structures in the studied magnetic fluids are schematically shown to the right.

The aggregation stability of magnetic fluids is directly determined by the behavior of stabilizants (surfactants or polymers) in the solvents [10-15]. Thus, in most cases as a first stage of the synthesis the concentrating solutions of fatty acids are added to the medium, where magnetic nanoparticles are produced. Already at this stage the aggregation of fatty acids themselves can affect the final stability of magnetic fluids. SANS makes it possible to reveal such aggregation and follow the transition from the isotropic solution to liquid crystalline state with the growth in the solute concentration, as it is shown in figure 2 for organic solutions of saturated mono-carboxylic acids with stiff linear structure. It is obtained that the critical concentration of this transition is inversely proportional to the length of surfactant molecules. Such aggregation decreases the number of free surfactant required for adsorption and further stabilization of magnetic nanoparticles, thus affecting the final stability of the magnetic fluid and explaining why longer saturated fatty acids are worse stabilizants as compared to shorter ones. After the SANS analysis of the interaction of acid molecules one comes to conclusion that the discussed transition goes easier in the presence of magnetic particles. In some cases the formation of micelle-type surfactant aggregates can appear in magnetic fluids. It is also reflected in the SANS curves, which makes it possible to describe quantitatively (in terms of micelle size and concentration) this process [14].

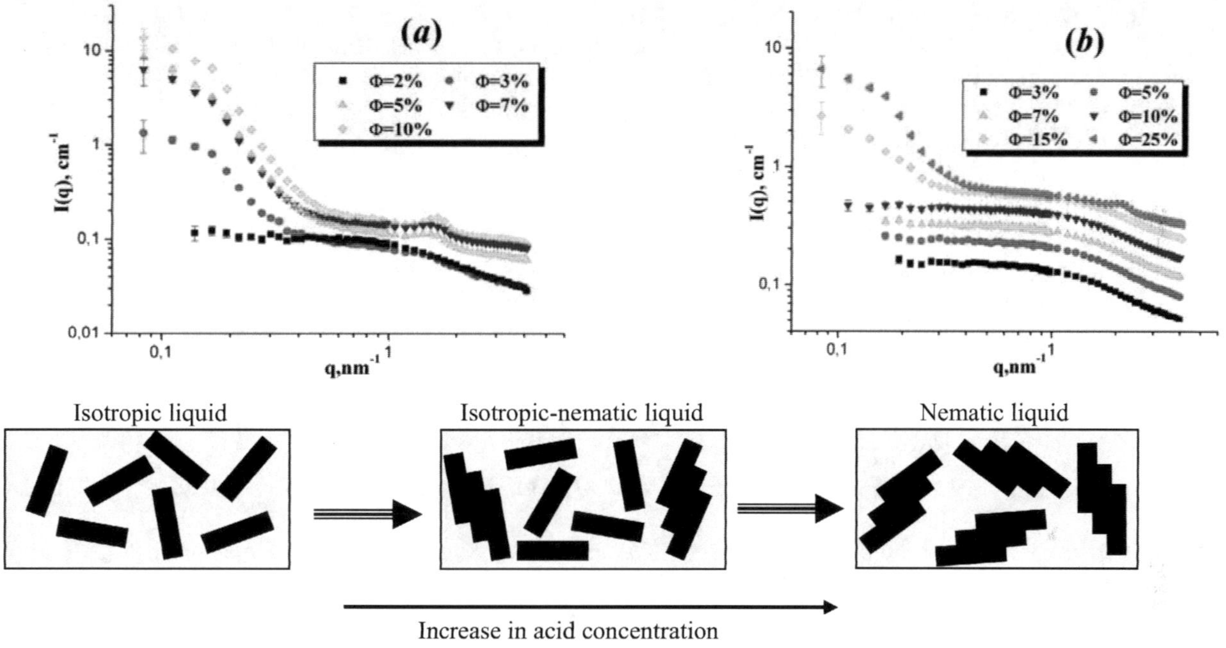

Figure 2. Formation of the liquid crystalline phase in bulk organic solutions of mono-carboxylic acids probed in stabilization of magnetic fluids. Experimental SANS curves from the solutions of stearic (C18) (a) and myristic (C14) (b) acids show the isotropic-nematic phase transition with the growth of the volume fraction of the acids in solutions. The scheme at the bottom illustrates the transition into the liquid crystalline state reflecting in the SANS curves.

4. Conclusions

Summarize we can deduce that SANS method can be successfully applied in the structure analysis of various types of magnetic fluids for biomedical applications. The interaction characteristics between surfactant/polymer molecules used in stabilization of magnetic fluids can be also obtained by SANS.

Structures of various types of ferrofluids were found and different rate of magnetic surface covering by stabilizants were observed. The formation of the found LC-phase in bulk solutions of mono-carboxylic acids is an important factor, which influences the stabilization efficiency of the studied acids in colloidal solutions of magnetic nanoparticles.

Acknowledgments. The work has been performed in the frame of the project Joint Research Groups, Helmholtz-RFBR (HRJRG-016). The program of JINR-HAS cooperation is also acknowledged.

References

[1] Vekas L, Avdeev M V and Bica D 2009 *Nanoscience in Biomedicine* (Ed. D Shi. Berlin: Springer-Verlag) p 650.
[2] Hafeli U and Zborowski M. (Ed.) 2009 *J. Magn. Magn. Mater.* **321**(10) p 1335.
[3] Rosensweig R E 1985 *Ferrohydrodynamics* (Cambridge: Cambridge University Press) p 344.
[4] Avdeev M V, Mucha B, Lamszus K, Vekas L, Garamus V M, Feoktystov A V, Marinica O, Turcu R and Willumeit R 2010 *Langmuir* **26** 8503.
[5] Avdeev M V and Aksenov V L 2010 *Phys. Uspekhi* **53** 971.
[6] Wignall G D and. Bates F S 1987 *J. Appl. Cryst.* **20** 28.
[7] Feoktystov A V, Avdeev M V, Aksenov V L, Petrenko V I, Bulavin L A, Bica D, Vekas L, Garamus V M and Willumeit R 2009 *Solid State Phenom.* **152-153** 186.
[8] Nagorny A V, Petrenko V I, Avdeev M V, Bulavin L A and Aksenov V L 2010 *J. Surf. Inv. X-ray Synchr. Neutron Tech.* **4**(6) 3.
[9] Petrenko V I, Aksenov V L, Avdeev M V, Bulavin L A, Rosta L, Vekas L, Garamus V M and Willumeit R 2010 *Phys. Solid State* **52** 974.
[10] Petrenko V I, Avdeev M V, Aksenov V L, Bulavin L A and Rosta L 2009 *J. Surf. Investigations. X-ray Synchr. Neutron Tech.* **3**(1) 161.
[11] Petrenko V I, Avdeev M V, Almásy L, Bulavin L A, Aksenov V L, Rosta L and Garamus V M 2009 *Coll. Surf. A* **337** 91.
[12] Petrenko V I, Avdeev M V, Aksenov V L, Bulavin L A and Rosta L 2009 *Solid State Phenom.* **152-153** 198.
[13] Avdeev M V, Bodnarchuk I A, Petrenko V I, Kholmurodov Kh T and Yaradaikin S P 2009 *Rus. J. Phys. Chem. A* **83**(7) 1129.
[14] Petrenko V I, Avdeev M V, Garamus V M, Bulavin L A, Aksenov V.L. and Rosta L. 2010 *Coll. Surf. A* **369** 160.
[15] Lancz G, Avdeev M V, Petrenko V I, Garamus V M, Koneracká M and Kopčanský P 2010 *Acta Physica Polonica A* **118**(5) 980.

Metal-oxide-based nanocomposites comprising advanced gas sensing properties

A A Ponomareva[1,2], V A Moshnikov[1], D Glöß[3], A Delan[2], A Kleiner[2], G Suchaneck[2]

[1]St. Petersburg State Electrotechnical University (LETI), 197376 St. Petersburg, Russia
[2]TU Dresden, Solid State Electronics Laboratory, 01062 Dresden, Germany
[3]Fraunhofer Institute for Electron Beam and Plasma Processing (FEP), 01277 Dresden, Germany

E-mail: ap_k@inbox.ru, Gunnar.Suchaneck@tu-dresden.de

Abstract. Using the sol-gel technique, nanocomposite SiO_2-SnO_2 thin films comprising a designed hierarchical pore structure were prepared on cost-efficient oxidized silicon substrates. These films exhibit advanced gas sensing properties both for reducing gases and oxygen due to an efficient analyte transport by mesoporores.

1. Introduction

Gas sensitivity of metal oxide gas sensors increases rapidly when the dimensions of oxide sensing materials become comparable or smaller than the typical thickness of the electron depletion layer [1]. Therefore, nanosized materials exhibit outstanding properties in sensor application, particularly low operation temperature and high sensitivity, attributed to their large surface area to volume ratio.

Hierarchical structures are higher dimensional structures composed of many, lower dimensional building blocks. Because they are generally larger than the individual nanostructures, their van der Waals attraction is relatively weak. On the other hand, hierarchical microstructures provide an effective gas diffusion path via well aligned mesoporous structures without sacrificing a high surface area [2]. Consequently, they exhibit high gas sensitivity and fast gas response. Hierarchical nanostructures for gas sensors were prepared by vapour phase growth, hydrothermal or hydrothermal/solvothermal reactions [2], by electrospinning and subsequent oxygen plasma etching [3], by ultrasonic spray pyrolysis [4], by flame-spray-pyrolysis [5, 6], and by pulsed laser deposition techniques [7]. Well defined hierarchical pore structures were prepared by sol-gel method in alkoxy-derived silica systems [8] as well as in aluminium salt solutions [9] utilizing a phase separation mechanism as a pore forming principle. Recently, one of our laboratories has applied this approach for the fabrication of SnO_2-SiO_2 composites comprising hierarchical macro-, meso- and micropores (according to IUPAC notation [10]). This minimises the build up of thermal stresses during sensor fabrication and, therefore, improves adhesion. Moreover, our technology is based on inexpensive precursors such as $SnCl_2$ and tetraethylorthosilicate (TOES). In the first case, advantages compared to alkoxides are besides the precursor costs a reduced sensitivity of the precursor to heat, moisture and light, and an improved time stability of the sol against the increase in viscosity. On the other hand, the presence of Cl^- increases the gelling time.

In this work, we investigate gas sensing properties of sol-gel deposited nanocrystalline SnO_2-SiO_2 composites in relation to their microstructure and demonstrate gas sensitivities at the state-of-art level.

2. Experimental

Nanocrystalline SnO_2-SiO_2 thin films comprising a hierarchical pore structure were fabricated using a polymeric sol-gel process. Tin (II) chloride dihydrate $SnCl_2{\bullet}2H_2O$ (M = 225.63 g/mol) was used as a tin precursor and tetraethylorthosilicate (TEOS) $C_8H_{20}O_4Si$ as a silica source (table 1). The solutions

Journal of Physics: Conference Series **345** (2012) 012029 doi:10.1088/1742-6596/345/1/012029

were aged 24 h at room temperature and spin-coated onto oxidized silicon wafers without an adhesion promoter. The films were calcined at about 550°C in air for 5 minutes by transferring the wafer into a hot furnace. Cooling down was performed in the furnace within 12 to 14 hours.

Table 1. Solvent and chemical reagent ratio used for sol-gel nanocomposite SiO_2-SnO_2 thin film deposition.

Composition* (Run number)	Solvent	Volume of solvent, ml	Weight of $SnCl_2{\cdot}2H_2O$, g	Volume of TEOS, μl
20%SiO_2-80%SnO_2	ethanol	10	0.242	60
20%SiO_2-80%SnO_2	*n*-butanol	10	0.154	28.7
10%SiO_2-90%SnO_2	*n*-butanol	10	0.173	19.1
100% SnO_2	*n*-butanol	10	0.193	-

* Percentages are in molar %

Composition analysis was performed by means of X-ray photoelectron spectroscopy (XPS) using MgK_α (1253.6 eV) radiation in the vacuum chamber of a Physical Electronics PHI 5702 electron spectrometer. The surface composition was evaluated by calculating the peak areas of high-resolution XP spectra. The signal was calibrated referring to the C1s hydrocarbon signal at a binding energy of 285 eV. In order to clean the sample from carbon residues and adsorbed at the surface gases, Ar^+ sputter cleaning (E_{kin} = 4 keV) was carried out removing an about 1.4 nm thick surface layer in two minutes. After argon ion bombardment, the C1s peak disappeared. In these cases XPS peaks were referred to the O1s signal at 531.15 eV. Microstructure of the deposited porous SnO_2-SiO_2 films was investigated by atomic force microscopy (AFM) using an NTEGRA Therma scanning probe microscope (NT-MDT, Zelenograd, Russia). Tapping mode imaging was performed in air using silicon cantilevers (Type NSG01, NDT-MDT, Moscow, Russia). AFM provides already a digitized image of 256x256 pixels suitable for numerical evaluation. Following an industrial standard for specifying the indication of surface texture of optical elements, especially its high spatial frequency errors (roughness) [11], an isotropic, one-dimensional power spectral density function $PSD(k)$ where k is the wavevector (the spatial frequency in reciprocal space), of the surface topography was calculated. The two-dimensional AFM image was averaged over the profiles traced in fast scan direction using the Open Source Software Gwyddion 2.21 [12]. This allows reducing noise due to instrument drift appearing more pronounced in the slow scan direction. Moreover, the averaging over 256 scans is approximately equivalent to the integration of the two-dimensional PSD over the slow scan direction [13]. The specific surface area (SSA) was determined by means of the Brunauer, Emmett and Teller (BET) method with nitrogen adsorption using a Sorbi-N4.1 instrument (ZAO META, Novosibirsk, Russia). UV-VIS optical properties of the films were investigated in the wavelength range of 300 to 2000 nm with steps of 2 nm by absolute reflectance and transmittance measurements in a two-beam spectrophotometer (Lambda 950, PerkinElmer, USA).

Sensor structures were prepared using film deposited onto oxidized silicon substrates with *n*-butanol as a solvent. Nickel-chrome and gold layer of a thickness of 10 and 35 nm, respectively, were evaporated onto the SnO_2-SiO_2 thin films. Thereby, the Ni-Cr layer serves for adhesion improvement. The patterned by a copper contact mask electrode layers formed 0.5 and 0.4 mm wide interdigital electrodes. Gas sensing experiments were carried out by using nitrogen (reference gas) and 1.5 vol.% hydrogen in nitrogen or oxygen provided by a computer controlled gas flow system. The purity of all gases was ≥ 99.999 Vol. %. The total gas flow was kept constant at 90 sccm. A gas sensor area of about 3.5 mm^2 on the silicon wafer was heated using a small platinum heater. The measurement setup is described in detail elsewhere [14]. Sensitivity to ethanol was determined in a computer controlled measurement setup described in [15]. Dry air dried was used as reference gas.

3. Results and discussion

XPS indicated a Si/(Si+Sn) ratio of about 0.27 at the surface (X-ray penetration depth about 5 nm) of both as-deposited and sputter-cleaned samples synthesized with a ratio Si/(Si+Sn) = 0.2. Carbon was present on the surface of as-deposited films by an amount of 22.5 %. Carbon disappeared after sputter cleaning. The higher silicon fraction obtained by XPS is probably caused by pores throughout the film with sizes larges than the film thickness of about 60 nm. Therefore, in the following Si/(Si+Sn) ratios based on the ratio in the solution are given. The $Sn3d_{5/2}$ peak appeared at 487.2 eV at slightly higher value than that of bulk SnO_2 (486.6 eV) indicating that tin is in the tetravalent state. The $Sn3d_{5/2}$ peak position did not change after sputter cleaning. The O1s peak was deconvoluted into two ones, a peak assigned to SiO_2 at 532.4 eV and a SnO_2 related one at 531.1 eV. Their peak area ratio was 0.2713 for the sputter cleaned sample in good agreement with the Si/(Si+Sn) determined from the XPS peak areas of the cations. In the $Cl2p_{3/2}$ region, no appreciable peak could be observed in the sample.

Optical properties of the SiO_2-SnO_2 nanocomposites were already reported in [16]. Films deposited using *n*-butanol as solvent exhibit an optical band gap of about 4 eV. The larger compared to bulk SnO_2 (3.62 eV) band gap is attributed to the contribution of the quantum size effect [17, 18, 19] observable also in ultrathin films [20]. In our case, the band gap value of about 4 eV corresponds to a crystallite size of about 5 nm [17]. The optical band gap of films using ethanol as solvent was about 3.4 eV. All films were strongly disordered with an Urbach tail characteristic energy of 0.50 to 0.75 eV making a reliable determination of the optical gap difficult. Therefore, we have also determined the isoabsorption gap E_{04} at an absorption coefficient of $\alpha = 10^4$ cm^{-1} yielding an optical gap of 4.51 and 3.95 eV for *n*-butanol and ethanol solvents, respectively.

The evaluation of the microstructure of hierarchical structures is challenging because of they span multiple length scales from the molecular to the macroscopic level. In our case, intrinsic film features are mostly pronounced in the range 10-100 nm making AFM a proper choice for microstructure characterization. The obtained by AFM surface topography can give qualitative insight into film formation processes such as cluster aggregation, pore formation, ordering etc.

Surface texture is characterized by the height of its peaks, the depths of its valleys and the distances that separate them. The calculation of the PSD function by Fourier decomposition of the digitized AFM image into spatial frequencies takes into account the statistical correlations between the different surface points and hence it gives a more general description than the RMS surface roughness alone. Thus, the PSD identifies wavelengths of features that contribute to the surface structure. The one-dimensional PSD considered in this work is representative of a homogeneous and isotropic surface.

Figure 1 shows the surface texture of sol-gel processed SnO_2 film and SiO_2-SnO_2 nanocomposites fabricated using *n*-butanol as solvent and the derived from AFM images PSDs. Practically no porosity was observed in the smooth films (RMS surface roughness about 0.9 nm) prepared without TEOS addition (Fig. 1a). The 10 at% SiO_2-90 at% SnO_2 sample exhibits the largest porosity with a three-dimensional film structure (Fig. 1b). This related to the fact that TEOS is known to favour the formation of three-dimensional network of linked oxide particles in the gel leading to a proper pore structure in the annealed composites [21]. The surface of 10 at% SiO_2-90 at% SnO_2 films is significantly roughened in the whole region of spatial wavelengths compared to SnO_2 films deposited by the same procedure without adding TEOS to the solution. The most pronounced feature size of 10 at% SiO_2-90 at% SnO_2 films equals to 1.25 μm (Fig. 1d). Note, that an increase of the PSD by two orders of magnitude corresponds to an increase of the RMS surface roughness R_{RMS} by one order of magnitude according to

$$R_{RMS} = \sqrt{\int_{k_1}^{k_2} PSD(k)dk} \tag{1}$$

where the reciprocal values of k_1 and k_2 determine the spatial wavelength bandwidth of the roughness measurement. The sample with 20 at% silica fraction had a decreased again roughness with circular pores of a broad size distribution (Fig. 1c). Spatial wavelengths below 250 nm are strongly attenuated giving evidence of a smoother surface at the nanometer level. The SSA of this film amounted to 87.21

m^2/g yielding an average pore size of about 10 nm. SSA and average pore size are in good agreement with the values of similar films investigated in [22]. On the other hand, it has to been mentioned that these nanopores known to enhance gas sensitivity [23] cannot be imaged by our AFM device.

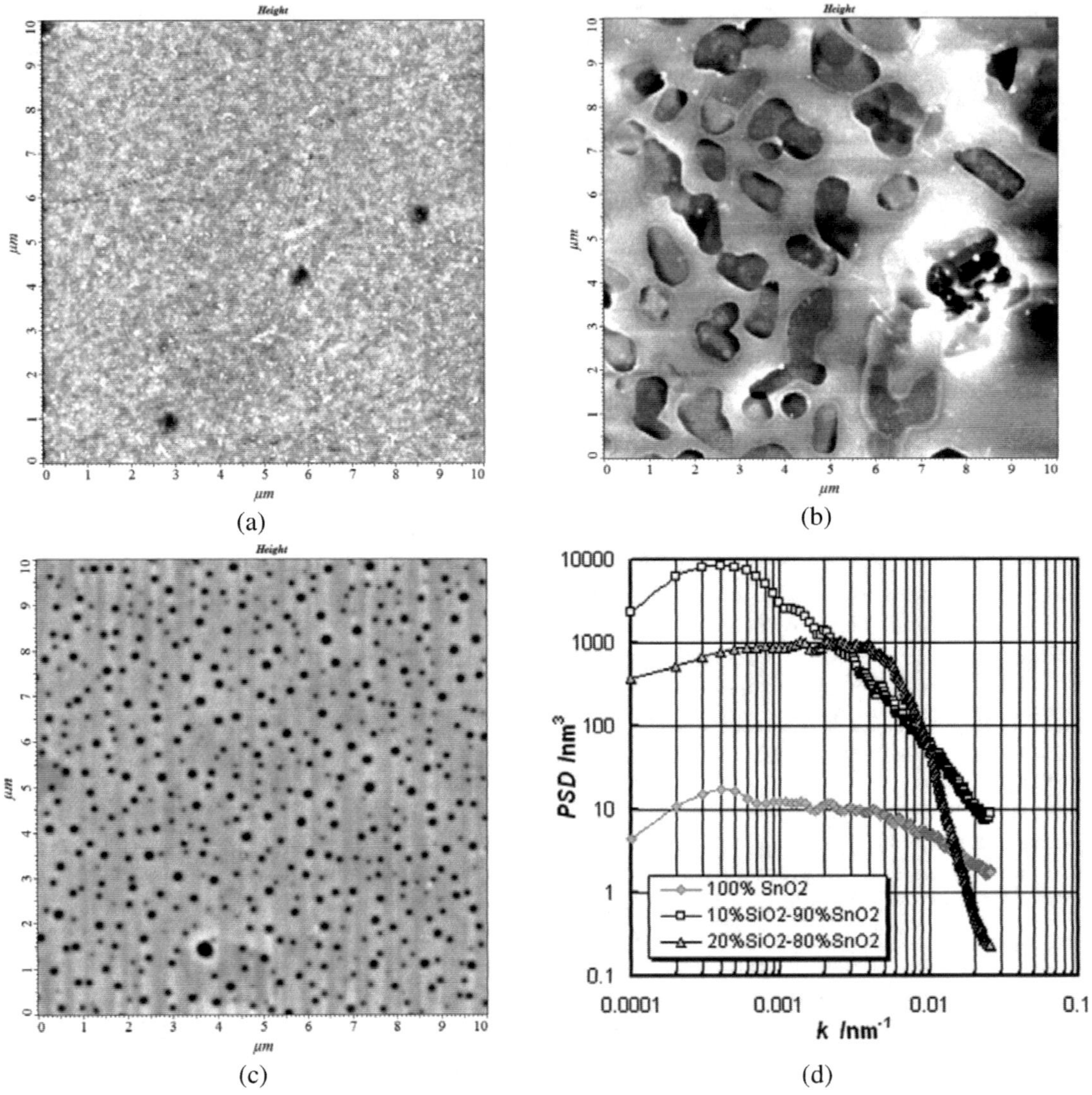

Figure 1. AFM-images of sol-gel layers: a – 100 at% SnO_2; b – 10 at% SiO_2 – 90 at% SnO_2; c – 20 at% SiO_2 – 80 at% SnO_2, d – power spectral density of the images in figures a-c.

Selected results of gas sensitivity measurements at 200°C are summarized in table 2. Preliminary values of hydrogen gas sensitivity were already reported in [24]. The highest gas sensitivities were achieved by the sensor with 10 at% silica fraction possessing the highest porosity.

Figure 2a shows the response time of hydrogen and ethanol sensors in dependence on nanocomposite silica fraction. The quickest response was obtained for 10 at% silica fraction providing a high three-dimensional porosity. Thereby, the impact of porosity on response time is more pronounced for larger gas molecules. This gives evidence that sensor response time is limited by the analyte transport in mesopores. The response time exhibits the typical dependence on the gas concentration, i.e., an increase of the response time with decreasing of gas concentration (not shown in figures) [24, 25, 26].

Table 2. Gas sensitivity of ($S=\Delta R/R_{air}$) of SiO_2-SnO_2 nanocomposite thin film sensors in dependence on analyte gas and gas concentration.

Nanocomposite	Gas sensitivity for the following gas concentration				
	H_2 4500 ppm	H_2 15000 ppm	C_2H_5OH 1000 ppm	O_2 61500 ppm	O_2 100000ppm
100 at% SnO_2	0.38	0.48	0.29	0.32	0.43
10 at% SiO_2-90 at% SnO_2	0.85	0.93	0.71	1.40	2.00
20 at% SiO_2-80 at% SnO_2	0.76	0.86	0.47	0.96	1.56

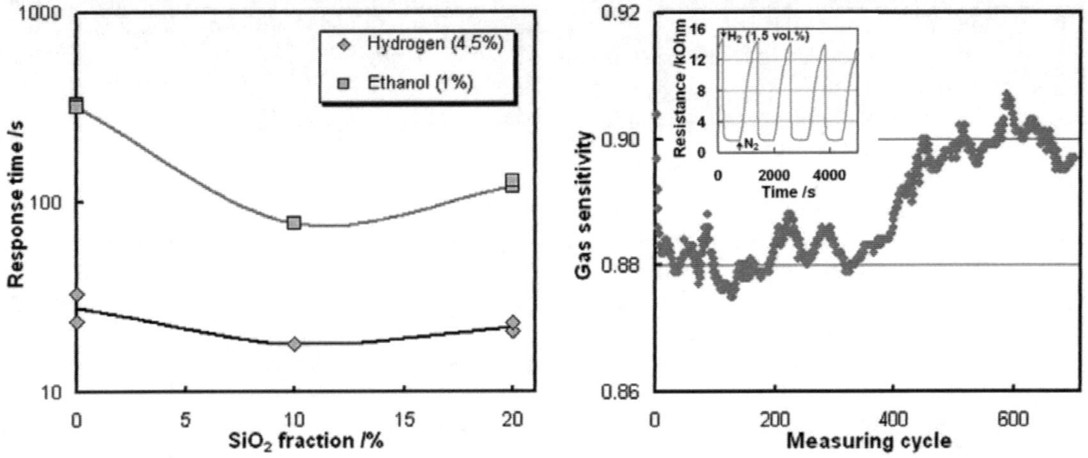

(a) (b)

Figure 2. Response time to hydrogen and ethanol in dependence on silica fraction of the nanocomposite (a) and gas sensitivity to hydrogen (1.5 vol.%) in dependence on measuring cycle number (b), all data measured at a temperature of 200°C.

In order to test long-term stability, seven hundred measuring cycles of hydrogen sensing were carried out. No degradation of hydrogen sensitivity was observed (Fig. 2b). Therefore, the fabricated in gas sensors are capable to overcome sensor drift and missing long-term stability known for state-of-art nanosized SnO_2 gas sensors [27, 28]. Longer term investigations are in progress now.

Conclusions

SiO_2-SnO_2 nanocomposites comprising a hierarchical pore structure were fabricated using tin(II) chloride dihydrate $SnCl_2•2H_2O$ as a tin source and TEOS as a silicon source. In order to achieve advanced gas sensing properties, the porous film structure was optimized by varying the silica fraction of the nanocomposite. For the optimized composition of 10 at%SiO_2-90 at%SnO_2, mesopores provide an efficient analyte transport decreasing sensor response and recovery times. These sol-gel deposited SiO_2-SnO_2 nanocomposites are promising materials for advanced metal oxide gas sensors.

Acknowledgements

This work was supported by contracts 16.740.11.0211, P399 and P2279 of the Federal Target Program "Scientific and Scientific-Pedagogical Personnel of Innovative Russia for the period 2009-2013", and by the Erasmus Mundus External Co-operation Window Programme of the European Union (A.P.).

References

[1] Xu Ch, Tamaki J, Miura N and Yamazoe N 1991 *Sensors and Actuators* B **3** 147
[2] Lee J-H 2009 *Sensors and Actuators* B **140** 319
[3] Zhang Y, Li J, An G and He X 2010 *Sensors and Actuators* B **144** 43
[4] Firooza A A, Hyodo T, Mahjoub A R, Khodadadi A A and ShimizuY 2010 *Sensors and Actuators* B **147** 554
[5] Tricoli A, Graf M and Pratsinis S E 2008 *Advanced Functional Materials* **18** 1969
[6] Qu Y, Xue T, Zhong X, Lin Y C, Liao L, Choi J and Duan X 2010 *Advanced Functional Materials* **20** 3005
[7] Chen Z, Pan D, Zhao B, Ding G, Jiao Z, Wu M, Shek C-H, Wu L C M and Lai J K L 2010 *ASNNano* **4** 1202
[8] Nakanishi K 1997 *Journal of Porous Materials* **4** 67
[9] TokudomeY, Fujita K, Nakanishi K, Miura K and Hirao K 2007 *Chem. Mater.* **19** 3393
[10] Rouquerol J, Avnir D, Fairbridge C W, Everett D H, Haynes J M, Pernicone N, Ramsay J D F, Sing K S W and Unger K K 1994 *Pure & Appl. Chem.* **66** 1739
[11] Draft DIN ISO 10110-8:2010, 2011-03-07
[12] Gwyddion – data analysis software: http://gwyddion.net
[13] Marx E, Malik I J, Strausser Y E, Bristow T, Poduje N and Stover J C 2002 *J. Vac. Sci. Technol.* B **20** 31
[14] Delan A, Karuppasamy A and Schultheiß E 2009 *Plasma Processes and Polymers* **6** S731
[15] Gracheva I E, Maksimov A I, Moshnikov V A and Plekh M E 2008 *Instruments and Experimental Techniques* **51** 462-465
[16] Ponomareva A A, Moshnikov V A and Suchaneck G 2011 *Technical Physics Letters* **37** 892-895
[17] Gu F, Wang S F, Lü M K, Zhou G J, Xu D and Yuan D R 2004 *J. Phys. Chem.* B **108** 8119
[18] Yu B, Zhu C and Gan F 1997 *Optical Materials* **7** 15
[19] Gu F, Wang S F, Lu M K, Cheng X F, Liu S W, Zhou G J, Xu D and Yuan D R 2004 *J. Cryst. Growth* **262** 182
[20] Gaidi M, Hajjaji A, Smirani R, Bessais B and El Khakani M A 2010 *J. Appl. Phys.* **108** 063537
[21] Blanco E, Esquivias L, R Litran R, Pinero M, Ramirez-del-Solar M and de la Rosa-Fox N 1999 Sonogels and Derived Materials *Appl. Organometal. Chem.* **13** 399
[22] Gracheva I E, Moshnikov V A, Karpova S S and Maraeva E V 2011 *J. Phys: Conf. Ser.* **291** 012017
[23] Yang J, Hidajat K and Kawi S 2008 *Materials Letters* **62** 1441
[24] Ponomareva A A, Moshnikov V A and Suchaneck G 2011 *Proc. E-MRS 2011 Fall meeting: Symp. K (Warsaw University of Technology, Poland, 19-23 September 2011)* p. 32
[25] Liewhiran C, Tamaekong N, Wisitsoraat A and Phanichphant S 2009 *Sensors* **9** 8996
[26] Cukrov L M, McCormick P G, Galatsis K and Wlodarski W 2001 *Sensors and Actuators* B **77** 491
[27] Korotcenkov G 2005 *Sensors and Actuators* B **170** 209
[28] Lu F., LiuY, Dong M and Wang X 2000 *Sensors and Actuators* B **66** 225

Synthesis and Characterization of Iron-Cobalt Nanoparticles

A.N. Popova
Institute of Coal-chemistry and Material science of the RAS SB
650000, Russia, Kemerovo, Sovetskiy av., 18

E-mail: zaharov@kemsu.ru

Abstract. Nanoparticles of iron-cobalt have attracted great interest in recent years because of their unique physical and chemical properties because of their industrial importance in preparing magnetic recording materials, pigments, catalysts, etc. With an objective to develop magnetic materials with high saturation magnetization Fe-Co nanosysytems were synthesized by method of chemical reduction reaction of salts aqua-solutions by hydrazine hydrate in alkaline medium. The synthesized powders were characterized by XRD, SEM, EDX and SQUID-magnetometry techniques. The steady-state grain size of 5-20 nm was calculated using XRD line broadening and SEM. The maximum saturation magnetization of 207 and 215 A•m2/kg was achieved at 300 and 5 K, respectively.

1. Introduction

Nanocrystylline materials, because of the very fine grain sizes, exhibit a variety of properties, which are different and often considerably improved in comparison with those of conventional coarse-grained polycrystalline materials [1]. Therefore, nanosized materials are the subject of a large number of studies in the last years from basic research and applications point of view. The high saturation magnetic moment, Curie temperature, and high permeability properties of Fe-Co alloys make its potential from magnetic recording head to drug delivery applications [2-4].

2. Materials and methods

Nanocrystalline Fe-Co alloys were synthesized by a chemical process involving the reduction of a mixture of iron sulphate and cobalt chloride with hydrazine-hydrate added slowly to a predetermined volume of water solution of the metal salt in reactor in the presence of magnetic field. The synthesized samples were characterized by X-ray powder diffraction (XRD) using a BRUKER D8-Advance X-ray diffractometer to identify the nanocrystalline state of the alloys and the phase of the crystalline grains. Microstructural evaluation was studied with Scanning Electron Microscopy (SEM) of JEOL model JSM-6390. Compositional analysis of the samples, carried out with an energy dispersive X-ray (EDX) analyzer (in conjunction with SEM) shows no significant impurity except a small trace of 2-5 at.% even smaller due to oxygen. The magnetic measurements were performed by a Quantum Design superconducting quantum interference device (SQUID) magnetometer of the type MPMS-XL5 with a maximum field of 2 T.

3. Results

Previously [5, 6], it was observed that for $x \leq 75$ a single bcc Fe(Co) solid solution was formed. Synthesis of Fe-Co alloys for Fe-rich region ($x \leq 25$) shows formation of a well-defined ferrite spinel

(Fe_2CoO_4) and the bcc Fe(Co) solid solution. Synthesis for Co-rich region ($x \geq 95$) were found leading to the coexistence of two phases (fcc + hcp) end products while the fcc Co(Fe) solid solution is observed for $x \geq 75$. Thus, the phase formation by chemical reduction of Fe-Co appears to be similar to what is observed in the well-known composition trend [7] of the bulk Fe-Co alloys at 650 K. Phase portrait of Fe-Co nanosystems was designed (Figure 1).

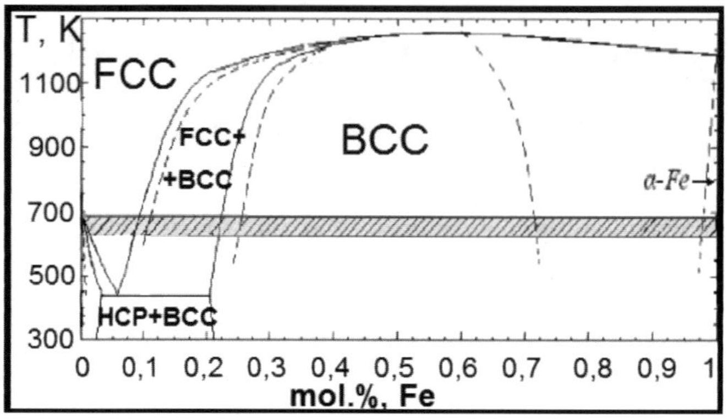

Figure 1. Phase diagrams of Fe-Co nanosystems.

The average grain size of Fe-Co crystalline particles were estimated from the XRD patterns by Scherrer analysis of the full widths at half maximum of the diffraction peaks. The grain size is between 5 and 20 nm. It was found that the samples are characterized by morphology with multilevel hierarchy. The crystalline particles form sphere-shaped ultrasonic disaggregation-resistant agglomerates. The Fe-Co agglomerates form long chain-like structures. Figure 2 shows the typical SEM images of Fe-Co samples.

Figure 2. SEM images of Fe-Co samples.

Figure 3 presents typical hysteresis loops of nanocrystalline Fe-Co alloys measured at 5 and 300 K. One can see the hysteresis loop features small changes in shape at these temperatures.

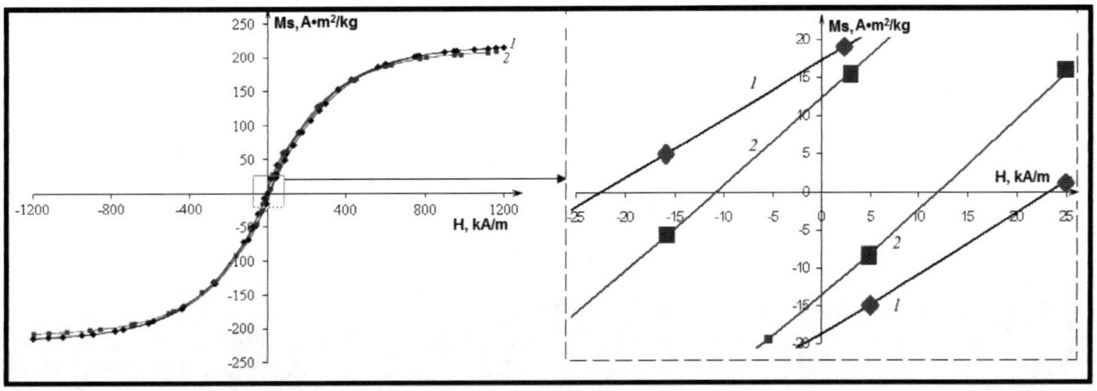

Figure 3. Magnetic characteristics of nanocrystalline Fe-Co alloys at 300 K (1) and 5 K (2).

Figure 4 shows the cobalt composition effect on the evolution of saturation magnetization. A peak value of saturation magnetization determined from magnetization curves is about 215 A•m2/kg.

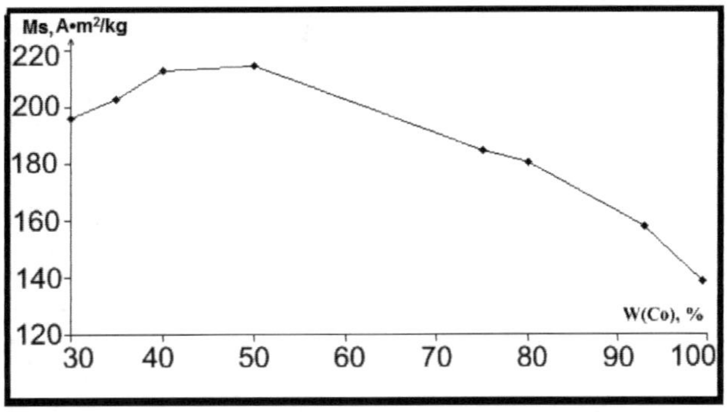

Figure 4. Magnetic characteristics of nanocrystalline Fe-Co alloys at 300 K (1) and 5 K (2).

Table 1 shows the maximum magnetization (MS) level and the coercive force (Hc) of the nanosized Fe-Co powders formed by reduction of salts mixture with hydrazine-hydrate. It was observed that nanocrystalline Fe-Co have high saturation magnetization and high coercivity.

Table 1. Magnetic properties of nanocrystalline Fe-Co alloys

Sample	T, K	Ms, A•m^2/kg	Hc, kA/m
0,5%Fe-99,5%Co	5	138,6	11,5
7%Fe-93%Co	5	157,9	15,1
20%Fe-80%Co	5	184,7	19,1
50%Fe-50%Co	5	214,7	8,3
60%Fe-40%Co	5	212,7	10,7
70%Fe-30%Co	5	196,1	9,1
50%Fe-50%Co	300	207,0	6,8
65%Fe-35%Co	300	203,1	7,0

4. Conclusions

With an objective to obtain alloys with highest possible MS value, the synthesis of nanocrystalline Fe-Co alloys was investigated by characterization of the powders using SEM, EDX, XRD, and SQUID-magnetometry. The results obtained show that the chemical synthesis by the reduction of a mixture of iron sulphate and cobalt chloride with hydrazine-hydrate can lead to obtain Fe-Co nanoparticles with chain-like structures, which present behaviour of hard magnetic materials.

References

[1] R. Birringer, H. Gleiter, H.P. Klein, P. Marquardt, 1984 *Phys. Lett.* **102**A 365.
2] G. Schmid, *Nanoparticles: From Theory to Application* (New York: Wiley Interscience), 2004, 443 p.
[3] Tartaj P., Morales M.P., Veintemillas-Verdaguer S., Gonsalez-Carreno T., Serna C.J., 2003 *J. Phys. D*, **36**, R182-R197
[4] Manuel Arruebo, Rodrigo Fernández-Pacheco, M. Ricardo Ibarra, and Jesús Santamaría 2007 *Nanotoday* **2 (3)** P. 22-32
[5] Yu. A. Zaharov, A. N. Popova, R. P. Kolmikov, V. M. Pugachev, V. G. Dodonov // 2008 *Perspective Materials* **6** 249-254
[6] Yu. A. Zaharov, V. M. Pugachev, V. G. Dodonov, A. N. Popova 2010 *Perspective Materials* **3** 60 – 72
[7] *Binary Alloy Phase Diagrams*, Massalski T. B., editor-in-chief (ASM INTERNATIONAL), 1986, 1086 p.

Acknowledgments

Author wishes to acknowledge assistance and encouragement from supervisor (Dr. Yuriy A. Zaharov) and colleagues (Valeriy Pugachev, Vadim Dodonov). The work was supported by The Integration Grant SB RAS.

Journal of Physics: Conference Series **345** (2012) 012034 doi:10.1088/1742-6596/345/1/012034

Synthesis of nanocomposites on basis of single-walled carbon nanotubes intercalated by manganese halogenides

M V Kharlamova, A A Eliseev, L V Yashina, A V Lukashin, Yu D Tretyakov

Lomonosov Moscow State University, 119992, Leninskie gory 1-3, Moscow, Russia

mv.kharlamova@gmail.com

Abstract. In this work $MnCl_2$@SWNT and $MnBr_2$@SWNT nanocomposites were prepared via capillary filling of single-walled carbon nanotube channels by melts of manganese halogenides with slow cooling to room temperature for better crystallization. Electronic properties of $MnHal_2$@SWNT nanocomposites have been investigated by X-ray photoelectron spectroscopy, optical absorption spectroscopy and Raman spectroscopy. It was found that the manganese halogenides show acceptor behavior and there is the charge transfer from the carbon nanotube walls to intercalated nanocrystals.

1. Introduction

Today single-walled carbon nanotubes (SWNTs) are an example of one of the most interesting nanomaterials, which attract the close attention of the research community due to their unusual electronic, chemical and structural properties accompanied by extraordinary mechanical behavior depending on the diameter and conformation of tubes [1-5]. However, the strong dependence that the functional properties have on the diameter and rolling vector of the nanotube, in combination with the practical difficulty of extracting of SWNTs of definite chirality, makes it necessary to develop methods for a directed change in their electron structure. Modifying of SWNT properties by introducing of various substances into the internal channels of nanotubes is of special interest. Moreover, single-walled carbon nanotubes have an extra small diameter and maximal anisotropy parameters, compared to other known nanostructures, which make them extremely attractive templates for forming of one-dimensional nanocrystals in the internal channels [6].

It should be noted that the filling of nanotube internal channels makes it possible to change the structure and properties of SWNTs without the necessity of separating them with respect to their chirality and electronic characteristics [7]. For instance, an intercalation of electron-donors (with Fermi level located in the conductance band of SWNTs) to the metallic tubes will lead to increase of electron density on carbon nanotube walls resulting in better conductance through the composite wire, while an intercalation of electron-acceptors (with Fermi level below the one of SWNTs) may cause the decrease of carrier density (and Fermi energy) in the system [7]. Moreover, the electronic properties of the obtained composites will be determined by the generalized electron density of the SWNT and the electron density on the one-dimensional crystal associated with it. Therefore, this approach, which is based on the charge transfer in X@SWNT system after introducing of one-dimensional nanocrystals of the electron-donor or electron-acceptor compounds into SWNT channels, makes it possible to control the electronic structure of single-walled carbon nanotubes and produce heterotransitions inside a single nanotube in the case of partially filled channels.

An extensive application of SWNT-based composites is unfortunately limited due to the absence of universal synthetic approaches for infilling of nanotube channels [7]. The traditional techniques include *in situ* filling of nanotubes by catalytic metals during CVD process and *ex situ* procedure (so-called capillary technique) via impregnation of pre-opened SWNTs [8, 9, 10, 11-16]. However, in many cases the formation of discrete clusters inside carbon nanotubes instead of the formation of continuous one-dimensional nanocrystals was observed, especially when *in situ* intercalation or wet-chemistry methods were applied [17]. Homogeneous filling of nanotube channels can be achieved when molten guest compounds are crystallized slowly inside SWNTs [18, 11, 19, 20]. This approach can be used if the surface tension of the guest compound is low (γ=130-170 mN/m) in order to enable the wetting of nanotube surface, as well as if it has a relatively low melting point (<800°C) to prevent the destruction of carbon nanotubes in the course of impregnation [15]. At present, a great number of studies report the filling of carbon nanotubes with halides, metals, oxides, hydroxides, chalcogenides, fullerenes and other materials [19-26]. However, in most of these works, most attention was paid to the studying of atomic structure of the obtained nanocomposites, whereas the electronic properties of intercalated nanotubes remained practically unknown. At the same time, these characteristics determine the practical application of nanomaterials on basis of single-walled carbon nanotubes. Therefore, in this work nanostructures on basis of SWNTs filled with $MnCl_2$ and $MnBr_2$ were synthesized and the influence of intercalated salts on the electronic structure of nanotubes was studied.

2. Experimental part

SWNTs were obtained by catalytic arc-discharge method using 0.8 cm diameter graphite rods with Y/Ni powder catalyst under the following conditions: helium pressure of 73.3 kPa and current of 100-110 A. Nanotubes were purified by multistage procedure consisting of oxygenation in air and rinsing with HCl to remove the catalyst. The purified samples with SWNTs content of 78 wt.% and catalyst content of 0.12 wt.% were pre-opened by heat treatment at 500ºC in dry air for 0.5 hour.

Pre-opened SWNTs (0.025 g) were grinded in agate mortar with $MnCl_2$ or $MnBr_2$ (Aldrich, 99,999 wt. %) in molar ratio of 1:1. In view of hygroscopic properties of these compounds this experimental stage was carried out in dry box. Then obtained mixture was evacuated in quartz ampoules at 1 Pa for 1 hour and sealed. The ampoules were heated at a rate of 1K/min up to the temperature of 100K above the melting point of $MnHal_2$. This temperature was maintained for 10 hours, whereupon the samples were slowly cooled at a rate of 0.1K/min. We believe that slow cooling provides better conditions for $MnHal_2$ crystallization inside the nanotube channels. The obtained samples were denoted by $MnCl_2$@SWNT and $MnBr_2$@SWNT, respectively.

The Raman spectra were acquired using a Renishaw InVia Raman microscope equipped with 20 mW 514 nm argon, 17 mW 633 nm HeNe and 300 mW 785 nm NIR diode lasers, variable power ND filters (power range 0.00005-100%) and near-excitation tunable (NeXT) filters. In Raman spectra the positions of all resonance lines were determined by least-square fitting of the experimental data with sets of Gaussian/Lorentzian convolution functions using "WiRE 2.0" software. Optical absorption spectra were acquired using Perkin-Elmer Lambda 950 spectrometer in the spectral range of 200-2500 nm with a spectral resolution of 1-4 nm. HRTEM was performed on JEOL 2100 microscope at 300 kV. For this study the samples were prepared by dispersing of nanotubes onto a carbon-coated copper grid. X-ray photoelectron spectra were obtained using monochromatic Al K_α irradiation at Axis Ultra (Kratos) spectrometer at pass energy of 5 eV. The samples were placed on metallic indium surface. The size of analyzed area equaled 300x700 microns.

3. Results and Discussion

The HRTEM images obtained for $MnCl_2$@SWNT and $MnBr_2$@SWNT samples are represented in Fig. 1. These data show a bundle of single-walled carbon nanotubes with an internal diameter of 1.4 nm. These data give evidence that nanotube channels are filled; moreover, the formation of one-dimensional crystals with an ordered structure is observed inside the channels in the case of $MnCl_2$. In the observed projection for $MnCl_2$@SWNT nanocomposite the motive of image is formed by two rows of contrast elements with the distance between them of 0.33-0.35 nm and periodicity of fragments repeating along SWNT axis of 0.30 nm.

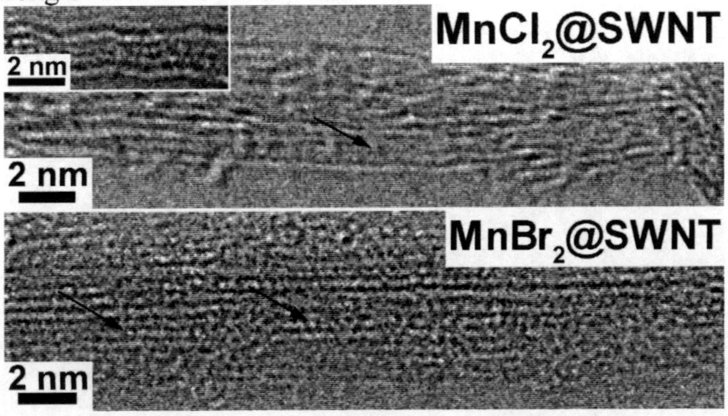

Figure 1. HRTEM images of $MnHal_2$@SWNT nanocomposites

The optical absorption spectra of SWNTs (Fig. 2) in the visible and near-IR regions illustrate the set of characteristic stripes corresponding to transitions between the van Hove singularities in semiconductor (S1–S2) and metal (M1) nanotubes with a diameter of 1.2–1.5 nm. In the spectra of $MnCl_2$@SWNT nanocomposite a partial quenching of optical transitions between the first van Hove singularities of semiconductor SWNTs takes place (S1 in Fig. 2). Such behavior of the system can be caused by the electron transfer between the nanotube and the crystal with either devastation or filling of the first van Hove singularities with electron density, which was observed earlier in our previous works [20, 21]. Also a strong shift (0.15-0.25 eV) of all peaks in composites towards lower energies takes place. This shift corresponds to the narrowing of energy gaps between the van Hove singularities in nanotubes after the filling of channels. Thus, the described changes in the spectra of composites (as opposed to unfilled SWNTs) evidence the charge transfer between the nanotube walls and the introduced nanocrystals, which leads to a significant change in the electronic structure of SWNTs.

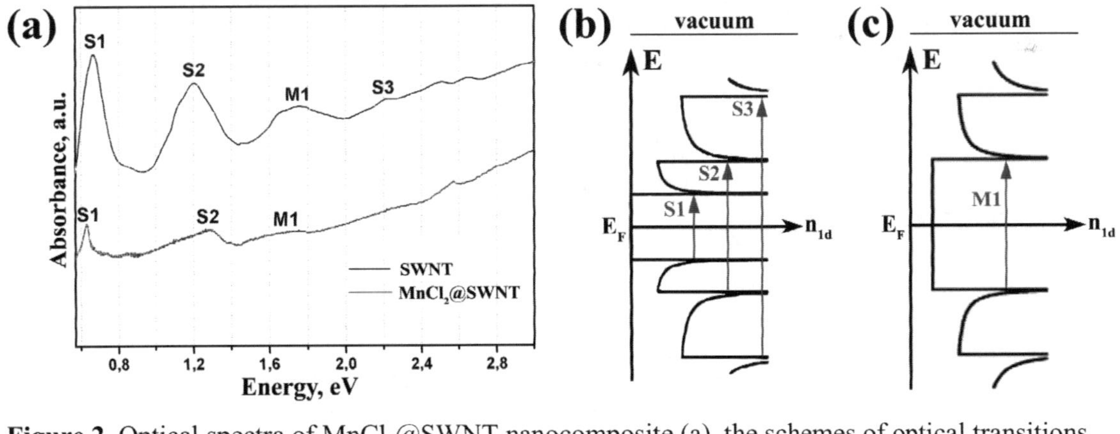

Figure 2. Optical spectra of $MnCl_2$@SWNT nanocomposite (a), the schemes of optical transitions between the van Hove singularities in semiconducting (b) and metallic SWNTs (c)

Figure 3 shows the characteristic RBM and G bands of Raman spectra of pristine SWNTs and $MnCl_2$@SWNT and $MnBr_2$@SWNT composites, which were obtained upon the excitation of samples by laser radiation with an energy of 1.96 eV and 1.58 eV (λ_{ex} = 633, and 785 nm, respectively) in the frequency region of 100–1700 cm^{-1}. Both RBM and G band positions for nanocomposites are up shifted relative to those of pristine SWNTs and it indicates the change in C-C bond energy. Also it may indicate the change of electronic structure of carbon nanotubes in course of intercalation. It should be noted that besides the significant shifts of RBM and G bands in Raman spectra of the $MnCl_2$@SWNT and $MnBr_2$@SWNT nanocomposites, which evidence the mechanical and electronic interaction between SWNT walls and intercalated nanocrystals, a substantial change in the profile of G band takes place. The profile of G band changes from typical for metallic nanotube form to the form that is typical for SWNTs of the semiconductor type. A possible explanation for the changes in G band profile is the transition of metallic SWNTs to a semiconductor state after the filling them with one-dimensional $MnCl_2$ and $MnBr_2$ crystals as a result of the charge transfer between SWNT walls and the intercalated nanocrystals.

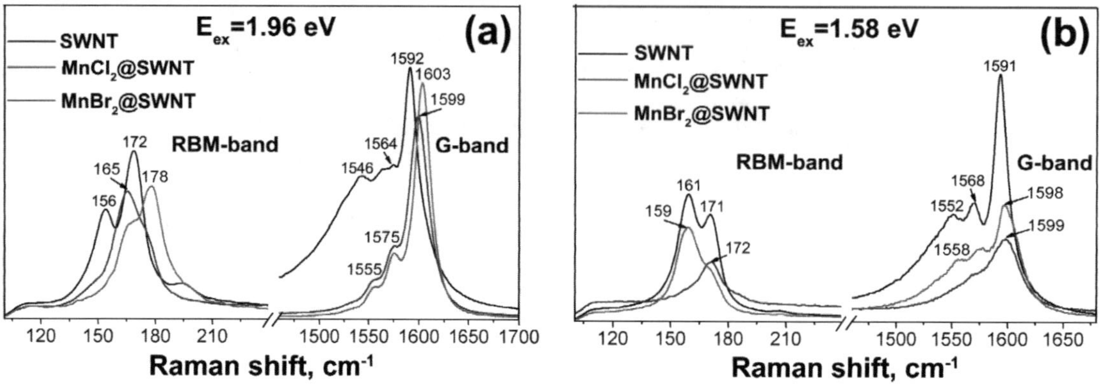

Figure 3. Raman spectra of $MnHal_2$@SWNT nanocomposites obtained at laser energies of 1.96 eV (a) and 1.58 eV (b)

Fig. 4 shows x-ray photoelectron spectra of pristine SWNTs and $MnBr_2$@SWNT sample. The C 1s spectrum of nanocomposite has a multicomponent structure. Component C I positioned at binding energy of 284.38 eV can be assigned to the pristine unfilled SWNT. Besides, new components C II and C III appear, with BE being lower for C II and higher for C III. Both components are evidently related to the filled nanotubes. The possible origin of C II component is the variation of SWNT work function as a result of Fermi level downshift and the corresponding shift of all peaks related to $MnBr_2$@SWNT to lower binding energies. Thus, the manganese halogenide show acceptor behavior. Moreover, this assumption is consistent with the whole dataset discussed above. The nature of the less intensive and broad component C III is evidently different. It is positioned at binding energy of 0.63 eV higher than C II. Probably it can be assigned to the local effects [20].

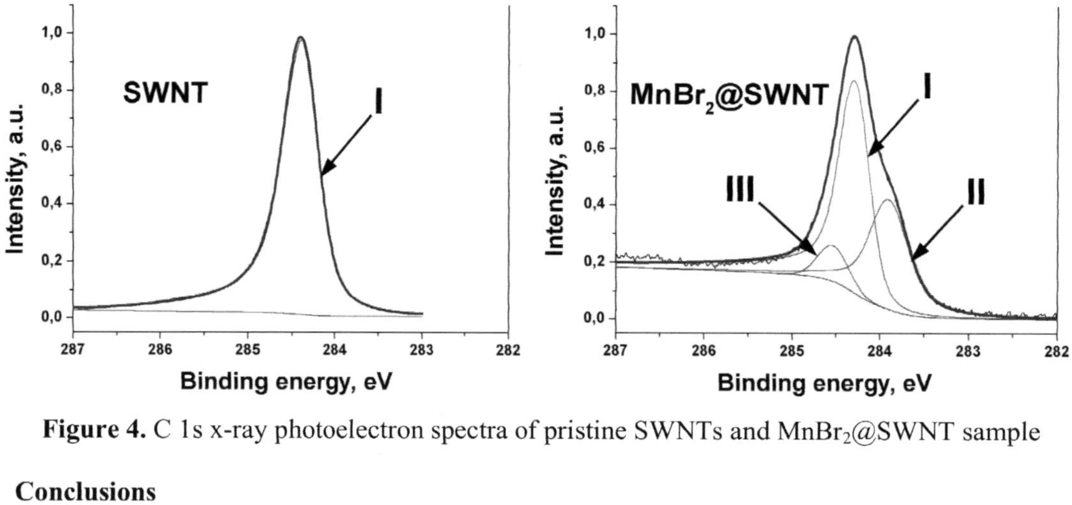

Figure 4. C 1s x-ray photoelectron spectra of pristine SWNTs and $MnBr_2$@SWNT sample

4. Conclusions

Thus, in present work $MnCl_2$@SWNT and $MnBr_2$@SWNT nanocomposites were obtained by the capillary filling of the internal channels of single-walled nanotubes with melts of manganese halogenides and subsequent slow cooling in order to obtain better crystallization of nanoparticles in

SWNT channels. It was found that the manganese halogenides show acceptor behavior and there is the charge transfer from the carbon nanotube walls to intercalated nanocrystals.

Acknowledgements
The authors thank Chan-Pu Liu, Chen-Yu Wang (National Cheng Kung University, Taiwan) and D.I. Petukhov (Lomonosov Moscow State University, Russia) for HRTEM measurements.

References

[1] Iijima S 1991 *Nature* 1991 **354** 56
[2] Odom T, Huang J, Kim P and Lieber C 1998 *Nature* **391** 62
[3] He R, Jin H, Zhu J, Yan Y and Chen X 1998 *Chemical Physics Letters* **298** 170
[4] Chen P, Wu X, Sun X, Lin J, Ji W and Tan K 1999 *Physical Review Letters* **82** 2548
[5] Treacy M, Ebbesen T and Gibson J 1996 *Nature* **381** 678
[6] Sloan J, Kirkland A, Hutchison J and Green M 2002 *Chem. Commun* 1319
[7] Eliseev A, Kharlamova M, Chernysheva M, Lukashin A, Tretyakov Yu, Kumskov A and Kiselev N 2009 *Russian Chemical Reviews* **78** 833
[8] Leonhardt A, Ritschel A, Kozhuharova R, Graff A, Muhl T, Huhle R, Monch I, Elefant D and Schneider C 2003 *Diamond and Related Materials* **12** 790
[9] Dujardin E, Ebbesen T, Hiura H and Tanigaki K 1994 *Science* **265** 1850
[10] Sloan J, Hammer J, Zwiefka-Sibley M and Green M 1998 *Chemical Communications* 347
[11] Flahaut E, Sloan J, Friedrichs S, Kirkland A, Coleman K, Williams V, Hanson N, Hutchison J and Green M 2006 *Chemistry of Materials* **18** 2059
[12] Kiang C, Choi J, Tran T and Bacher A 1999 *Journal of Physical Chemistry B* **103** 7449
[13] Chancolon J, Archaimbault F, Pineau A and Bonnamy S 2006 *Journal of Nanoscience and Nanotechnology* **6** 82
[14] Brown G, Bailey S, Sloan J, Xu C, Friedrichs S, Flahaut E, Coleman K, Hutchison J, Dunin-Borkowski R and Green M 2001 *Chemical Communications* 845
[15] Ebbesen T 1996 *Journal of Physics and Chemistry of Solids* **57** 951
[16] Mittal J, Monthioux M, Allouche H and Stephan O 2001 *Chemical Physics Letters* **339** 311
[17] O'Connell M, Bachilo S, Huffman C, Moore V, Strano M, Haroz E, Rialon K, Boul P, Noon W, Kittrell C, Ma J, Hauge R, Weisman R and Smalley R 2002 *Science* **297** 593
[18] Sloan J, Novotny M, Bailey S, Brown G, Xu C, Williams V, Friedrichs S, Flahaut E, Callender R, York A, Coleman K, Green M, Dunin-Borkowski R and Hutchison J 2000 *Chemical Physics Letters* **329** 61
[19] Xu C, Sloan J, Brown G, Bailey S, Williams V, Friedrichs S, Coleman K, Flahaut E, Hutchison J, Dunin-Borkowski R and Green M 2000 *Chemical Communications* 2427
[20] Eliseev A, Yashina L, Brzhezinskaya M, Chernysheva M, Kharlamova M, Verbitsky N, Lukashin A, Kiselev N, Kumskov A, Zakalyukin R, Hutchison J, Freitag B and Vinogradov A 2010 *Carbon* **48** 2708
[21] Kharlamova M, Eliseev A, Yashina L, Petukhov D, Liu C, Wang C, Semenenko D, Belogorohov A 2010 *JETP Letters* **91** 196
[22] Sloan J, Friedrichs S, Meyer R, Kirkland A, Hutchison J and Green M 2002 *Inorganica Chimica Acta* **330** 1
[23] Kirkland A, Meyer R, Sloan J and Hutchison J 2005 *Microsc. Microanal* **11** 401
[24] Yashina L, Eliseev A, Kharlamova M, Volykhov A, Egorov A, Savilov S, Lukashin A, Puttner R and Belogorokhov A 2011 *J. Phys. Chem. C* **115** 3578
[25] Thamavaranukup N, Höppe H, Ruiz-Gonzalez L, Costa P, Sloan J, Kirkland A and Green M 2004 *Chem . Commun.* 1686
[26] Brown G, Bailey S, Novotny M, Carter R, Flahaut E, Coleman K, Hutchison J, Green M and Sloan J 2003 *Appl. Phys. A.* **76** 457

Journal of Physics: Conference Series **345** (2012) 012035 doi:10.1088/1742-6596/345/1/012035

The impact of thermal treatment conditions on the formation of crystalline structure of Ce-Zr-oxide composite obtained by a modified sol-gel technique

E A Trusova, A A Khrushcheva, L I Shvorneva
A.A. Baikov Institute of Metallurgy and Material Science of RAS, 119991 Leninskii pr. 49, Moscow, Russia

E-mail addresses: trusova@imet.ac.ru (E.A. Trusova), anastasiya.xr@gmail.com (A.A. Khrushcheva)

Abstract. We present the results of the modified sol-gel synthesis of ultrafine ceria-doped zirconia powder for medical ceramics (implants) and catalytic purposes (environmental catalysis and petrochemistry). Special attention has been paid to study the influence of thermal treatment on crystallite size and crystal lattice parameters of zirconia doped by ceria. Zirconyl chloride and cerium nitrate were used as metal sources, and tetraethylammonium hydroxide (TEAH) was used as a sol stabilizer at molar ratio TEAH/Σ (Ce + Zr) equal to 0.5. It was proved that zirconium and cerium practically completely were included in the obtained solid solutions, since their phase compositions fully correspond to initial quantities of cerium and zirconium in reaction mixture. It was shown that average crystallite size of the obtained powders did not exceed 75Å, and the powders were resistant to thermal treatment. It was established that stabilization of the crystal lattice of ZrO_2 occurs through formation of a cubic ceria sublattice.

1. Introduction

Ultrafine powders of zirconia doped by ceria, are promising materials with a wide application range in a different innovative technologies of solid oxide fuel cells, gas sensors, oxygen-conducting materials, a fine-grained composite ceramics with grain size less than 2 microns for medical purposes and in catalysis [1]. During last 10-15 years, solid solutions CeO_2-ZrO_2 attracted attention of researchers in the fields of physical chemistry, technology of ceramics and nanotechnology. It is known that solid solutions CeO_2-ZrO_2 have high oxygen capacity, high mobility of lattice oxygen and high thermal stability [1]. Increasing demand of modern Hi-tech industry in the new materials with high oxygen capacity need in development of novel commercially scaled techniques for synthesis of ultra-dispersed powders [2].

Modified sol-gel technique is one of the technologically acceptable methods, which allows constructing of the structure of materials with new properties at the atomic and molecular level. Thermal treatment is important step of a structure formation, because this stage conditions determine not only a form of definite structure, but also ensure high dispersion of the material formed.

Previously, we reported about modified sol-gel synthesis of ultrafine $Ce_xZr_{1-x}O_2$ powders with average crystallite size of 8 - 11 nm [3]. The aim of this work is to develop modified sol-gel synthesis of ultrafine zirconia doped by ceria for medical ceramics (implants) and catalytic purposes

Journal of Physics: Conference Series **345** (2012) 012035 doi:10.1088/1742-6596/345/1/012035

(environmental catalysis and petrochemistry). Special attention is paid to study the influence of thermal treatment on crystallite size and crystal lattice parameters of ZrO_2, stabilized by CeO_2.

2. Experiment

2.1. Preparation

Zirconyl chloride and cerium nitrate were used as metal sources, and tetraethylammonium hydroxide (TEAH) was used as sol stabilizer at a molar ratio TEAH/Σ (Ce + Zr) equal to 0.5. Sol-gel synthesis was carried out by stirring with a magnetic stirrer at a temperature of 80-90°C in the course of 40 min. Then the sol was evaporated and the product was calcined at a temperature of 500-550°C for 1-21 hours.

2.2. Characterization

The phase composition was study by X-ray diffraction on a DRON-3 and Shimadzu XRD-6000 diffractometeres with monochromatic radiation CuK_α, and identification of crystalline phases was carried out by using a database JCPDS 2003. The crystallite size was calculated by Scherrer formula: $D = 0,94\ \lambda/\cos\theta\ (b - b_i)$, where λ - wavelength; b - broadening (full width at half maximum) of X-ray lines; b_i-instrumental broadening; θ - angle reflection.

3. Results and discussion

According to X-ray diffraction (Figure 1) obtained powders were single-phase systems with general formula $Ce_{0.06}Zr_{0.94}O_2$. All of them were nanopowders, as the crystallite size did not exceed 75 Å. It was proved that zirconium and cerium practically completely included in the obtained solid solutions, since their phase compositions fully correspond to initial quantities of cerium and zirconium in reaction mixture. This phenomenon was confirmed by yield of powders from theoretical around 96-99%. The absence of characteristic for amorphous phase halo on the diffractograms of the samples also proves the full inclusion of initial amounts of cerium and zirconium in the composite. The appearance of a small halo visible in all diffraction patterns may be due to lithol (industrial petroleum vaseline) which was used for attachment the powders in the cell of diffractometer. Comparison of diffraction patterns of composites $Ce_{0.06}Zr_{0.94}O_2$ (Figure 1) and pure ZrO_2 shows that they are almost identical.

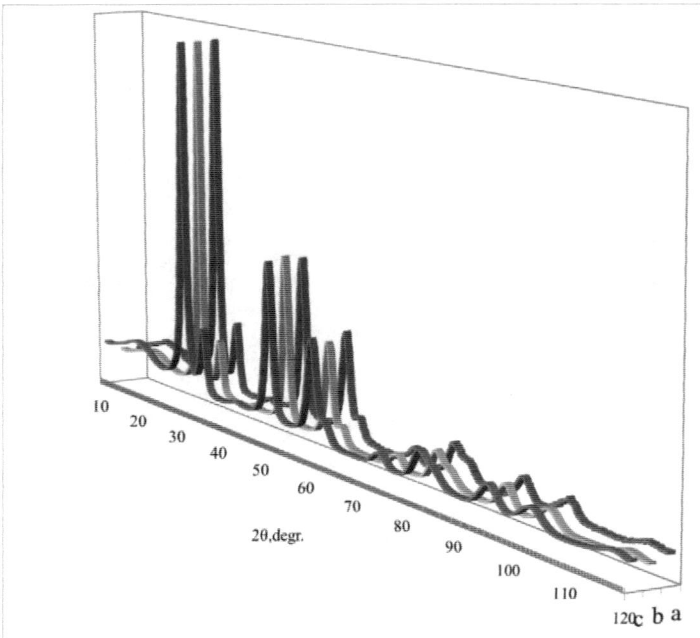

Figure 1. XRD patterns of samples: *a*- $Ce_{0.06}Zr_{0.94}O_2$ calcined at a 500°C for 1 hour, *b* - $Ce_{0.06}Zr_{0.94}O_2$ calcined at 550°C for 21 hours, *c* – ZrO_2

Table 1. Average crystallite size (D), microdeformation (e), and lattice parameters (a and c) of composites $Ce_{0.06}Zr_{0.94}O_2$, obtained in different modes of thermal treatment.

Thermal treatment conditions		D,Å	e,%	a, Å	c, Å
Temperature, °C	Duration, h				
500	1	63	0.98	3.618	5.146
500	21	73	0.79	3.608	5.169
550	1	75	0.77	3.607	5.179

Apparently, the characteristic for ceria cubic crystal was formed as a sublattice in the structure of ZrO_2. Table 1 shows the results of X-ray diffraction of composites $Ce_{0.06}Zr_{0.94}O_2$, obtained under different conditions of thermal treatment. It is shown, that duration and treatment temperature weakly affect the dispersion of composite systems, stabilized by ceria. Thus, prolongation of calcination at 500°C from 1 to 21 hs leads to the crystallite growth from 63 to 73Å. At the same time the amount of microdeformations reduced significantly, and the parameters correspond to the tetragonal crystal lattice of ceria doped zirconia [4]. If the calcination temperature increases from 500 to 550°C for the duration of 1 h, the crystallite size increases from 63 to 75Å. The microdeformation amount reduced significantly, and the parameters correspond to the tetragonal crystal lattice of ceria doped zirconia too [4].

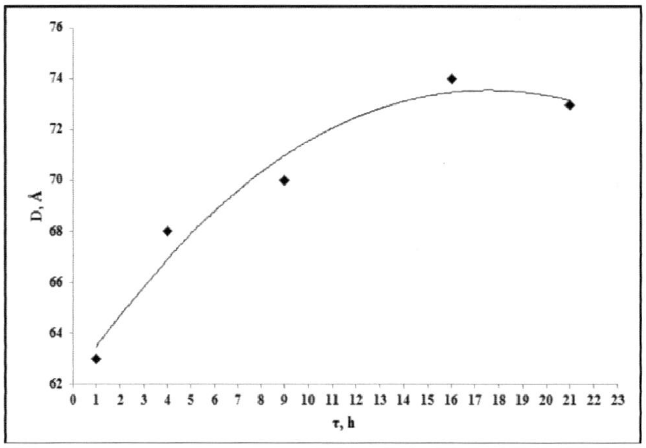

Figure 2. Dependence of the average $Ce_{0.06}Zr_{0.94}O_2$ crystallite size (D) of the duration of calcination (τ) at 500°C

Figure 2 shows the plot of average crystallite size of the duration of the composite $Ce_{0.06}Zr_{0.94}O_2$ calcination at 500°C. When the thermal treatment duration is 16 or more hours crystallite growth became slower and did not exceed 75Å.

4. Conclusions

- Method of obtaining the single-phase composites $Ce_{0.06}Zr_{0.94}O_2$ by modified sol-gel synthesis with using of TEAH was developed.
- It was shown that obtained powders have average crystallite size don't exceed 75Å, and are resistant to thermal treatment.
- It is established that the stabilization of the crystal lattice of ZrO_2 occurs through the formation of a cubic CeO_2 sublattice.

Acknowledgements

This work was supported by grant RFBR № 09-08-00917-a, Program of Basic Research № 22 of the Presidium of RAS, and Program of Department of Chemistry and Material Sciences № 7 of RAS.

References

[1] Cuif J P, Blanchard G, Touret O, Marczi M, Que´mere´ E 1996 *Warrendale, Society of Automotive Engineers*; PA: **73**

[2] Boaro M, Trovareli A, Hwang J.H, and Mason T O 2002 *Solid State Ionics*; **147**: 85-95

[3] Trusova E A, Khrushcheva A A, Vokhmintcev K V 2011 Sol-gel synthesis and phase composition of ultrafine ceria-doped zirconia powders for functional ceramics *J. Eur. Ceram. Soc.*, *Elsevier*, (*Preprint* JECS8522)

[4] Mineral crystal structure database of IEM RAS, Inf. card № 3206. *WWW-MINCRYST* (http://database.iem.ac.ru/mincryst/rus/s_full.php)

Journal of Physics: Conference Series **345** (2012) 012047 doi:10.1088/1742-6596/345/1/012047

Manipulation by optical properties of luminescent ordered organic nanoclusters via exciton-phonon coupling

AV Sorokin, I I Fylymonova, S L Yefimova, Yu V Malyukin
Institute for Scintillation Materials, STC "Institute for Single Crystals", NAS of Ukraine, 60 Lenin ave., 61001 Kharkov, Ukraine

E-mail: sorokin@isma.kharkov.ua

Abstract. The way to manipulate by spectral properties of cyanine dye luminescent ordered nanoclusters (J-aggregates) has been demonstrated. It has been observed a surfactant shell formation around the J-aggregates of four different dyes. This leads to exciton delocalization length increasing which causes an exciton self-trapping suppression. The latter results in spectral changes observed.

Molecular aggregates, called J-aggregates, are well-ordered luminescent organic nanoclusters of non-covalently coupled luminophores such as cyanines, porphyrines and some other dyes [1-3]. High degree of dye molecules ordering in J-aggregate chains results in the appearance of an excitonic narrow absorption band (so-called J-band) bathochromically shifted with respect to a monomer band. Spectral properties of J-aggregates are governed by the exciton delocalization length, which is usually equal to up to tens monomers, rather than the physical length of aggregates [3]. Due to excitonic nature of electronic excitations, J-aggregates reveal a number of unique spectral properties such as very narrow for organic molecules spectral lines (tens of cm^{-1} at liquid helium temperatures), large extinction coefficients (hundreds of thousands of $cm^{-1}\cdot M^{-1}$), giant third-order optical nonlinearities up to 10^{-5} esu and so on [1-3]. Such optical properties do the J-aggregates a very promising object for many applications, e.g. spectral sensitization in photovoltaics, nonlinear optical devices, luminescent probes in biology and medicine and so on [1-3].

Usually exciton-phonon interaction in J-aggregates suggested being week so exciton self-trapping in J-aggregates is not taken into consideration [3]. But there is one representative of J-aggregates family (amphi-PIC J-aggregates) for which exciton ST has been obviously demonstrated [4]. It is clear evidence of strong exciton-phonon interaction in this type of J-aggregates. Moreover, exciton-phonon coupling constant g measurement for merocyanine J-aggregates formed in Langmuir-Blodgett films demonstrates significant dependence of exciton-phonon interaction on J-aggregate preparation conditions [5]. Authors shown that for some cases exciton-phonon coupling is strong enough with $g >$ 1 and exciton self-trapping should be suggested [5].

The aim of present report is to demonstrate importance of exciton-phonon interaction taking into account and to show dependence of exciton properties like luminescence quantum yield on exciton-phonon coupling for J-aggregates of 4 different cyanine dyes (figure 1).

Sample preparation and experimental details could be found in [6-8]. Luminescence decay times for L-21 and amphi-PIC J-aggregates which was appeared to be impossible to measure using our previous setup [6] have been measured using picosecond spectrofluorimeter FluoTime 200 (PicoQuant, Germany). This lifetime spectrometer was supplied by MCP-PMT P3809U-50 (Hamamatsu, Japan) and laser head LDH-P-FA-530 with λ_{em} = 531 nm (PicoQuant, Germany). The FWHM of instrument response function (IRF) was 96 ps.

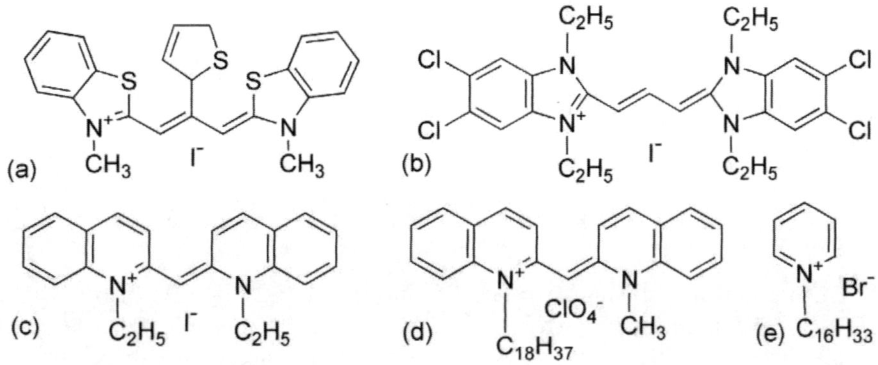

Figure 1. Molecular structures of the dyes and the surfactant investigated: a) L-21, b) JC-1, c) PIC, d) amphi-PIC and e) CPB

Recently we have shown significant spectral changes for J-aggregates of different dyes (figure 1) caused by surfactant – J-aggregate interaction [6,7]. In particular, exciton delocalization length growth, luminescence quantum yield increase up to 15 times and lifetime decrease has been observed (Table 1).

Table 1. Exciton delocalization lengths (N_{del}), luminescent quantum yields (φ), radiative lifetimes (τ_0) and exciton-phonon coupling constants (g) for cyanine dyes J-aggregates in absence and presence of surfactant CPB

	N_{del}		φ, %		τ_0, ns		g	
		CPB shell		CPB shell		CPB shell		CPB shell
L-21	< 1	28	2.5	17.5	6	1.95	2.2	0.6
JC-1	4	9	1.5	19	16.65	4.2	1.4	0.85
PIC	86	93	2.5	38	16	1.85	0.75	0.55
amphi-PIC	4	11	0.3	0.5	11.7	9	1.1	0.9

Figure 2 demonstrates the most obvious spectral changes for JC-1 J-aggregates at interaction with surfactant CPB. One could see J-band (λ_{max} = 680 nm) narrowing and growth, hypsochromically shifted band appearing (λ_{max} = 530 nm, so called H-band) in absorption spectrum (figure 2a). Luminescence spectrum (figure 2b) reveals significant exciton emission band (λ_{max} = 682 nm) growth with simultaneous monomer emission band (λ_{max} = 585 nm) decreasing.

On the base of anisotropy decay analysis it was suggested that the surfactant molecules form a kind of shell around J-aggregates [6]. To check this assumption, TEM images have been obtained for amphi-PIC J-aggregates in the absence and presence of CPB surfactant (figure 3). Amphi-PIC J-aggregates are revealed to be cylinders with a diameter of ~ 4 nm and 50–100 nm length (figure 3a). In the presence of CPB, elongated shapeless objects with a mean diameter of ~ 30 nm and 50–100 nm length are observed (figure 3b). Such a picture confirms the J-aggregates-surfactant complex formation and supports our idea concerning a surfactant shell creation.

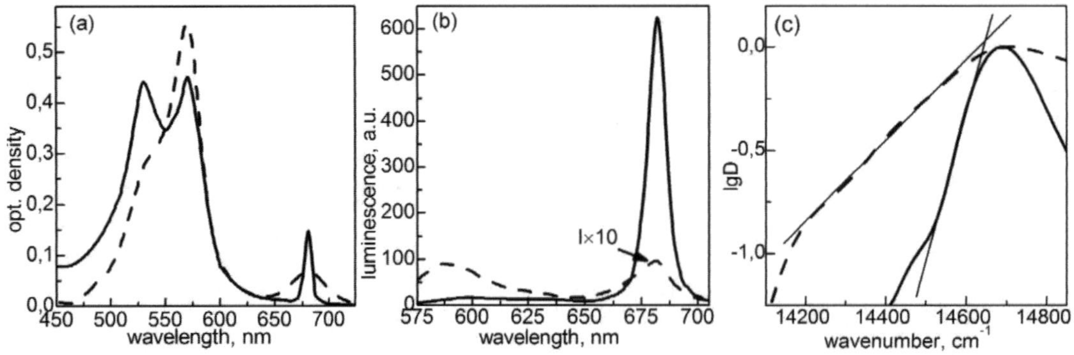

Figure 2. Absorption (a) and luminescence (b) spectra of L-21 dye (C = 10^{-4} M) in a binary solution DMF / water buffer Tris-HCl (pH = 8.0) (1:19) in the absence (dashed lines) and presence (solid lines) of CPB (C = 10^{-3} M). (c) – semilogarithmic plot of J-bands from (a).

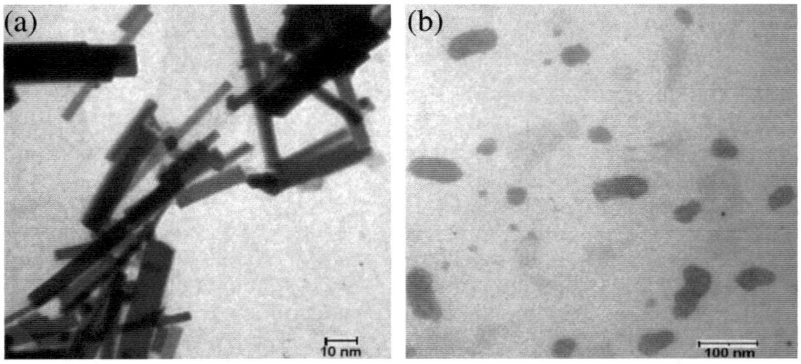

Figure 3. TEM images of amphi-PIC J-aggregates: (a) in absence and (b) in presence of CPB surfactant

It was supposed that luminescence quantum yield increasing occurs due to exciton-phonon interaction weakening and exciton self-trapping suppression as a result [7,8]. Indeed, for amphi-PIC J-aggregates exciton ST suppression at interaction with CPB surfactant could be observed directly at low temperatures (figure 4) [8].

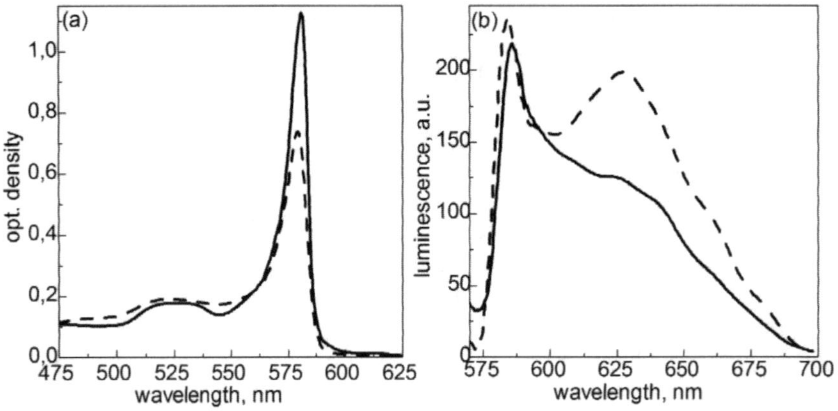

Figure 4. Absorption (a) and luminescence (b) spectra of amphi-PIC J-aggregates in the presence (solid line) and absence (dashed line) of the surfactant CPB at T = 80 K.

At low temperatures in luminescence spectrum of amphi-PIC J-aggregates (figure 4b) two different bands are observed: narrow band (λ_{max} = 585 nm), which was ascribed to free exciton emission, and wide band (λ_{max} = 630 nm), which was ascribed to self-trapped exciton emission [4]. It could be seen that CPB cause the same effect on amphi-PIC J-band at low temperatures as on J-bands of other J-aggregates at room temperatures (figure 2a), namely leading to J-band (λ_{max} = 580 nm) increasing and narrowing (figure 4a). At luminescence spectrum both emission bands decreasing could be observed (figure 4b). But if the free excitons band with λ_{max} = 585 nm reveals only slight decrease in intensity, the self-trapped excitons band with λ_{max} = 630 nm decreases significantly. Moreover, the first band decreasing could be ascribed rather to ST excitons contribution reducing than the free excitons luminescence suppression (figure 4b).

Unfortunately, ST excitons luminescence for other types of J-aggregates hasn't been found. But it is possible to find the exciton-lattice coupling constant from a slope of long-wavelength tail of the excitonic band by using the Urbach rule [8]:

$$\alpha(E) = \alpha_0 \exp\left(-\sigma \frac{E_0 - E}{kT}\right), \tag{1}$$

where $\alpha(E)$ is the absorption coefficient,

$$\sigma = \sigma_0 \frac{2kT}{\hbar\omega_p} \tanh\left(\frac{\hbar\omega_p}{2kT}\right), \tag{2}$$

where $\hbar\omega_p$ is an energy of the principal interacting phonon mode and σ_0 is the limit of σ at high temperatures. σ_0 and g have a simple reciprocal relationship:

$$g = \frac{s}{\sigma_0} \tag{3}$$

For 3D simple cubic lattices s = 1.5, for 2d square lattices s = 1.24 and for 1D chains s is found to be temperature dependent, scaling as $T^{1/3}$ [8].

The application of Urbach-Martienssen rule for amphi-PIC is demonstrated on figure 5.

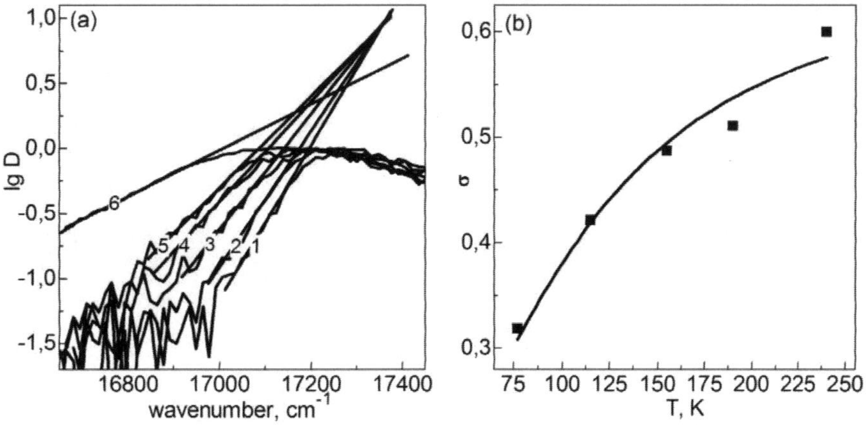

Figure 5. a) Temperature dependence of Urbach tails for amphi-PIC J-band in the absence of CPB. Temperatures designation: 1 – 80 K, 2 – 105 K, 3 – 155 K, 4 – 190 K, 5 – 220 K, 6 – 290 K; (b) Temperature dependence of low temperature Urbach slopes and its approximation by equation (2).

So, we have found such parameters: σ_0 = 0.65 and ω_p = 220 cm^{-1} and σ_0^{CPB} = 0.8 and ω_p^{CPB} = 225 cm^{-1}, i.e. $\sigma_0 < \sigma_0^{CPB}$. For estimation, we choose s = 1 as limit of s(T) dependences for linear 1D chain at the temperatures under consideration [8]. It gives us g = 1.55 and g^{CPB} = 1.25 thus confirms exciton-phonon coupling weakening.

But comparing Urbach tails for J-band at room (figure 5a, curve 6) and low (figure 5a, curves 1-5) temperatures one could see a significant difference. Indeed, J-aggregates obviously affects by different surrounding in solutions and glass-like matrices. However, semilogarithmic plot of J-bands at room temperature reveals steeper slope in CPB presence (figure 2c), i.e. $\sigma < \sigma^{CPB}$. Analyzing literature data [1-5] we have found that interacting phonon modes are laid within range 50–300 cm^{-1}, that is corresponded to σ/σ_0 = 0.85–0.99 at T = 290 K according to equation (2). So, averaging we obtain $\sigma/\sigma_0 \sim 0.92$ which is corresponded to $\omega \sim 205$ cm^{-1} in close correlation with data obtained for amphi-PIC J-aggregates in low temperature cases. So, finally we could to determine the exciton-phonon coupling g constants for all J-aggregates under investigation and to confirm the exciton-phonon weakening at the surfactant shell formation for all cases (Table 1).

We could to conclude that solvate shell changing for J-aggregates via surfactant addition results in multiple effects on exciton properties such as exciton delocalization increasing, exciton-phonon interaction weakening and luminescence quantum yield enhancement. The results obtained could be used for J-aggregates luminescence properties control for different applications.

[1] Shapiro B I 2006 *Russ. Chem. Rev.* **75** 433
[2] Wurthner F, Kaiser T E and Saha-Muller Ch R 2011 *Angew Chem. Int. Ed.* **50** 3376
[3] Knoester J, Agranovich V M in: Agranovich V M, Bassani G F (Eds.). 2003 *Thin Films and Nanostructures: Electronic Excitations in Organic Based Nanostructures*, vol. 31. Elsevier, Amsterdam, Oxford
[4] Malyukin Yu 2006 *Phys. Stat. Sol. C* **3** 3386
[5] Yamaguchi A, Kometani N, Yonezawa Y 2006 *Thin Solid Films*, **513** 125
[6] Guralchuk G Ya, Katrunov I K, Grynyov R S et al. 2008 *J. Phys. Chem. C.* **112** 14762
[7] Katrunov I K, Yefimova S L, Sorokin A V, Malyukin Yu V 2010 *Funct. Mat.* **17** 52
[8] Katrunov I K, Sorokin A V, Yefimova S L, Malyukin Yu V 2011 *Mol. Cryst. Liq. Cryst.* **57** 535

Journal of Physics: Conference Series **345** (2012) 012037 doi:10.1088/1742-6596/345/1/012037

Substantiation of using the method of electrophysical effects on the dispersion medium in devices for a highly effective filtration of gas mediums from nanosized aerosolsof various origin

Andrey Grishin, Ivan Yagodkin, Andrey Posagennikov, Valeriy Melnikov

Institute for Physics and Power Engineering, Russia, square Bondarenko 1, Obninsk, 249033

Grishin.andrew@gmail.com

Abstract. The problem of gas purification at nuclear power plants, steam generating stations, radiochemical, instrument-making and other industries is one of the most important directions in the field of technologies for environmental protection from aerosols of various origin including radioactive and toxic ones. It is of great interest worldwide. An increase in the power of nuclear power plants and higher requirements to the cleanup of gas emissions of enterprises necessitate the continuous improvement of techniques and methods of gas purification.

The ecological safety including radiation safety related to using equipment of enterprises of radiochemical, nuclear, instrument-making, and other industries widely depends on the state of the air medium. That is why the quality of filter equipment, especially aerosol filters, which enable a highly effective cleanup of air mediums from nanosized aerosols with minimum expenses, has acquired importance.

It is known, that the finer aerosols are, the deeper they penetrate into the respiratory airway and the more dangerous are for human beings. The protective ability of the human organism appears to be insufficient when very fine particles with diameter of up to 1 micron penetrate into the lungs and deposit mainly in the alveolar tissues. Up to 70 % of aerosols, which have been accumulated in the human lungs affected by silicosis, consist of such particles [Khulkvist B., 1959].

Nanoparticles can be in the air as suspension for a long time and be moved by airflows at long distances.

Toxicological investigations have shown that nanoparticles can potentially influence on human health [Oberdörster G., 2000 Phil. Trans. Roy. Soc. London Ser. A 3582719].

In this connection it is necessary to create highly effective systems for air purification from nanosized aerosol particles.

The development of filter materials has reached its limit. It has been proceeded from decreasing the size of filter fibers (Fig. 1). Today, further decrease in the size of fibers is an economically unprofitable expensive measure. Therefore, it is very important to search for other ways of developing the filtration basing not on the decrease in the size of filter material fibers but on change of properties of the medium to be filtered, while influencing on it by various methods.

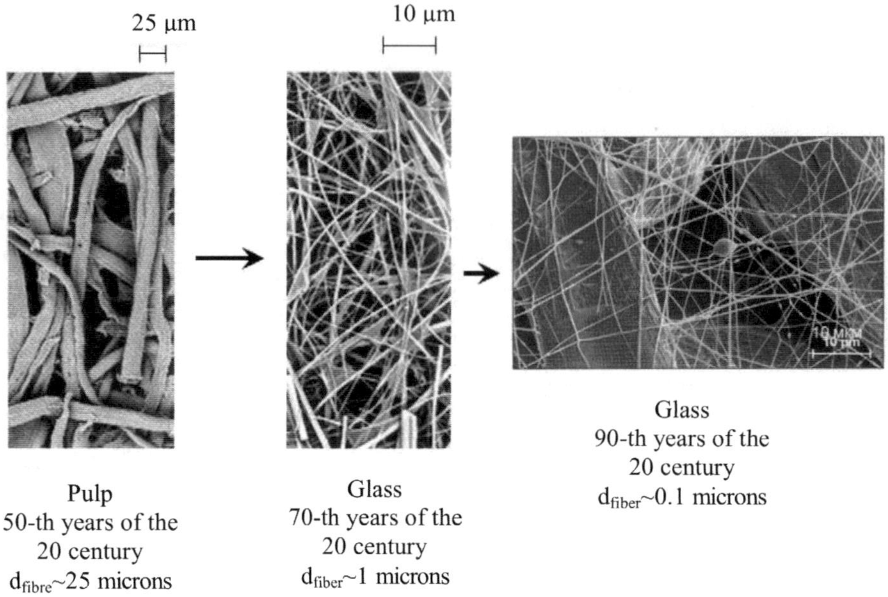

Fig. 1 The stages of development of filter materials

There are a lot of methods for aerosols purification using various effects on the medium to be cleaned up. However, none of conventional methods is suited to capture nanosized aerosol particles. For the capture of nanosized aerosol particles, the method of the electrophysical effect on the disperse medium followed by capturing nanosized aerosol particles on highly effective filter material has been proposed.

The method of the electrophysical effect on the disperse medium is based on charging particles to provide their subsequent more effective capture by the highly porous filter medium.

Physical processes that accompany the formation of charged aerosols and their interaction with other materials are diversiform and complex. These processes include the charge of aerosol particles of different size and composition, motion of charged particles in natural and exterior electric fields, interaction of charged aerosol particles among themselves and with other aerosol particles (processes of electrical coagulation), deposition and interaction of charged aerosol particles with various objects etc. Of special interest is the study of possible physical mechanisms, conditions and tendencies of the effect of aeroionization on aerosol particles to facilitate their effective capture in various filter systems, which could find a promising application in some technological processes and devices.

The maximum efficiency is provided due to the filtration system consisting of polyester filter material, metal strainer and metal felt. In this system, the consecutive arrangement of dielectric and conductive elements enables to involve the most physical mechanisms of capturing charged aerosol particles. Among the considered electrode systems, the maximum efficiency is achieved with use of sawtooth electrodes. For effective capturing aerosol particles, optimal distances between electrodes and a range of feeding high voltage were found out.

Our experiments indicated the positive results of aeroionization, which enabled to increase the efficiency of capturing aerosol particles with diameter of 0.1 microns on "rough" cheap filter material (class G2 in accordance with GOST R 51215-99) from 3 % up to 95 % (Fig. 2).

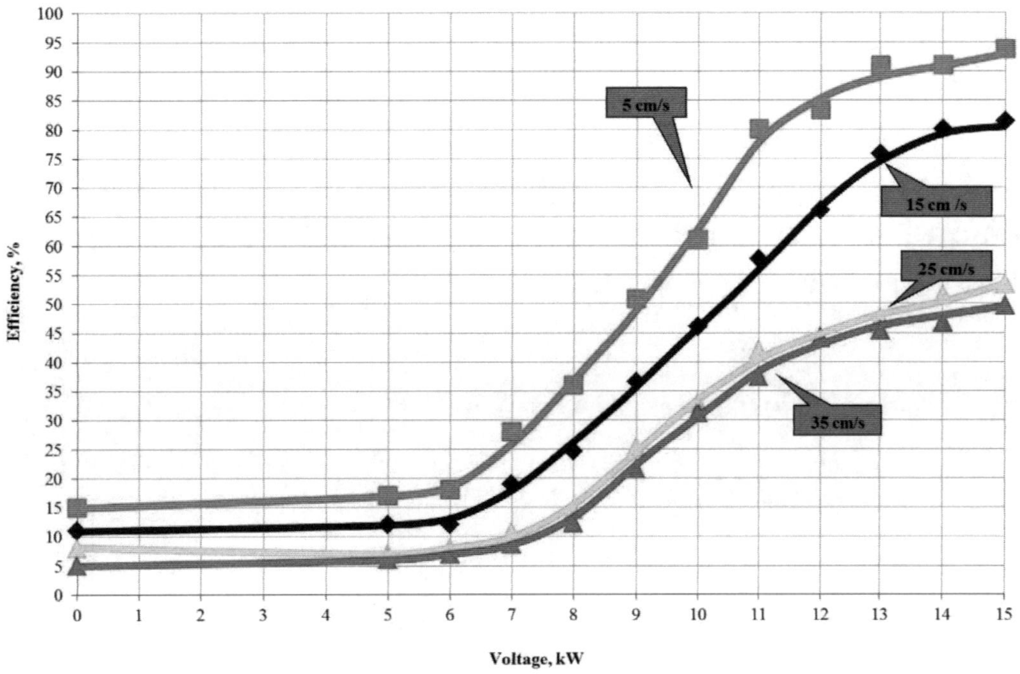

Fig. 2. The dependence of the efficiency of the capture of aerosol diameter of 0.1 microns for different speeds of the air stream from voltage

The selection of optimum parameters influencing on the efficiency of capturing aerosols from the airflow and aeroionization can enable to capture nanosized aerosol particles from the airflow.

The cost efficiency of this device will allow to prolong active life of expensive highly effective aerosol filters at least by 2 times and reduce the expenses related to their premature replacement.

Social and economic effects from the introduction of the developed product lie in higher safety of enterprises of nuclear, radiochemical, instrument-making, medical, nanotechnological and other industries, in the maintenance of human life and health, improvement of ecological situation as well as reduction of material and power consumption of manufacture.Start your abstract here...

References

[1] Khulkvist B., Ionizing natural radiation sources. - M.: Divisions of the characters., 1959.-199c.
[2] Oberdörster G., 2000 Phil. Trans. Roy. Soc. London Ser. A 3582719.

Journal of Physics: Conference Series **345** (2012) 012038 doi:10.1088/1742-6596/345/1/012038

The research of the technological parameters effect on the shape and structure of the disperse dye flow of the inkjet printers produced in the electric field.

M.G. Verdiyev, Sh.Sh. Nabiev, N.A. Kamnev

Dagestan State Technical University, Kalinin Prospekt, a post office №30, Makhachkala, Dagestan Republic 367030, Russian Federation.

E-mail: mailbox@lab112.ru, shihnabi@gmail.com

Annotation. The study is devoted to the results of experimental searches of the disperse dye flows of inkjet drawn on paper in an electric field that can be used to make improvements to technology of improving the quality and the print admissible capacity of various printed materials in inkjet technology.

1. Introduction

It became necessary to save the results of calculations by recording when the first electronic computer was invented in the 50s the last century. In order to do it, people were specially taught to print received information at typewriters. It suggested the researchers and designers the idea of switching the typewrites directly to computer. And Remington-Rand Corporation invented the first printer for the computer UNIVAC (Universal Automatic Computer), named UNIPRINTER in 1953. Since then, the technologies of printing of inkjet printers, developed the last century, are not subjected to the substantial changes. Of course, the technical process is perfecting, the number of dots per inch is increasing, the quality and the speed are improving, but no significant changes in the technology of drawing ink on paper aren't happening.

One of the important units of inkjet printer is the print head, it consists of a large number of nozzles, to which the ink leads up to. The ink is supplied to the nozzles with capillary properties and is kept from leaking out owing to the surface tension powers of its liquid components. There are special pellicate elements in one of the structures in the print head, integrated to create and draw the microscopic particles of ink on paper. There are different types of print heads, depending on the operating principle of the formed mechanism and drawing of the ink particles on the media - paper [1]:

piezoelectric (used by companies Epson and Brother) [2];
gas bubbles (Canon) [3];
drop-on-demand (Hewlett-Packard) [4].

The first method of drawing the ink particles on paper is based on the physical phenomenon, which consists in the deformation of the piezoelectric element in an electric field, and in a sudden increase of pressure in the capacity for the ink and the second and third are on the effect of pressure increasing in the nozzle of the print head by pulse heating of ink above the point of its boil. The disadvantages of existing methods are: heterogeneous form (in the form of smudge) and a large spread of sizes of ink particles throwing out the print head nozzles [5]. To obtain uniform in size and shape (spherical) of ink nanosized particles a lot of researches are conducted, a large part of which is aimed at changing structural parameters of print head, however, it has not resulted in a practical significant improvement in quality of print.

One of the effective methods of obtaining spherical in shape, homogeneous, nanosized particles is the method of the ink dispersing in an electric field [6-10]. In this method of dye drawing the sizes of spherical regular forms of ink particles have the order of units and tens of nanometers. The lower limit of particle sizes in this technology is determined by minimal sizes of dye particles, contained in the ink structure. The lower limit of the particle sizes of the particle flow is $2 \div 12$ nanometers for pure liquids. In famous technologies [1-4], the_volume of ink droplets is 3 : 6 pikolitres, that corresponds to the sizes of order 10^{-5} m, i.e. at the level of hundreds of micrometers [1]. In our case, the particle diameter of dispersed flow, measured with atomic power microscopy, amounts to 10^{-6}, and more times less pikoliter, therefore, the powers of medium resistance – air or any gas due to the motion of particles from the nozzle to the media are not sufficient to change their ball shape. The power of the medium resistance are reduced by reducing the relative speed (in regard to the medium entrained with flow) of the medium [11]. In the known technologies particles have the form of "smudge" [1-5]. Thus, in an area, coated with the media pikolitr drops in famous technologies [5] will be placed about 10^2 or more particles produced by the proposed technology. Certainly, it will increase the admissible capacity, contrast, and quality characteristics of inkjet technology. The study is devoted to experimental research of the influence of technical parameters on the shape and structure of disperse dye flow of inkjet printers produced in the parameters of electric field of inkjet printer dispersed flow of the structure of the electric field in the area of flow formation.

2. Description of the experimental setup

Schematic section of the experimental setup is shown on the figure 1. Inside the rigid metal frame 1, made of rectangular steel profile symmetrically on two bearings fixed frame 3 providing the possibility of rotation around the horizontal axis. On the inner frame 3, fluoroplastic electrical isolation, adjustably installed metallic capillary 4 and the second electrode - working table 5. The second electrode - work-table is grounded. After installing the optimum position of the capillary required the distance between its end and the table is regulated by a screw of micromanipulator ST-3. Investigated liquid and a high potential are supplied to the capillary by fluoroplastic pipes, laid through the pipe 6, the bearing sleeve 2 and the frame profile 3. To measure power transfer of dispersed particles flow, microammeter 7 serves, through which the work-table 5 is grounded- the second electrode. Dispersed flow of liquid 8, the working table 5, and the capillary 4 are hermetically isolated from the medium, by a composite framework made of two cylindrical halves, made of transparent Plexiglas and attached to the inner frame 3 (not shown). Structure regulation of the electric field in the formation of particulate flow is performed by changing the diameter and position of the metallic disk 9 installed on the end of the capillary 4. Capillary is supplied with adjustable in the range 0 - 50 kV potential of a DC voltage source one electrode which is grounded. Power source is not shown. Capillary by fluoroplastic pipe is connected to the capacitance (capacitance is not shown) for the research liquid. Fluid consumption through the capillary is changed by changing the position of the capacity concerning to the end of the capillary or by regulation the pressure in it. The method of the change of the ink consumption through the capillary by changing the position of capacitance achieves the highest degree of accuracy, so as for regulation of pressure dropin with such precision the authors haven't had more accurate measuring instruments and pressure gauges.

For substance dispersing with a fusing temperature above medium fluid capacity is installed directly on the capillary and heated with electromagnetic radiation or ohmic heating. In the latter case, capillary with heating device must be thoroughly electrically isolated from the installation and the insulated from the medium. At the dispersion of conductive liquids capacity must also be thoroughly electrically isolated from the casing installation.

The inner frame of the set freely rotates on bearings around a horizontal axis at 180 degrees clockwise or counterclockwise. Free rotation of the inner frame is designed for experimental evaluation of the shape and size of dispersed flow particles on its magnitude deviations on the media

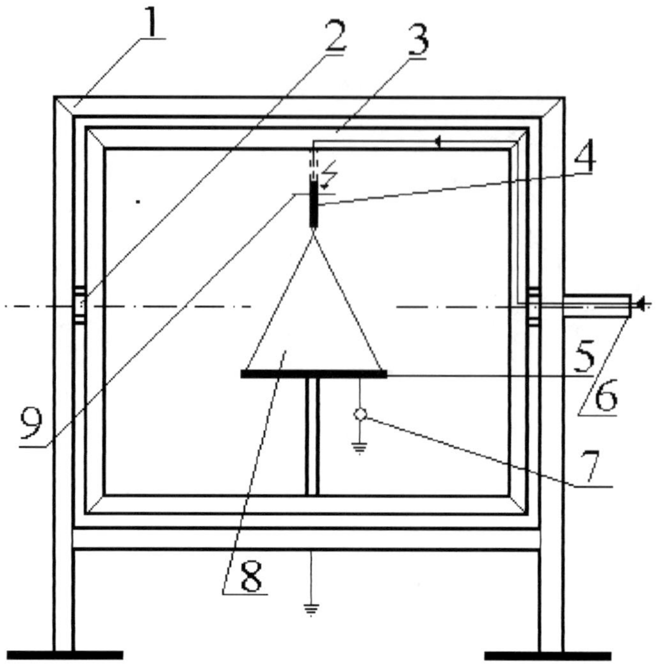

Figure 1. The schematic drawing of the experimental setup.

- a piece of paper pinned on the work table in the vertical plane, when you turn it in the gravitational field at 180 [7]. For the experimental evaluation of the average particle size of dispersed flow first axis is oriented vertically and the on the media - paper receive the image of section, then the system of the inner frame with fixed positions of the capillary and paper rotates at 90 and again get the image of the flow on the paper . After that, the system of inner frame rotated in the opposite direction at 180 degrees and fix he position of the center flow, and then measure the magnitude of the mixing center of the image on a sheet of paper.

3. The results of experimental searches.

New ink mark «HAMELEON» was selected for research, developed with German dyes and components have been tested at the request of the Japanese Company «Revcol» [12] National Institute for Chemistry Science, Tokio, Japan. At various installation options were researched the modes of dispersed ink flow. The ink dispersion is made with capillaries of diameter d in $3*10^{-4}$, $4*10^{-4}$, $4,5*10^{-4}$ m on coverslips and paper, fixed on the desktop setup. Dependence of the structure and cross-sectional area of ink dispersed flow on a variety of installation options and technological conditions of the process was determined by the images of dispersed flows received on paper or the cover glass. The researches were conducted at variable: the level of ink in the reservoir concerning the end of the capillary, the capillary drive position, the distance between the electrodes for different diameter disks, fixed on the capillary.

At optimal technological regimes liquid ink cone elongates from capillary and, with the top facing down, which begins with dispersed flow at corresponding to its potential, as shown in the photographs shown in figure 2.

Fluid consumption through the capillary is changed by way of capacity position changing concerning to the end of capillary dispersion or by adjusting the pressure in it. Searchers were conducted at liquid levels above the end of the capillary in the range $0 \pm 0{,}01$ m, i.e. at zero, an excess (100 Pa) and negative (100 Pa) pressures in the working volume. At zero level of the liquid under an electric field from the end of the capillary is extended drop in a cone shape. When the potential increases frequency of breakdown drops from the end of the capillary rises and subsequently is transformed into a continuous thin stream of fluid starting at the top of the cone. The increase of potential up to 500 ÷ 1000 V leads to rupture of the jet and the formation of non-uniform cross section of the conical section of the dispersed flow, and in the center is still very thin stream of fluid. The Further growth of potential leads to the disintegration of the central stream, and dispersed flow becomes uniform. The photo 2 shows stationary finely divided stream of jet printer ink for the set modes of dispersion.

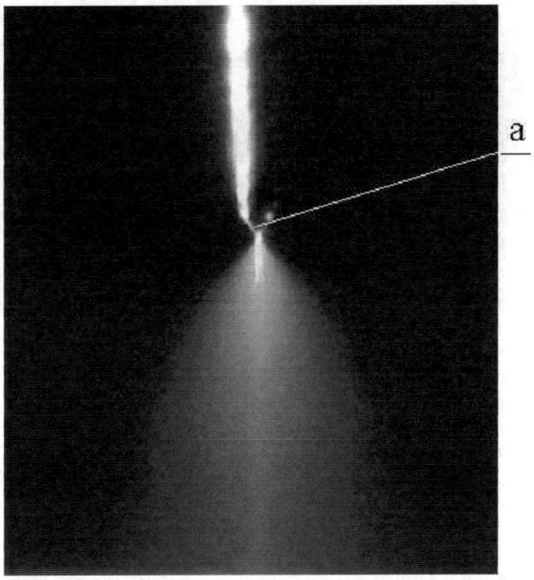

Figure 2. The photo of steady dispersed flow from the top of inverted cone, jet printer ink . At a potential up to 5 kV, depending on the distance between the electrodes, there is a uniform dispersion of particulate flow, with a cone on top of the capillary.

The optimal value of the potential steady dispersed flow depends on the distances between the electrodes, the structure of the electric field at the end of the capillary (in the formation zone of dispersed flow) and the properties of the . When the values of the capillary potential are more than 5 kV, cone at the end of the capillary disappears and the dispersion process becomes unstable regime. An unstable dispersion is initially in the form of two streams, then with the growth of the potential number of dispersed flows is increased to four, as shown in figure 3. Figure 3 shows images of cross sections of flow depending on the capacity of the capillary at a constant distance between the electrodes. In our opinion, the instability of dispersed flow can be explained by the potential growth and by not a sufficient flow of fluid through the capillary, resulting in a lack of fluids - a charge media is the destabilization of dispersed flows.

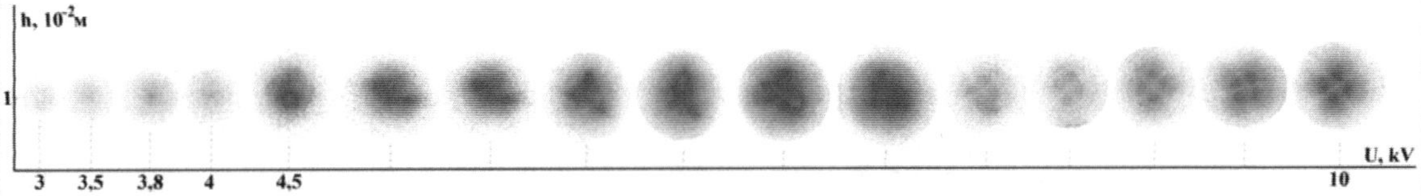

Figure 3. Dynamics of changes in cross section and the structure of ink dispersed flow of inkjet printers, depending on the capacity of dispersion.

The main interest for researches is the stationary mode of dispersion when the dispersion of sizes and density and distribution of particles in the stream of flow are minimal. Optimization of modes of ink dispersed flow in the process is done by studying - volt-ampere characteristics of dispersed liquid flows and the dependence of the cross sections structure: the potential of the dispersion, the distance between the electrodes, the structure of the electric field in the area of the stream formation, consumption of substances and their physical properties.

Figure 4 shows the volt-ampere characteristics for different levels of ink in a reservoir above the end of the capillary (because of absence of accurate data on the of the ink density in the paper gives the height of the level of ink above of the end of the capillary instead of hydrostatic pressure).

During the researches the level of dye over the end of the capillary was changed in the range from -10^{-2} to $+ 10^{-2}$ m. As we can see from the figure the increase in level of ink over the end of the capillary does not affect the curves of volt-ampere characteristic of the dispersed flow to a potential value equal to 7 kV and at potentials more than 7 kV there is a discrepancy of curves by increasing flow instability and current oscillations of transfer. As It's known from experimental data at negative values of differential pressure (level), the potential of dispersion increases with increasing its value. With the growth of a negative differential pressure, the potential of dispersion increases, in our opinion, owing to the growth of work done with electric field at the capillary rise of ink in the tube to the point of disintegration of the jet. Therefore, the potential growth does not increase the current transfer with dispersed flow. Maximum current transfer of dispersed flow reaches at the zero level drop of ink in the reservoir concerning the point of dispersion. The dependence of flow structure on: the potential dispersion of the ink and the height of their level over the end of the capillary are shown in figure 5. Figure 5 shows how a cross section structure of flow and density of its particles over the cross section when the altitude level and the potential of disintegration change.

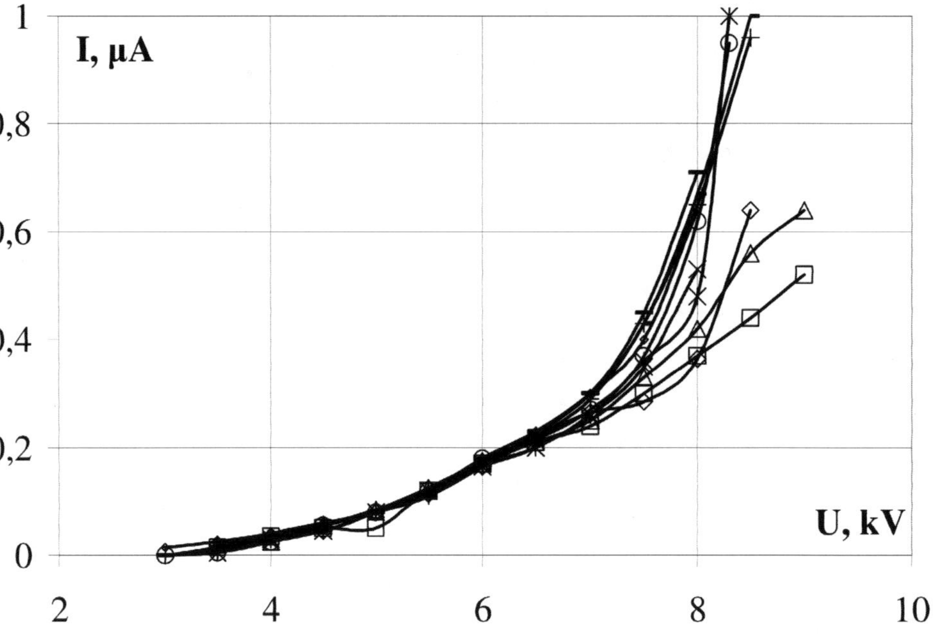

Figure 4. The effect of the level h2 dye in the reservoir above the end of the capillary on current-voltage characteristic of dispersed flow (◇ - -10^{-2}m , □ - -7×10^{-3}m, △ - -5×10^{-3}m, × - -3×10^{-3}m, ✳ - -2×10^{-3}m, ○ - -10^{-3}m, + - 3 $\times10^{-3}$m, ‾ - 5 $\times10^{-3}$m, – - 7 $\times10^{-3}$m).

When the ink level over the end of the capillary is zero, the homogeneous dispersion flows are observed up to 4 kV, then the density of particles in the cross section of dispersed flow becomes non-uniform and as a result of it a dark spot is in the center of image. The increase in the potential up to 5kV and more results in an unstable mode of dispersion, increasing the area of dark spot, and its displacement from the center. With further increase in potential spots become darker, and dispersed flow splits into two or more.

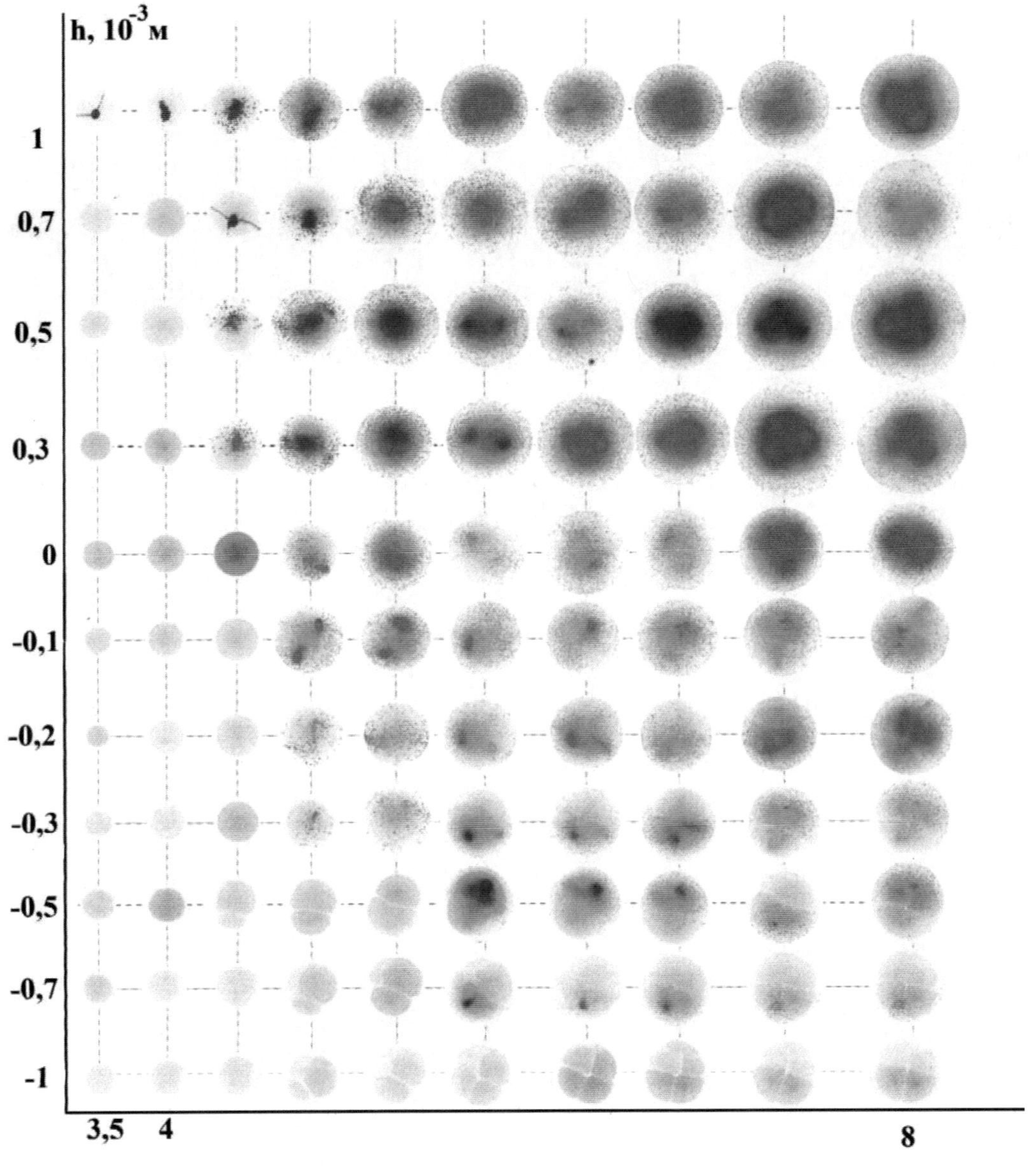

Figure 5. Dynamics of change in flow structure depending on the level of ink over the end of the capillary (in the figure some errors are caused by errors : the holding time of flow and installation of dispersion capacity, but the overall figure of the process is clearly seen, as noted in the literature [6]).

Increasing of the ink level over the end of the capillary in the range from 0 to – 10 m results in a shift of the unstable dispersion toward lower potentials. And when the ink level equal to 10 m it's impossible to find the sustainable mode of dispersion, due to the large consumption of it. Apparently, caused by that charges do not have time to drain to the entire stream and in the middle of flow is the stream of fluid, which is either not charged or not charged up to the required capacity to divide. This phenomenon may be caused by the contents of the different components in the ink, as result of it the regime of stable dispersion for one component may coincide with the unstable mode dispersion of the other component.

If we analyze the current-voltage characteristics shown in figure 4 and the data in figure 5, we'll see that increasing of the ink level over the end of the capillary does not resulting in significant change of the current, while the flow consumption through the capillary increases. This can be explained by the fact that the charges do not flock on the entire flow of ink and that's why the central part of the flow is keen on peripheral due to electrical forces in a form of very thin trickle what we see on the stream and as the result of it the darkening is observed in the center flow .

When the level of ink drops from the end of the capillary in the range from 0 to -10^{-2} m ink consumption is reduced through the capillary and electrical charges have time to flow down, that's explains the uniformity of the structure of the flow cross-section even at non-stationary mode dispersion.

In this case, up to 4,5 kV, we observe a steady mode dispersion, and then decay of the flow into multiple parts. This figure also shows that an increase of ink consumption the cone base radius of flow is growing, as the consumption is increases with the growth of level altitude (in the figure the consumption is increasing from bottom to top).

Figure 6 shows the results of researches of the effect of capacity dispersion on the ink consumption at different levels of ink over the end of the capillary. By increasing the capacity of the capillary and the ink level from 0 to 35×10^{-3} m, within the error of measurement of time and changes in level, consumption increases linearly, as can be seen from the figures 6 and 7.

The maximum potential value, for which a dispersion is observed above is limited with phenomenon of the breakdown.

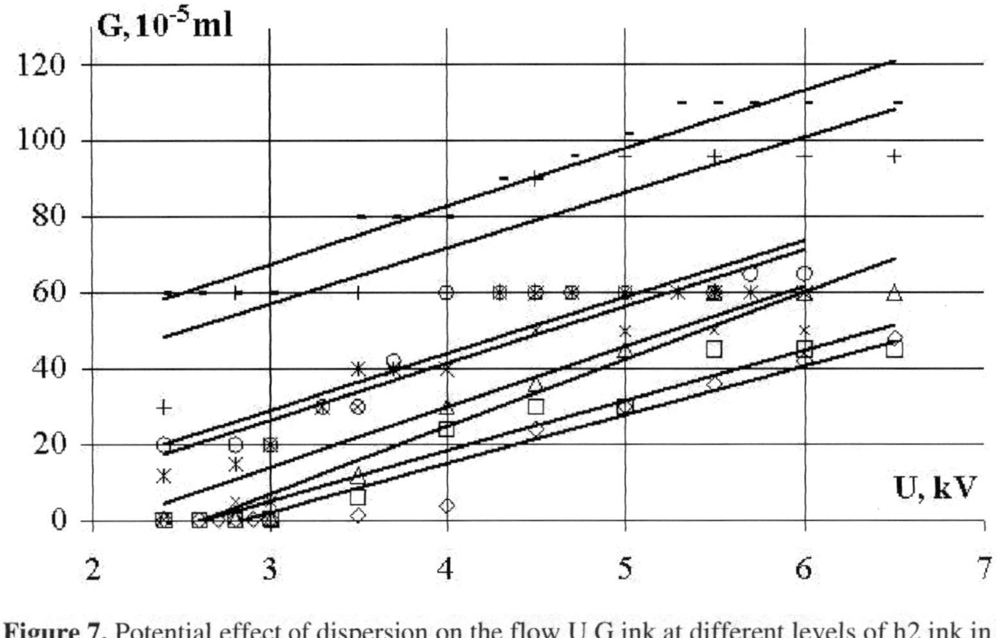

Figure 7. Potential effect of dispersion on the flow U G ink at different levels of h2 ink in the reservoir above the end of the capillary (one division is 10^{-5} liters, ◇ - 0m, □ - 3×10^{-3}m, △ - 5×10^{-3}m, × - 7×10^{-3}m, ⋇ - 10^{-2}m, ○ - 15×10^{-3}m, + - 25 $\times10^{-3}$m, ⁻ - 35 $\times10^{-3}$m)

The next stage is devoted to the research of distance effect between the electrodes on the current-voltage characteristics and the structure of dispersed flows. Figure 9 shows current-voltage characteristics for different distances between the electrodes. With increasing potential strength of the current grows polynomially.

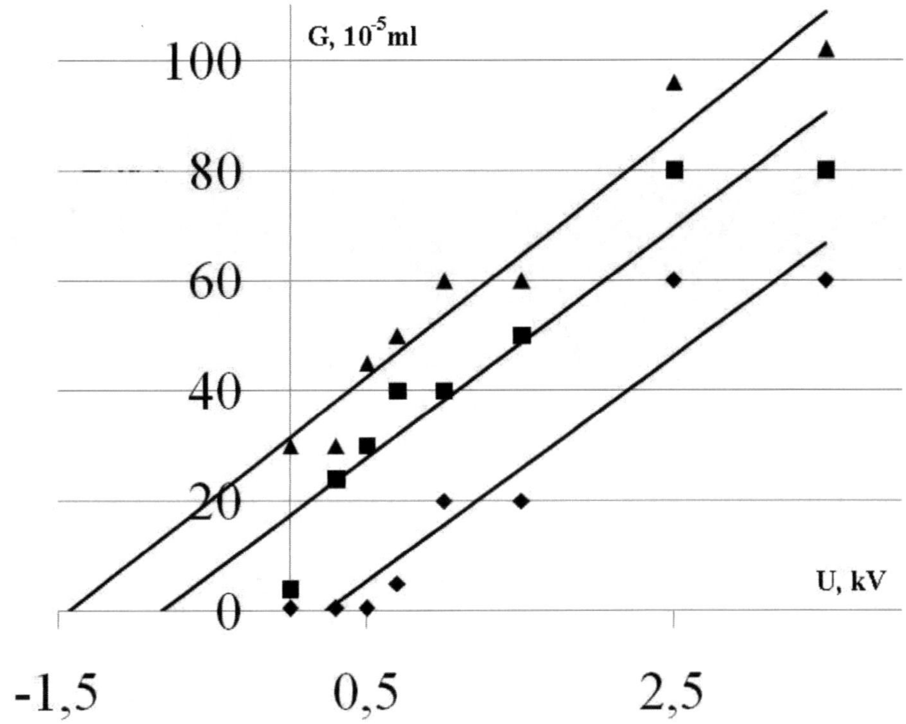

Figure 8. The effect of the dye level in a reservoir above the end of the capillary on its consumption at different dispersion potentials (◆ - 3kV, ■ - 4kV, ▲ - 5kV).

The increase of the distance between the electrodes in the range of 5×10^{-3}m to 10^{-2}m brings to a polynomial growth of the current. As can be seen from the graph when the distance between the electrodes increases the offset of curves voltage characteristics is observed toward higher potentials, as the potential formation of a stable dispersion flow is shifted to the right. When the distance between the electrodes 3×10^{-3} m the area of minimum cross section of the base of the free cone flow is 0,5024 $\times10^{-6}$ m^2, and at a distance of 10^{-2} m area the base flow is equal to 6,1575 $\times$ 10^{-6} m^2 With increasing of distance between the electrodes the diameter of the conical flow base grows linearly, as well as the potential beginning of disintegration.

Figure 10 shows the cross-sectional area of dispersed flows from the distance h between the electrodes at the altitude of the ink over the end of the capillary is zero. The size and shape of dispersed flow can be changed by adjusting the structure of the electric field or by the influence of a magnetic field in the flux of particles to the second electrode. For all cases of free dispersed flow in a steady state the cross section of basis are coincided in the form to the ideal circles. More homogeneous dispersed flows in the cross section are observed from the apex of ink cone in the modes of the beginning of the jet decay formed at the end of the capillary. With further grow capacity of capillary increases the consumption of ink through the capillary, as evidenced by higher saturation flux images on the media paper or a slide in unit time. At this mode, the experimentally determined dependence of the ink on the time of change in change of the rate level in the reservoir at different potentials of the capillary, and as a reservoir used gostirovanny vessel with a capacity of 1milliliter and scale divisions of 100 units, i.e. with the scale of 10^{-5} liters / division.

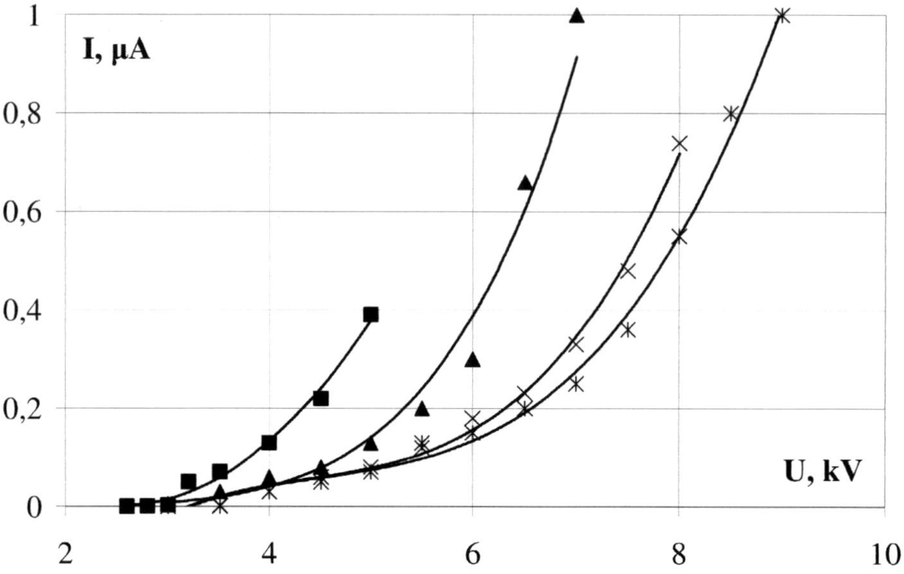

Figure 9. The dependence of current-voltage characteristics of dispersed ink flow of jet printers at distance h_1 between the electrodes (■ - 5×10^{-3}м, ▲ - 7×10^{-3}м, × - 9×10^{-3}м, ⋇ - 10^{-2}м).

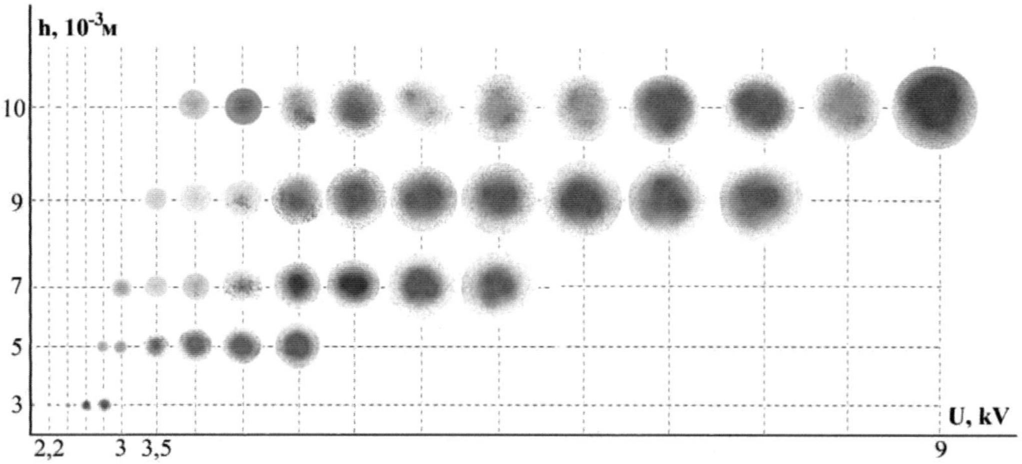

Figure 10. Dynamics of changes in cross-sectional area and the structure of dispersed ink flows on the second electrode, depending on the capacity of the capillary at different distances between the electrodes.

4. Conclusion

On the basis of numerous experimental searches are shown that the structure of the dispersed ink flow of jet printers and their dispersion mode in electric field depends on: the installation options, the potential dispersion, the ink consumption through the capillary and its electrical conductivity. During the research the optimal mode dispersion at different installation options and the modes of uniform on section of stationary dispersed flow are set. This technology can be successfully used for drawing of ink films on basis of nanopowder CIGS semiconductor for producing of flexible (roll) solar energy converters, in ink and 3D printers.

Acknowledgement
The work was performed under financial support of foundations and START and UMNIK.

References

[1]. http://www.micpocxema.ru/pech3.html
[2]. How is an inkjet printer constructed? http://www.colorit. ukrbiz.net/Pages/technology-of-printing2.htm
[3]. A. Asai Three-Dimensional Calculation of Bubble Growth and Drop Ejection in a Bubble Printer. Trans. ASME. J. Fluid. Eng. 1992. V. 114. P. 638-641.
[4]. P.H. Chen, W.-C. Chen, S.-H. Chang Bubble growth and ink ejection process of a thermal ink jet print head//Int. J. Mech. Sci. 1994. V. 262. P. 205-221.
[5]. VI Artemov, AF Ginevsky, NI Pavitsky. Numerical simulation of thermal ink head processes of ink jet printer. / / Proceedings of the Third Russian National Conference on Heat Transfer. 2002. T.4. Moscow: Publishing House of MEI. S. 221-224.
[6]. M.G. Verdiyev, O. Ivanov, S.A. Ninalalov The study of liquid dispersion in an electric field. Journal of Engineering Physics. - 1989. - V.57. Number 2. - S. 275 - 280.
[7]. M.G. Verdiyev, T.A. Ismailov, M.Sh. Abidov. The method for determining the characteristic parameters of the fluid particles dispersing in an electric field. RF patent № 2372613, 2009.
[8]. M.G. Verdiyev, T.A. Ismailov, M.Sh. Abidov. Technology for producing monodisperse nano powders of various substances. Advanced technology and equipment for materials micro - and nano-electronics.: Works IV Russian-Japanese Seminar. - M.: MGIU, 2006. S. 211 - 220.
[9]. M.G. Verdiyev, T.A. Ismailov, I.M. Isaev, M.SH. Abidova. Problems of development of the device splash of ink for inkjet printers (thesis). Sat mes. Reports. XXIV Final Scientific and Technical Conference: faculty, staff and students DGTU. - Makhachkala DGTU, 2003.- S. 629.
[10]. M.G. Verdiyev Obtaining the thin films of liquids by dispersing them in an electric field. Electronic processing of materials. - 1991. - № 4. - S. 36 - 41.
[11]. L.D. Landau, E.M. Lifshitz. Theoretical Physics. Study Guide. In 10 volumes, T. VI. Hydrodynamics. 3rd edition, revised. Moscow: Nauka, 1986, 736.
[12]. http://www.revcol.ru

Journal of Physics: Conference Series **345** (2012) 012039 doi:10.1088/1742-6596/345/1/012039

Supracrystalline analogues of carbon nanomaterials for the hydrogen storage

R A Brazhe and I S Olenin
Department of Physics, Ulyanovsk State Technical University,
32 Severny Venetz Street, Ulyanovsk, Russian Federation

E-mail: i.olenin@ulstu.ru

Abstract. It is shown that, in contrast to the known two-dimensional carbon nanoallotropes graphene and graphane their supracrystalline analogues have a greater specific surface area and, consequently, a greater sorption capacity. It allows to recommend them as promising hydrogen storage for the hydrogen energetics.

1. Introduction

In connection with the exhaustion of hydrocarbon energy sources and the pollution of the environment when they are used, and taken into account the potential danger of nuclear power engineering many experts believe that the future for hydrogen energetics. Hydrogen is a highly effective, ecologically clean and wide spread in the nature of energy source. However, the current tack of reliable ways to store and reversible hydrogen sorption at the reasonable pressures and temperatures is an obstacle to its use, particularly in vehicles. The International Energy Agency has established the following requirements to be met by current hydrogen storage systems: the mass content not less than 5 wt.%, the dehydrogenation not greater than 373 K, and efficient cyclic stability [1,2].

As it is shown in Ref. [2,3], the carbon nanofibers and nanotubes can be the convenient hydrogen sorbents. The physical sorption in these structures is possible at the room temperature, but it requires the high pressures (~10 MPa). However, in Ref. [4] was shown that the accumulation of hydrogen in an amount of 0.5-7.0 wt.% is possible in single-walled carbon nanotubes at the room temperature and the pressure of 0.1 MPa. The getting of graphene [5,6], and then graphane [7], drew attention of the researchers to these materials as a promising hydrogen storage media [8,9].

2. Discussion

Appeal from the classical 2D crystals (graphene and graphane) to the 2D supracrystals [10-11] is a further step in hydrogen storage systems search. Supracrystals represent the quasi infinite two-dimensional crystalline structures, where in the nodes of crystalline lattice atoms or ions are replaced by their symmetrically organized atomic complexes. As the result, the specific surface area and sorption capacity increase.

From the geometrical point of view 2D supracrystals represent Kepler nets, i.e. the flat mosaics consist of the regular polygons. There are eleven varieties of such structures. Indices behind the parentheses denote the following: first index determines the type of unit cell (4 – square, 6 - hexagonal), a next index determines the type of enclosing cell (Figure 1). At first, the number of sides for the nodal cell id indicated, then (in parentheses) the type of polygon in central of supracell is indicated to distinguish the structures $(C)_{63(6)}$ and $(C)_{63(12)}$.

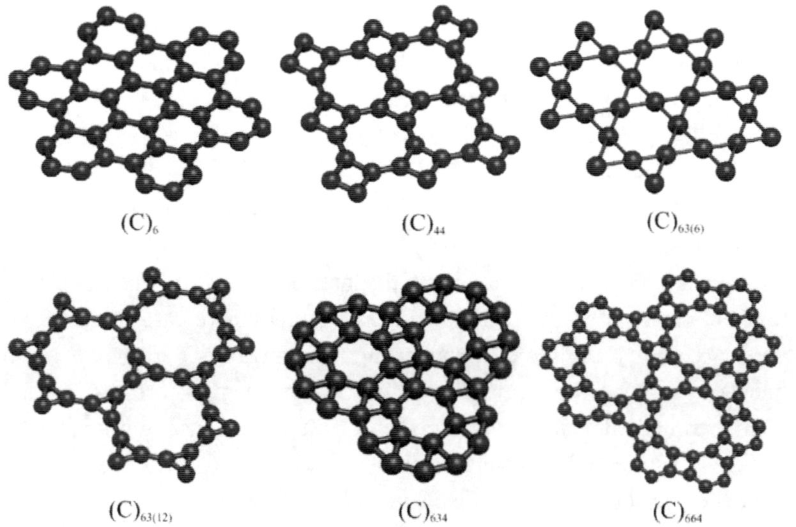

Figure 1. Fragments of the graphene and the five possible types of carbon 2D supracrystals.

The calculated top values of specific surface area for the carbon 2D supracrystals in comparison with graphene and represent in Table 1. Both sides of the sheet were taken into account. The calculations of the bond length l and the energy per one atom for the equilibrium state have been provided on the basis of density functional theory [13] by using of software package ABINIT-5.8.4 [14]. The calculation of specific surface area we have been carried out using the following formulas wich can be easily deduced from the cell's geometry.

$(C)_6$: $$s = \frac{3\sqrt{3}}{2}\frac{N_A}{\mu}l^2,$$

$(C)_{44}$: $$s = \frac{1}{2}\left(1+\sqrt{2}\right)^2\frac{N_A}{\mu}l^2,$$

$(C)_{63(6)}$: $$s = \frac{4\sqrt{3}}{3}\frac{N_A}{\mu}l^2,$$

$(C)_{63(12)}$: $$s = \frac{\sqrt{3}}{6}\left(2+\sqrt{2}\right)^2\frac{N_A}{\mu}l^2,$$

$(C)_{664}$: $$s = \frac{\sqrt{3}}{12}\left(3+\sqrt{3}\right)^2\frac{N_A}{\mu}l^2,$$

$(C)_{634}$: $$s = \frac{\sqrt{3}}{6}\left(1+\sqrt{3}\right)^2\frac{N_A}{\mu}l^2.$$

Here N_A is Avogadro number, μ is molar mass (for the carbon μ = 12 g/mol), and l is the bond length.

Table 1. Structural, energetic and sorption characteristics of 2D supracrystals.

Parameter	$(C)_6$	$(C)_{44}$	$(C)_{63(6)}$	$(C)_{63(12)}$	$(C)_{664}$	$(C)_{634}$
Hybridization	sp^2	sp^2	sp^3	sp^2	sp^2	sp^3
l, Angstrom	1.42	1.43	1.86	1.71	1.56	2.17
$\|E_{atom}\|$, eV	13.5	12.9	14.2	14.7	11.3	13.3
Specific surface area, m^2/g	2630	2630	7100	5790	3950	1320

 doi:10.1088/1742-6596/345/1/012039

From Table 1 you can see that the structure $(C)_{63(12)}$ has the most depth of potential pit were atoms are located and theoretically must to be more stable than graphene. Other structures have some less thermal stability, but they also must exist at the room temperature. The specific surface area of same 2D supracrystalline carbon nanoallotropes is greater than for graphene: $(C)_{63(6)}$ in 2.7 times, $(C)_{63(12)}$ in 2.2 times, and $(C)_{664}$ in 1.5 times.

Supracrystalline nanoribbons [15] can be rolled into nanotubes [16] wich have a greater specific surface area than carbon ones and higher sorption capacity.

Graphane is obtained by the hydrogenation of the graphene, i.e. by the chemical joining of the hydrogen atoms to the carbon ones (Figure 2). 2D supracrystals where carbon atoms are in the state of sp^2-hybridization, i.e. have only one free electron, and have supracells with only even number of sides in the polygon behave itself by the same way. There are the structures $(C)_{44}$ and $(C)_{664}$. These structures have been searched its stability in the plane by their energy minimization using software package GROMACS. As graphene this structures are able to absorb (by chemisorption) up to 7.7 wt.% of hydrogen and have the greater specific surface area than graphene.

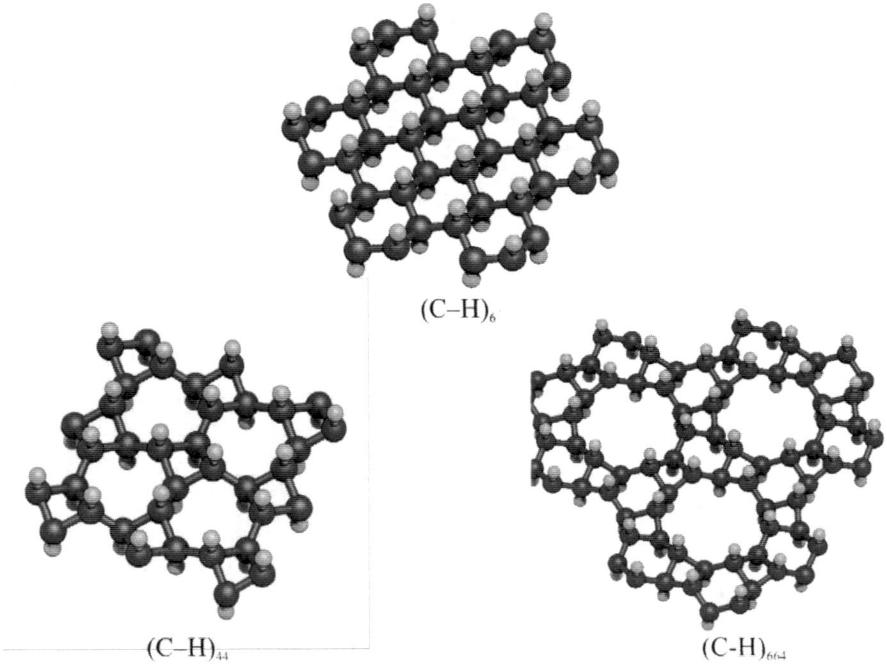

Figure 2. Graphane and supracrystalline analogues of graphane.

3. Conclusion

1) The specific surface area of graphene like and graphane like 2D supracrystals may be in some times more than for graphene and their sorption capacity is, accordingly, greater.

2) Both planar and nanotubular 2D supracrystalline carbon nanoallotropes have a commensurable with the graphene thermal stability and may be exist at the room temperature.

3) The proposed supracrystalline materials can be used in the hydrogen storage systems for the hydrogen energetics.

References

[1] Azhazha V M 2006 *Voprosy atomnoi nauki. Ser.: Vakuum, chistye materialy, sverhprovodniki* **1** 145-152 (Russian)
[2] Tarasov B P and Goldshleger N F 2002 *International scientific journal for alternative energy and ecology* **3** 20-39
[3] Nechaev YU S 2006 *Physics-Uspekhi* **176** 582-610
[4] Dillon A C and Heben M J 2000 *Int. Symp. Metal–hydrogen System: Abstr* **2** 2–10
[5] Novoselov K S, Geim A K, Morozov S V, Jiang D, Zhang Y, Dubonos S V, Grigorieva I V and Firsov A A 2004 *Science* **306** 666–669
[6] Novoselov K S, Jiang D, Schedin F, Booth T J, Khotkevich V V, Morozov S V and Geim A K 2005 *Proc. Nat. Acad. Sci* **102** 10451–10453
[7] Elias D C, Nair R R, Mohiuddin T M G, Morozov S V, Blake P, Halsall M P, Ferrari A C, Boukhvalov D W, Katsnelson M I, Geim A K and Novoselov K S 2009 *Science* **323** 610–613
[8] URL: http://www.forbes.ru/tehno/budushchee/13405-grafen-materiya-tolshchinoi-v-atom
[9] URL: http://www.nanonewsnet.ru/news/2009/grafan-novyi-material-na-osnove-grafena
[10] Brazhe R A and Karenin A A 2009 *Matematicheskoe modelirovanie fizicheskih, ekonomicheskih, tehnicheskih, sotsialnyh sistem i protsessov: mat. Mezhd. konf.* 51-52 (Russian)
[11] Brazhe R A, Karenin A A and Olenin I S 2010 *Radioelektronnaya tehnika: mezhvuzovskii sbornik nauchnyh trudov* 156-161 (Russian)
[12] Brazhe R A, Kochaev A A, Aref' eva P A, Karenin A A and Olenin I S 2010 *Prilozh. k zhurnalu «Fizicheskoe obrazovanie v vuzah»* **17** 115 (Russian)
[13] Satanin A M 2009 *Nizhnii Novgorod: NGGU im. N. I. Lobachevskogo (Natsional' nii issledovatel' skii universitet)* 64 (Rassian)
[14] URL: http://www.abinit.org.
[15] Aref' eva P A, Brazhe R A and Karenin A A 2010 *Radioelektronnaya tehnika: mezhvuzovskii sbornik nauchnyh trudov* 141-147 (Russian)
[16] Brazhe R A and Karenin A A 2010 *Radioelektronnaya tehnika: mezhvuzovskii sbornik nauchnyh trudov* 148-155 (Russian)

Journal of Physics: Conference Series **345** (2012) 012041 doi:10.1088/1742-6596/345/1/012041

Oligonucleotide microarray for subtyping of influenza A viruses

Klotchenko S.A.[1], Vasin A.V.[1], Sandybaev N.T.[2], Plotnikova M.A.[1], Chervyakova O.V.[2], Smirnova E.A.[1], Kushnareva E.V.[1], Strochkov V.M.[2], Taylakova E.T.[2], Egorov V.V.[1], Koshemetov J.K.[2], O.I. Kiselev[1], A.R. Sansyzbay[2]

[1] Research Institute of Influenza of the Ministry of Health and Social Development of the Russian Federation, 15/17 Prof. Popova St., St. Petersburg, Russia,
[2]Research Research Institute for Biological Safety Problems of the RK NBC/SC ME&S RK, Gvardeiskiy, Republic of Kazakhstan

E-mail: vasin@influenza.spb.ru

Abstract. Influenza is one of the most widespread respiratory viral diseases, infecting humans, horses, pigs, poultry and some other animal populations. Influenza A viruses (IAV) are classified into subtypes on the basis of the surface hemagglutinin (H1 to H16) and neuraminidase (N1 to N9) glycoproteins. The correct determination of IAV subtype is necessary for clinical and epidemiological studies. In this article we propose an oligonucleotide microarray for subtyping of IAV using universal one-step multisegment RT-PCR fluorescent labeling of viral gene segments. It showed to be an advanced approach for fast detection and identification of IAV.

1. Introduction

Influenza is an acute respiratory disease brining on annual seasonal epidemics and irregular hardly predictable pandemic outbreaks. The etiological agents of influenza are the viruses of the same name, the members of *Orthomyxoviridae* family. Among the three known types (A, B and C) of influenza viruses the most serious infectious threats arise from influenza A viruses (IAV) due to their genetic variability and capacity for a rapid and global spread. IAV are classified into subtypes on the basis of the serological differences of two surface glycoproteins: hemagglutinin (HA) and neuraminidase (NA). IAV infect humans, birds, pigs, horses and marine mammals, the natural reservoirs of IAV are waterfowls. 16 subtypes of HA and 9 subtypes of NA have been described so far. In human population only three HA (H1, H2, H3) and two NA (N1, N2) subtypes predominate, but the infections of humans with high pathogenic IAV strains, such as H5N1, H7N7 and H9N2, have been recently detected. Moreover, the reassortment of IAV segments can lead to the emergence of new virus variants, such as for example «Spanish flu» virus in 1918 or swine-origin A/H1N1 virus in 2009.

Since the pandemic potential of IAV is associated with the spread of new atypical virus subtypes, the development of rapid and precise diagnostic methods of IAV detection and subtyping is an extremely urgent problem of influenza surveillance and epidemiology. One of the most promising technologies for IAV subtyping is microarray approach, the number of which have been developed recently [1-7]. Earlier we offered the method of IAV genomic RNA amplification using universal primers for application in influenza subtyping microarrays. This pair of primers to the conservative terminal regions of all IAV segments greatly unifies influenza microarray systems making them independent from the complex set of multiplex primers specific for certain IAV subtypes. In current paper we propose universal IAV subtyping oligonucleotide microarray using the approach described above.

2. Materials and methods

IAV strains were obtained from the collection of Research Institute of Influenza of the Ministry of Health and Social Development of the Russian Federation.

2.1. RNA isolation.

Total RNA was isolated from the allantoic fluid using TRIZol reagent (Invitrogen, USA). RNA concentration and integrity were measured using NanoDrop ND-1000 spectrophotometer (NanoDrop Technologies, USA).

2.2. Microarray design.

Oligonucleotide probes, containing 5'-aminolinker, were synthesized and HPLC purified by «DNA-synthesis» (Russia). Probes (300 ng/μl) in 3×SSC were immobilized on Vantage Aldehyde Slides (CEL Associates, Estonia) using SpotBot 3 Personal Microarrayer (ArrayIt, USA). After printing slides were treated with UV (0.09 J/cm^2). Before hybridization slides were boiled in miliQ H_2O for 1 min.

2.3. Fluorescent labelling of IAV genomic segments.

Fluorescent labelling of IAV cDNA was performed in the process of the direct incorporation of Cy3-dCTP by multisegment RT-PCR using SuperScript III One-Step Quantitative RT-PCR System (Invitrogen, USA) with universal MBTuni-12 primers [8,9] as described earlier [10]. The reaction mixture additionally contained 20 μl 1mM Cy5-dCTP. Amplification conditions for Rotor-Gene 6000 thermocycler (Corbett Research, Australia) were as follows: 1) 45°C for 60 min; 2) 94°C for 2 min; 3) 5 cycles at 94°C for 30 sec, 45°C for 30 sec and 68°C for 3 min; 4) 31 cycles at 94°C for 30 sec, 57°C for 30 sec and 68°C for 3 min.

2.4. Hybridization.

30 μl of fluorescently labeled cDNA was incubated at 99°C for 2 min, chilled on ice for 2 min, afterwards 10 μl of 99% formamide and 10 μl of 5× hybridization buffer were added. Hybridization was performed in 50 μl using «Whatman» frames at 37°C for 2 hours with agitation at 250 rpm. After hybridization slides were washed in the mixture of 1×SSC buffer and 0.1% SDS.

2.5. Microarray scanning.

After hybridization slides were scanned using ScanArray Express ("PerkinElmer", USA) with 5 μm resolution. Data analysis was made using ScanArray software.

3. Results and discussion

The selection of IAV subtype-specific oligonucleotide probes was made by the multiple alignments of more than 15 000 HA and 16 000 NA sequences deposited in GenBank. On the basis of the alignments, made using MAFFT software (http://mafft.cbrc.jp/alignment/server/), sequences were divided into separate groups, for which specific oligonucleotide probes were designed using OligoWiz software (See tables 1A,B,C). It was not always possible to select the unique probe for each HA or NA subtype, for this reason several probes were selected for some subtypes. Oligonucleotide probes were synthesized and immobilized on microarray slides using contact printing method. IAVmicroarray consisted of 16 identical subarrays, containing 20 rows, corresponding to 16 probes for HA (H1-H16), 9 probes for NA (N1-N9) and 7 control probes for NP and M segments (See figure 1). The diameter of each spot was 200 nm, all probes were immobilized in 3 repeats.

Table 1A. Oligonucleotide probes for IAV hemagglutinin subtyping.

Probe	Oligonucleotide sequence (5` → 3`)[a]	Length, n.
H1-1	GAAGGGAGAATGAACTATTACTGGACACTAGTAGAGCCGGGAGACAA	47
H1-2	GTCTCCCTGGGGGCAATCAGYTTCTGGATGTGYTCYAATGGG	42
H1-3	ACAGGACTAAGGAACATCCCATCCATTCAATCCAGAGGTTTGTTTGG	47
H2-1	TAAGGAATGTTCCCCAGATTGAATCAAGAGGATTGTTTGGGGCAAT	46
H2-2	GTCACCGTGACTCATGCCAAGGACATTCTTGAGAAAACGCATAATGG	47
H2-3	TATGCTACAGTAGCAGGYTCCCTGTCACTGGCAATCATGATA	42
H2-4	TGGGATGTCATAAATTTTGAGAGCACTGGTAATTTAATTGCACCAGAATA	50
H3-1	ATGTGGGCCTGCCARARAGGCAACATTAGGTGCAACATTTGC	42
H3-2	ATGTGGGCTTGCCAAAAAGGCAACATCAGATGCAACATTTGC	42
H3-3	CAGCAACTGTTACCCTTATGATGTGCCGGATTATGYCTCCCT	42
H4-1	GCACTRCTTTTAGCCTTTATTTTGTGGGCTTGTCAGAATGGAAACAT	47
H4-2	ATTTTGTGGGCTTGTCAGAAYGGAAACATCCGGTGCCAGATTTG	44
H4-3	TTCCATATCATGCTTYTTGCTCGTTGCACTRCTTTTAGCCTT	42
H5-1	TTGGGACATCAACACTAAACCAGAGATTGGTACCAARAATAGCTACTAGA	50
H5-2	TTATTCAACAGTGGCRAGTTCCCTAGCACTGGCAATCATGGT	42
H5-3	CCCAACAATAAAGAGRAGYTACAATAATACCAACCAAGAAGATCTTTTGG	50
H5-4	CCCAACAATAAAGAGGACCTAYAACAACACCAATGTAGAAGACCTTTTAA	50
H5-5	TTTATAGAGGGAGGATGGCAGGGAATGGTAGATGGTTGGTATGG	44
H5-6	CTAGATGTCTGGACTTATAATGCTGAACTTCTGGTTCTCATGGAAAATGA	50
H6-1	CTTGGTGTGTATCAAATTCTTGCYATTTATAGTACGGTATCGAGCAG	47
H6-2	CAAATCCTTGCYATTTATAGTACGGTATCGAGCAGTCTRGTTTTGG	46
H6-3	GCAATGGGTCTTTGGATGTGTTCAAATGGTTCAATGCARTGCA	43
H7-1	ATGGGATTGGTTTTCATTTGCATAAAGAATGGAAACATGCRGTGCACTAT	50
H7-2	CTTCGGGGCATCATGTTTCATACTTCTGGCCATTGCAATGGG	42
H7-3	CATCAAAATGCACAAGGAGARGGAACTGCAGCTGACTACAAA	42
H7-4	TGGTTTAGCTTCGGGGCATCATGCTTCCTWCTTCTTGCCATTGCAATGGG	50
H8	ATTTACAGTACAGTGGCGGCCAGTCTYTGCTTGGCAATCCTG	42
H9-1	CTTACAAAATCCTYACCATTTATTCGACTGTCGCCTCATCTCTTGT	46
H9-2	GCAATGGGGTTTGCTGCCTTCYTRTTCTGGGCCATGTCCAAT	42
H9-3	TTCTGGGCCATGTCYAATGGATCTTGCAGATGCAACATTTGTAT	44
H10-1	CTTTTGGCTGTCATCATGGGGCTTGTTTTCTTCTGTYTGAAAAATGGAAA	50
H10-2	GTCATCAATTGGACYAAGGATTCAATAACCGACATCTGGACTTATCARG	49
H11-1	GATCTCCATGATTCTAATGTTCGAAACCTCCATGAAAAGGTCAGACGAAT	50
H11-2	TGGGCGTGCAGYAATGGATCATGTAGATGTACCATTTGCATT	42
H12	TACTGCTCATGATTATTGGGGGTTTCATTTTCGGRTGTCAAAATGGAAAT	50
H13-1	AGTGTTGTGYTAGTAGGACTCATACTCTCTTTCATCATGTGGGCC	45
H13-2	ATAAATATGCTTGCAGACAGAATAGATGAYGCTGTAACTGATGTA	45
H14	TGCATCACCCATCAAGCGATAATGAGCAAACGGATCTCTACAAGG	45
H15	GCTGATCTGATAATAGAAAGAAGAAATTCAAGTGACATCTGTTACCCAGG	50
H16-1	TAATGCCATTGATGAAGGAGATGGTTGCTTCAATCTTCTTCACAA	45
H16-2	ATTCGAAATGGGACATATAATCATGAGGACTACAAAGAAGAGTCACAA	48

Table 1B. Oligonucleotide probes for IAV neuraminidase subtyping

Probe	Oligonucleotide sequence (5` → 3`)[a]	Length, n.
N1-1	GGGTTGGTCTTGGCCAGACGGTGCTGAGTTGCCATTYACCATT	43
N1-2	TCCTAATGGATGGACARATACCGACAGTGATTTCTCAGTGAAACAGGATG	50
N1-3	TGGTCTTGGCCAGACGGTGCTGAGTTGCCVTTCACCATTGAC	42
N1-4	CAAGAGTCTGAATGTGCATGTGTAAATGGYTCTTGCTTTACTGTAATGAC	50
N2-1	CAAGTGTGYATAGCATGGTCCAGCTCAAGTTGTCACGATGGAAA	44
N2-2	TTTGGSRACCAAACAAGTGTGCATAGCATGGTCCAGCTCAAG	42
N2-3	TTTTGTGGCACTTCAGGYACHTATGGAACAGGCTCATGGCCT	42
N2-4	TGTGGHACYTCAGGTACATATGGAACAGGCTCATGGCCTGAT	42
N3	AGYAATAGTATAGTTACTTTCTGTGGAYTAGACAATGAACCTGGATCGGG	50
N4	TGTGGTGTTAATTCTGATACCACAGGTTGGTCATGGCCYGATGGC	45
N5	TTTTGTGGTGTTTCMAGTGAGGTCCCAGGRTGGTCCTGGGATGATGG	47
N6-1	TCATGCCATGACGGCATCTCAAGAATGTCRATCTGCATGTCA	42
N6-2	GAGCGATTRGGATCATGGTCCTGGCATGATGGTGCTGARATC	42
N6-3	GAGCGATTGGGATCKTGGTCATGGCATGATGGGGCTGAAATC	42
N7-1	GTTGAAGGATGGGTAGTGGTGGCYAAGGACAATGCCATAAGATT	44
N7-2	CAGTTGGGTCCGGTTCCTTCCCCGATGGGGCACARATCCAAT	42
N8-1	ATATGGACCTCWAGYAGCTCCATTGTGATGTGTGGAGTAGAYCAT	45
N8-2	TGAGTGTAGAAATAGGGCAATCACCCAATGTGTACCAGGCAAGGT	45
N8-3	TGGTCRTGGCACGATGGAGCTATTCTTCCYTTTGACATCGAT	42
N9	ATGTGTTCCAGYACAGAATTCCTKGGACAATGGAACTGGCCTGAT	45

Table 1C. Oligonucleotide probes for IAV typing

Probe	Oligonucleotide sequence (5` → 3`)[a]	Length, n.
NP-1	ACGAAAAGGCAACGAACCCGATCGTGCCTTCCTTTGACATGA	42
NP-2	ATGAGTAATGAAGGDTCTTATTTCTTCGGAGACAATGCAGARGAG	45
M2-1	GCAGARTGCTGTGGATGTTGACGATRGTCATTTTGTCAACATAG	44
M2-2	CCTATCAGAAACGAATGGGGGTGCAGATGCAACGATTCAAGTGA	44
M2-3	CCTTCTACGGAAGGAGTRCCWGAGTCTATGAGGGAAGAATATCG	44

[a]Apart from the four (A, G, C, T) bases, abbreviation codes are used: Y denotes C or T; R – A or G; W – A or T; S – G or C; M – A or C; K – G or T; V – A or C or G; H –A or C or T.

For the validation of IAV subtyping microarray we used different IAV strains. Total RNA from IAV samples was isolated using TRIZol reagent. Full-length cDNAs, corresponding to IAV genes, were amplified using universal MBTuni-12 primers in the process of the RT-PCR with Cy3-dCTP as suggested and described earlier [10]. The Cy3-labeled IAV cDNA samples were denaturated and further used for hybridization with microarray slides during 2 hours at 37°C.

First of all we used entirely characterized strains A/Vic/3/75(H3N2), A/chicken/Kurgan/5/05 (H5N1), A/Brisbane/59(H1N1), A/swine/1976/31(H1N1), A/Brisbane/10/07(H3N2), A/California/07/09(H1N1). Data presented on figures 2 and 3 showed the specific binding of viral cDNAs with the probes, immobilized on the microarray slide. All IAV strains were subtyped correctly.

After that we used two partly characterized IAV strains of H2N2 and H1N1 subtype and four IAV strains, for which only HA subtype was determined (A/H6_51c, A/H3_402, A/H3_111 and A/H9_121). Significant levels of fluorescence were observed only for probes, specific to the corresponding IAV subtypes, for all other oligonucleotides fluorescence intensities were on the approximately equal level and comparable with standard error (See figure 3). NA subtypes were unambiguously determined: N1 for A/H6_51c strain, N8 for A/H3_402 and A/H3_111 strains and N7 for A/H9_121.

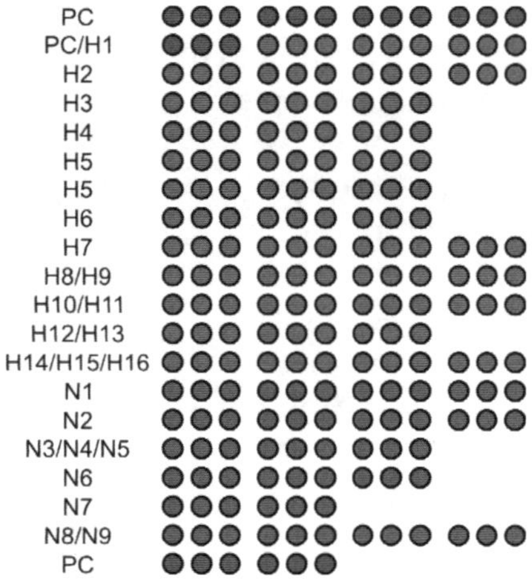

Figure 1. Layout of IAV subtyping microarray. H – hemagglutinin, N – neuraminidase, PC – positive control

Finally two IAV strains of the unknown subtype were analyzed. The subtype of these IAV strains could not be determined by the traditional methods. We easily determined these strains as A/H5N1 and A/H3N8 using IAV subtyping microarray.

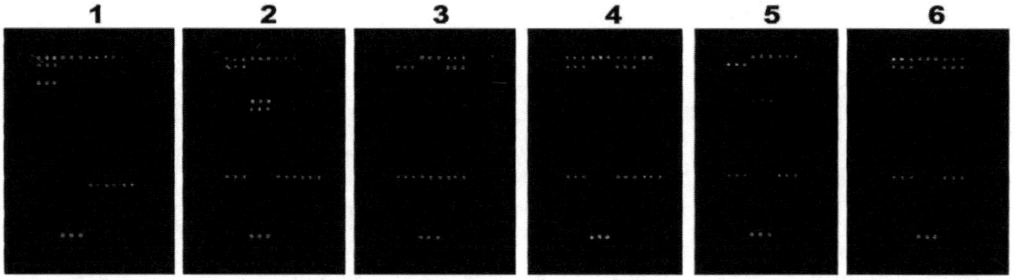

Figure 2. The fluorescence intensities of IAV subtyping microarray after hybridization with Cy5-cDNA, obtained from the (1) A/Vic/3/75(H3N2), (2) A/chicken/Kurgan/5/05(H5N1), (3) A/Brisbane/59(H1N1), (4) A/swine/1976/31(H1N1), (5) A/Brisbane/10/07(H3N2) and (6) A/California/07/09(H1N1) virus strains.

The developed method for identification and subtyping of IAV via hybridization on a microarray slide showed good results. All the studied samples were correctly classified into subtypes according to the fluorescent signals from capture probes. We did not observe significant cross-reactivity by capture probes. Currently developed oligonucleotide IAV subtyping microarray in combination with multisegment RT-PCR method for amplification of fluorescently labeled cDNA of all possible IAV subtypes with a single pair of primers is a very promising approach for rapid and accurate identification of all subtypes of IAV. Moreover, it allows molecular screening not only of HA and NA, but of all other IAV genes.

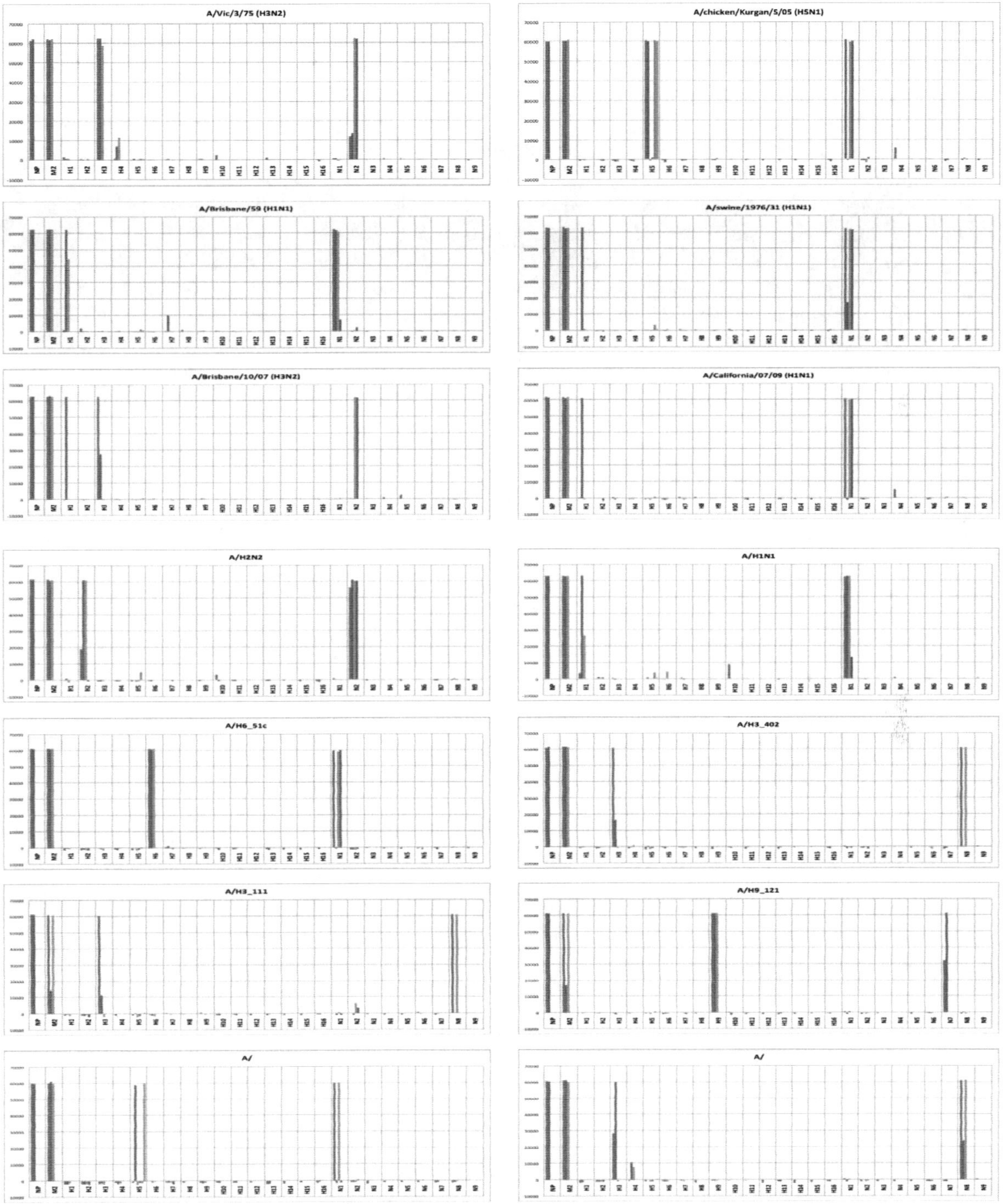

Figure 3. Relative values of the spot fluorescence intensities, derived after hybridization with reference IAV cDNA samples. Each column corresponds to the 3 spots mean intensity

4. References

[1] Dawson E D, Moore C L, Smagala J A, Dankbar D M, Mehlmann M, Townsend M B, Smith C B, Cox N J, Kuchta R D, and Rowlen K L 2006 *Anal. Chem.* **78** 7610–7615.

[2] Fesenko E E, Kireyev D E, Gryadunov D A, Mikhailovich V M, Grebennikova T V, L'vov D K, Zasedatelev A S. 2007 *Infl. Other Resp. Viruses*. **1**(3) 121-129.

[3] Ghedin E, Sengamalay N A, Shumway M, Zaborsky J, Feldblyum T, Subbu V, Spiro D J, Sitz J, Koo H, Bolotov P, Dernovoy D, Tatusova T, Bao Y, St.George K, Taylor J, Lipman D J, Fraser C M, Taubenberger J K, and Salzberg S L 2005 *Nature* **437** 1162–1166.

[4] Han X, Lin X, Liu B, Hou Y, Huang J, Wu S, Liu J, Mei L, Jia G, Zhu Q. 2008. J. *Virol. Methods* **152**(1-2) 117-121.

[5] Kessler N, Ferraris O, Palmer K, Marsh W, Steel A, 2004 *J. Clin. Microbiol.* **42** 2173–2185.

[6] Li J, Chen S, Evans H, 2001 J. *Clin. Microbiol.* **39** 696–704

[7] Sengupta S, Onodera K, Lai A, and Melcher U 2003 *Clin. Microbiol.* **41** 4542–4550.

[8] Hoffmann E, Stech J, Guan Y, Webster RG, Perez DR. 2001 *Arch Virol.* **146**(12) 2275-89.

[9] Zhou B, Donnelly M E, Scholes D T, St George K, Hatta M, Kawaoka Y, Wentworth D 2009 *J Virol.* **83**(19) 10309-13.

[10] Vasin A V, Sandybaev N T, Plotnikova M A, Klotchenko S A, Chervyakova O V, Strochkov V M, Taylakova E T, Elpaeva E A, Komissarov A B, Egorov V V, Koshemetov J K, Kiselev O I, Mamadaliev S M 2011 *Journal of Physics: Conference Series*. **291** 012006

Journal of Physics: Conference Series **345** (2012) 012040 doi:10.1088/1742-6596/345/1/012040

Atomic force microscopy as a tool to study *Xenopus laevis* embryo

E A Pukhlyakova[1],Yu M Efremov[2] , D V Bagrov[2], N N Luchinskaya[3], D O Kiryukhin[3], L. V. Belousov[1] and K V Shaitan[2]

[1] M.V. Lomonosov Moscow State University, Faculty of Biology, Department of Embryology, 111992, Leninskie Gory, 1/12, Moscow, Russia

[2]M.V. Lomonosov Moscow State University, Faculty of Biology, Department of Bioengineering, 111991, Leninskie Gory, 1/73, Moscow, Russia

[3]Engelhardt Institute of Molecular Biology, Russian Academy of Sciences, 117984, Moscow, Russia

E-mail: yu.efremov@gmail.com

Abstract. Atomic force microscopy (AFM) has become a powerful tool for imaging biological structures (from single molecules to living cells) and carrying out measurements of their mechanical properties. AFM provides three-dimensional high-resolution images of the studied biological objects in physiological environment. However there are only few AFM investigations of fresh tissue explants and virtually no such research on a whole organism, since most researchers work with cell cultures. In the current work AFM was used to observe the surface of living and fixed embryos and to measure mechanical properties of naive embryos and embryos with overexpression of guanine nucleotide-binding protein G-alpha-13.

1. Introduction

Cell- and tissue-scale mechanics integrates numerous processes within the multicellular organism from local cell movements to global tissue sculpting during morphogenesis. However, so far only few quantitative analyses of tissue and embryo biomechanics have been carried out [1, 2]. The challenge is to learn how gene products actually generate or modulate forces and provide cells and tissues with mechanical properties to transmit these forces in appropriate patterns. Recently many methods for measuring forces at molecular and cellular levels have been developed and atomic force microscopy-based indentation is one of them [3, 4, 5]. This method has a good resolution and requires rather simple sample preparation procedure.

The embryos of *Xenopus laevis*, a South African clawed toad, are frequently used in biological research because of the large embryo size and simplicity of manipulation [6]. We used AFM to study morphology of embryo surface and to carry out force measurements.

2. Materials and methods

In our previous work [7] we studied both fixed and living *Xenopus laevis* embryos with AFM, in order to visualize different stages of its development and to observe dynamic structure changes. We have developed a special sample preparation protocol for the observations. Embryos were placed and tightly held in the holes made in agarose layer in such a way that the region of interest (e.g. animal or vegetal pole) was turned up. Such immobilization method did not affect the embryo viability, since the embryos could develop normally after prolonged (5–6 hours) experiments. Imaging was performed in contact mode in buffer solution or cultivation medium.

For the force measurements special cantilevers modified with 9 μm microspheres were used (figure 1). Such modification led to precise control of probe geometry and at the same time provided an estimate of the mechanical properties averaged over a relatively large contact area [5]. This setup allowed us to carry out measurements similar to those made with NanoNewton force measurement device described in [8].

The elastic (Young's) modulus of embryo was determined using Hertzian contact mechanics. For the naive embryos the measurements were carried out from blastula to neurula stages and for the embryos with overexpressed G-alpha-13 protein – only at gastrula stage.

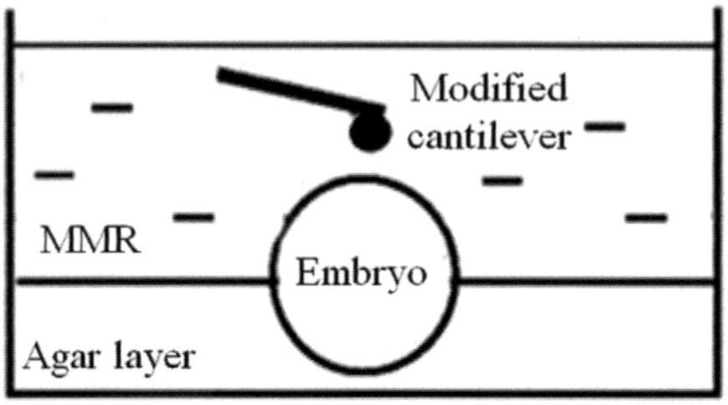

Figure 1. Scheme of experimental setup for force measurements of *Xenopus Laevis* embryos.

3. Results and discussions

3.1. Imaging the surface of *Xenopus Laevis* embryos

As described earlier [7], AFM can be used to image the embryos surface at various stages of development. AFM data were compared with the data from scanning electron microscopy (SEM). AFM investigation of fixed embryos allowed us to obtain data similar to SEM. We have also shown that AFM can be applied to study living embryos and visualize cell movements on the embryo surface *in vivo* in real time. The measured rate of elongation or shortening of the cell boundary fragments was ~0.5-1.5 μm/min.

The changes of microvilli quantity and distribution on the cell surface and the cytoskeleton remodelling were also observed (figure 2). In the area, indicated by box 1, cytoskeleton is clearly visible as network on the surface. The area, indicated by box 2, is rather smooth on the first image and some microvilli-like protrusions appear on the second image. Since AFM can be used to study living embryos, it can also be used to monitor the impact of bioactive compounds on embryogenesis.

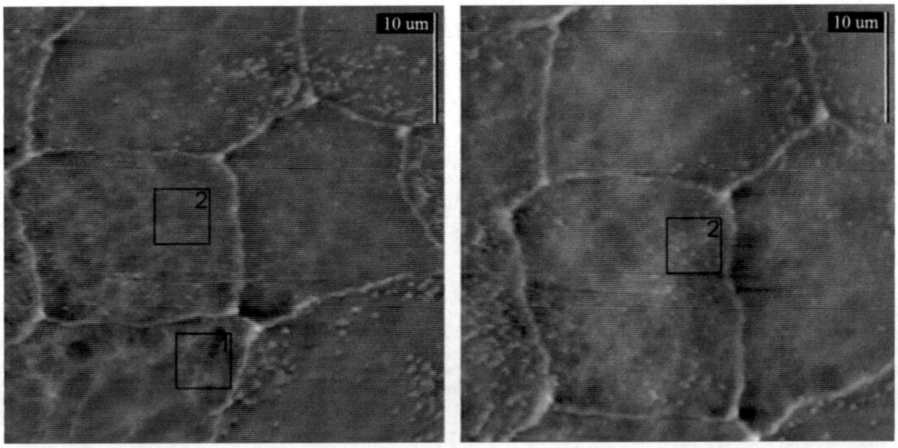

Figure 2. Two consecutive contact error images (~5 min on each scan) of living embryo surface (gastrula stage). The microvilli and cytoskeleton rearrangements are visible. In the area, indicated by box 1, cytoskeleton is clearly visible as network on the surface. In the area, indicated by box 2, some microvilli-like protrusions appeared on the second image.

3.2. Force measurements of *Xenopus Laevis* embryos.

Apart from imaging of the embryo surface AFM allowed us to measure the Young's modulus of the living embryo. The force measurements were carried out on living embryos during development from blastula to neurula stages. Our measurements of more than five embryos of each stage showed that there is a strong natural variability in stiffness among the embryos collected from the same or different females (1–200 kPa). Using our technique of immobilization of living embryos we managed to measure the stiffness properties in different embryo regions. On the gastrula-stage embryos we found that Young's modulus on the animal pole was more than two times higher than on the vegetal pole. Our data show the increasing of the embryos' elastic modulus (Young's modulus) on the animal pole during gastrulation. These phenomena can be explained by the fact of turgor blastocoel pressure growth, which stretches the blastocoel roof and may lead to rise of actin polymerization [9].

AFM is a convenient tool to study pathways of actin polymerization and regulation of cell stiffness. In our experimental model we measured the Young's modulus of G-alfa-13 overexpressing embryos under lysophosphatidic acid (LPA) treatment. LPA is a ligand of G-protein-coupled receptor [10]. For the downstream signal transduction LPA receptor uses heterotrimeric G proteins. LPA receptor activation leads to the structural change of the G protein and to the dissociation of its alpha subunit. It was showed that proteins of G-alpha-12/13 family can regulate actin cytoskeleton reorganization and cell motility. G-alpha-13 protein through RGS-RhoGEF complexes, such as LARG, p115Rho-GEF, PDZ-RhoGEF, leads to small GTFase RhoA activation which controls the actin polymerization. Thus, the signal cascade G13– RGS-RhoGEF– RhoA regulates actin cytoskeleton state [11, 12].

Numerous studies have proved the actin cytoskeleton to be the major contributor to the mechanical properties of single cells [13], however little is understood about the contribution of the actin cytoskeleton into mechanical properties of tissues. We found a significant growth of Young's modulus (about 100%) of the embryos with G-alpha-13 overexpression after addition of LPA into the cultivation medium (Figure 3). Apparently, LPA-activated signal cascade G13 – RGS-RhoGEF – RhoA induces the RhoA activation in the cell cortex of Xenopus embryos and increases its stiffness. Addition of LPA to ordinary embryos (without G13 overexpression) caused an increase of Young's modulus which did not exceed 15%.

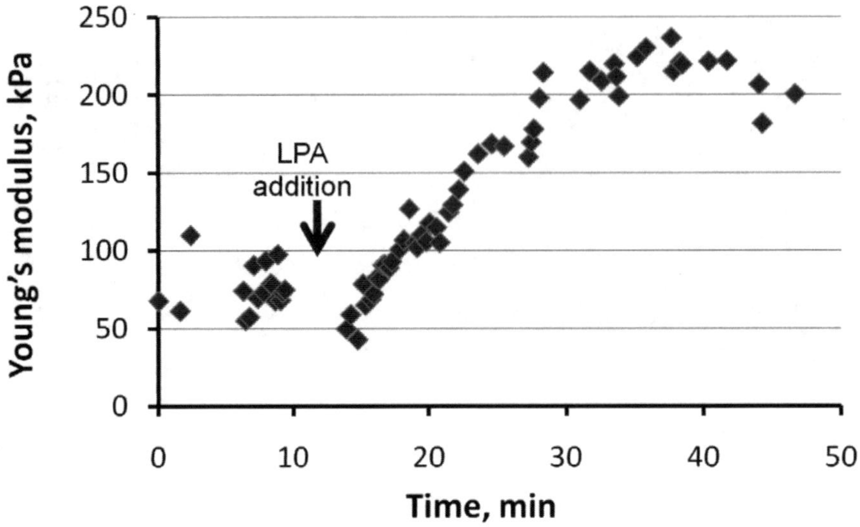

Figure 3. Growth of the Young's modulus of embryo with overexpression of G-alpha-13 protein after LPA addition.

4. Conclusions

We have shown that AFM is a powerful and convenient tool to study living embryos. We have detected an increase of Young's modulus upon impact of LPA on embryos with G-alpha-13 overexpression. Our findings allow us to recommend AFM as a method for studying morphology and stiffness properties of the different regions of the embryo during development.

Acknowledgments

The work was supported by the federal target program "Scientific and educational research personnel to innovative Russia" and Russian Fond for Fundamental Investigations (RFFI); Grant number: 05-04-48681.

References

[1] Zhou J, Kim H Y, Wang J H, Davidson L A 2010 *Development* **137** 2785-2794
[2] Zhou J, Kim H Y, Davidson L A 2009 *Development* **136** 677-688
[3] Efremov Yu M, Bagrov D V, Dubrovin E V, Shaitan K V, Yaminskii I V 2011 *Biophysics* **57** 257-267
[4] Kuznetsova T G, Starodubtseva M N, Yegorenkov N I, Chizhikc S A, Zhdanov R I 2007 *Micron* **38** 824–833
[5] Rico F, Wojocikiewicz E P, Moy V T 2008 *Applied Scanning Probe Methods IX* ed M Tomitori, B Bhushan, H Fuchs (Berlin: Springer Berlin Heidelberg) p 89-109
[6] Gurdon J B, Hopwood N2000 *Int. J. Dev. Biol.* 2000 **44** 43-50
[7] Efremov Yu M, Pukhlyakova E A, Bagrov D V, K.V. Shaitan K V 2011 *Micron* **42** 840-852
[8] Davidson L, Keller R 2007 *Methods Cell Biol.* **83** 425-439
[9] Beloussov L V, Luchinskaya N N, Ermakov A S and Glagoleva N S 2006 *Int. J. Dev. Biol.* **50** 113-122
[10] Lloyd B, Tao Q H, Lang S and Wylie C 2005 Lysophosphatidic acid signaling controls cortical actin assembly and cytoarchitecture in Xenopus embryos. *Development* **132** 805-816
[11] Kelly P, Casey P J and Meigs T E 2005 Biologic Functions of the G12 Subfamily of Heterotrimeric G-Proteins: Growth, Migration, and Metastasis. *Biochemistry* **46** 6677-6687
[12] Sternweis P C, Carter A M, Chen Z, Danesh S M, Hsiung Y F, Singer W D 2007 Regulation of Rho guanine nucleotide exchange factors by G proteins. *Adv. Protein Chem.* **74** 189-228
[13] Luby-Phelps K 2000 *Int. Rev. Cytol.* **192** 189-221

Journal of Physics: Conference Series **345** (2012) 012042 doi:10.1088/1742-6596/345/1/012042

Design of a sensor for the blood AB0 group antibodies detection

D.V. Kolesov[1], G.A. Kiselev[1,2], M.A. Moiseev[3], A.A. Kudrinskiy[3], I.V. Yaminskiy[1,3]

[1] OOO "Biosensor Academy", Moscow, Russia

[2] A.N.Frumkin Institute of physical chemistry and electrochemistry RAS, Moscow, Russia

[3] M.V. Lomonosov Moscow State University, Moscow, Russia

kolesov@nanoscopy.ru

Abstract. Control the content of the blood group antibodies in the plasma of the recipient is an important task in modern transplantation. In this paper we proposed to use micromechanical cantilever sensors for detection of the low concentrations of AB0 blood group antibodies in serum. The technique of chemical modification of cantilever surface to create the receptor layer was developed. The apparatus, which provides data acquisition from multiple microconsoles simultaneously was created. We carried out experiments by the detection in a solution the β antibodies with a concentration of 300 times less than the native content of antibodies in the blood. Change in surface stress due to formation of antigen-antibody complexes on the cantilever surface was 0.075 N/m. The resulting lateral strain, apparently, induced by repulsion between the complexes during the sorption of antibodies in layer of antigens, immobilized on the surface. The possibility of regeneration of sensory layer for repeated measurements was shown.

1. Introduction

A huge number of patients around the world await suitable for transplants. AB0-incompatible organ transplantation may be the solution to this problem. The human blood groups are determined by a set of antigenic characteristics of red blood cells - erythrocytes. The ABO blood group system is the most important blood type system in transplantation. It is characterized by presence on the red blood cells of two types of antigens A and B and associated with them alpha (anti-A) and beta (anti-B) antibodies in the blood plasma. A number of studies [1] have shown that removal of blood group anti-A and anti-B antibodies from patients' blood can reduce the risk of hyperacute organ rejection in ABO-incompatible transplantation.

For the successful solution of blood purification problem, the method of monitoring the presence of antibodies in the serum should be developed. One of the best candidates for the sensor is microcantilever gauges. This technology is based on the conversion of energy of the complementary interaction to the mechanical deformation of the cantilever and has been actively developing in the last two decades. This type of sensor has high sensitivity and selectivity, but also allows monitoring in real time [2].

2. Experimental and results

Experiments by the antibody determination were carried out on the Atomic Balance BioScan instrument (Biosensor Academy, Moscow). The device can simultaneously record small deflections of the two cantilevers. We have developed a technique of chemical modification of gold surface of the cantilever with antigens type A and B. The investigation object was the solution of monoclonal antibody beta released from mouse serum. Concentration of antibody in buffer solution was 300 times less than the natural concentration of antibodies in human blood.

During the experiments, two cantilevers were modified with antigens A and B, respectively, and then placed in the flow liquid cell of the device. After the equilibrium establishment, the buffer solution containing antibodies was injected into the liquid cell. There was a complementary binding of antibodies from the solution to antigens on the surface of one of the cantilevers. As a result, forces of intermolecular interaction in the sensory layer change during the formation of antigen-antibody complexes. The target signal is the difference between the two cantilevers deflections. After achieving the plateau, the total difference of surface tension caused by the interaction of antigen-antibody complexes on the surface of the cantilever was 0.75 N/m (Fig.1).

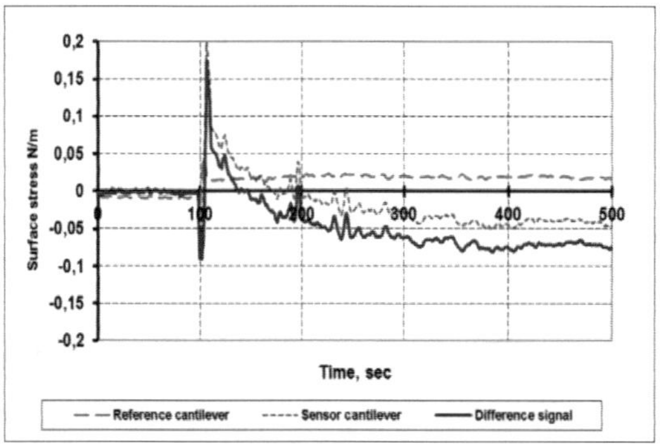

Z

Figure 1 Development of the forces of surface tension during the formation of AG-AT complexes on the surface of the cantilever.

Then the recovery of sensory layer with a 10% urea solution was conducted, and the investigated solution re-injected. During the second part of experiment, the difference of the forces of surface tension amounted to 0.65 N/m. The time to plateau has increased from six to ten minutes (Fig.2). This is evidenced by the partial decomposition of the complexes formed in the first stage.

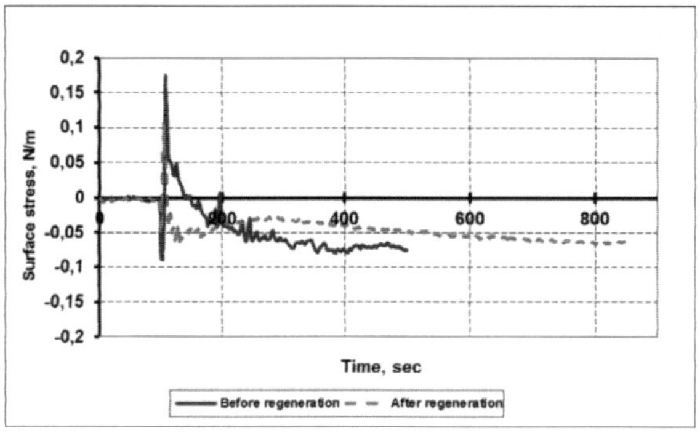

Figure 2. The dependence of surface stress from time during the formation of AG-AT complexes on the cantilever surface before and after the regeneration of sensory layer

3. Conclusions

In this study we demonstrated the method for determination of markers of blood groups using microcantilever sensors with receptor layer which contains antigens to antibodies of blood groups. It has been demonstrated to detect the β antibodies with concentration 300 times smaller than native content in human blood plasma. The resulting difference in surface stress due to formation of a layer of antigen-antibody complexes on the surface of the cantilever was 0.075 N / m. It was shown the possibility of regeneration of sensory layer of micromechanical sensors for the purpose of repeatable analysis. Thus, the results suggest the potential applicability of cantilever sensors for the challenges of modern transplantation, specifically, control of the content of antibodies in the serum.

Acknowledgement
This work was partially supported by the Ministry of Education and Science of the Russian Federation (project no. 16.513.11.3021)

References
[1] Squifflet J.P. et al. Lessons learned from ABO-incompatible living donor kidney transplantation: 20 years later. Exp Clin Transplant 2004, 2:208–213
[2] N.V. Lavrik, M.J. Sepaniak, P.G. Datskos. Cantilever transducers as a platform for chemical and biological sensors // Rev. of Sci. Ins., – 2004, – Vol.75, – No.7, – pp.2229-2253

IV Nanotechnology International Forum (Rusnanotech 2011) IOP Publishing
Journal of Physics: Conference Series **345** (2012) 012043 doi:10.1088/1742-6596/345/1/012043

Morphological changes in the kidney, liver and spleen during prolonged administration of iron nanoparticles

N.A. Navolokin[1], G.N. Maslyakova[1], A.B. Bucharskya[1], X.M. Kong[1], V.V. Zuev, B.A. Medvedev [2], A.A. Ignatiev [2], T.V. Bochkaryeva[2]

[1]SEI HPE Saratov SMU n.a.V.I. Razumovsky

[2]Saratov State University n.a. N.G. Chernyshevsky

E-mail: navolokin1@rambler.ru

Abstract. We determined the cytotoxic effect of iron nanoparticles of 70 nm, with a single per oral administration in an experiment on white outbred mice. Morphological changes were evaluated in the internal organs.
Thus, changes depend on the concentration of nanoparticles at long-term per oral exposure: identified violations of the structure of the liver, kidneys and spleen as venous plethora and degeneration of cells at 250 and 500 mkg / kg dose of nanoparticles are reversible, changes in the organs were pronounced with a dosage of 1000 mkg / kg.

1. Introduction

Nanotechnologies have become a revolutionary step in the various fields of science and promising area for scientific research in coming years, as these studies are based on the application and use of substances at the molecular level, and structures of one nanometer (one billionth of a meter) and 100 nm. The possibilities of using nanoparticles in biomedical purposes were a rapidly developing field of nanotechnology, which allows to open new possibilities in the diagnostics and treatment of various diseases.

At present, nanostructures were established for biomedical applications in various fields, including visualization of various biological objects and delivering drugs to cells, tissues and organs. The unique features of nanoparticles, such as high surface activity, the stability of biomolecular absorption, changes in physical and chemical properties under the influence of physical fields, their small size, comparable to biomolecules, the severity of the magnetic properties and biocompatibility, open wide prospects for creation of new drugs based on nanopreparations, that can be used in the treatment of various diseases, including cancer.

It should be noted that only few studies in the field of magnetic nanoparticle delivery and their application for MRI and CT scans were carried with using of morphological methods. In addition, we found no published data of comparative analysis of structural and functional changes that occur in the kidneys, liver and spleen of laboratory animals at peroral administration of different doses of magnetic nanoparticles.

Objective: to study the morphological changes in the kidneys, liver and spleen of laboratory animals at peroral administration of Fe nanoparticles in experiment.

Materials and methods: Fe nanoparticles of 70 nm ± 10 nm, 28 white outbred male mice, organometric and morphological methods.

The design of the experiment: the animals were divided into three experimental groups of seven mice in each group, during 5 days they received administered per orally solution of iron nanoparticles at a concentration 250 mkg/kg, 500 mkg/kg, 1000 mkg/kg. Each mouse of control group received

Published under licence by IOP Publishing Ltd

administered per orally saline during 5 days . Animals were derived from the experiment by decapitation the next day.

2. Results and their discussion

Organometric study.

The coefficient K was introduced for increasing the objectivity of evaluation of organometric study, which was calculated as the ratio of the mass of organ to the weight of the nimal body. Median was used in the statistical analysis due to the heterogeneity of the samples .

Table 1. Indicators coefficient (K) for kidney

Series of experiment	Coefficient (K) Fe	Control
250 мкг/кг mkg/kg,	103,4±3,35	122,5 ± 8,9
500 мкг/кг mkg/kg,	115,6±13,184	122,5 ± 8,9
1000 мкг/кг mkg/kg,	113,2±14,33	122,5 ± 8,9

Findings were the following: reduction in the K (Tab. 1) was noted in mice compared with controls, suggesting an increase in weight of kidneys.

Pathological changes in organs.

The following changes have been identified due repeated per oral administration of iron nanoparticles: in the kidney - swelling of the kidney convoluted tubules was marked at a dosage of 250 mg /kg. Changes were represented by small granular dystrophy and swelling of the epithelium, the poorly defined plethora of vessels of the brain substance with the introduction of iron nanoparticles at a concentration of 500mkg /kg. Degeneration of convoluted tubules and a more pronounced hyperemia of vessels of the cortex and the glomeruli were observed at the introduction of iron nanoparticles in the maximum concentration of 1000 mkg/kg . Other authors who have studied the morphology of the internal organs of rats with intravenous injection of Fe_3O_4, also note the venous plethora of vessels of the brain substance, but in our experiment there was no expansion of Shymlansky-Bowman capsule, which is possible due to other dimensions and method of introducing nanoparticles [4,7].

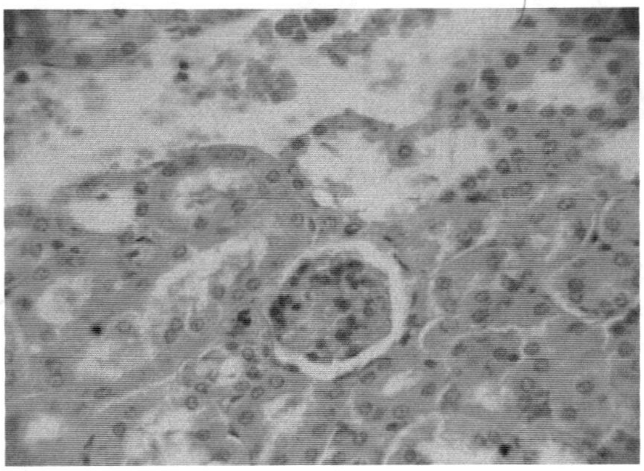

Figure 1. Morphological changes in the kidneys with the introduction of iron nanoparticles at a concentration of 1000 mg/kg, which are represented by degeneration of convoluted tubules and expressed plethora of vessels of the cortex and glomeruli.

In the liver - a moderately pronounced hyperemia of blood vessels and degeneration of hepatocytes were indicated with a dosage of 250 mkg / kg. The severity of plethora dystrophy was increasing with a dosage of 500 mkg / kg and a large number of the black granules were observed in the blood and between hepatocytes. There were a very marked hyperemia of blood vessels and degeneration of cells with a dosage of 1000 mkg / kg.

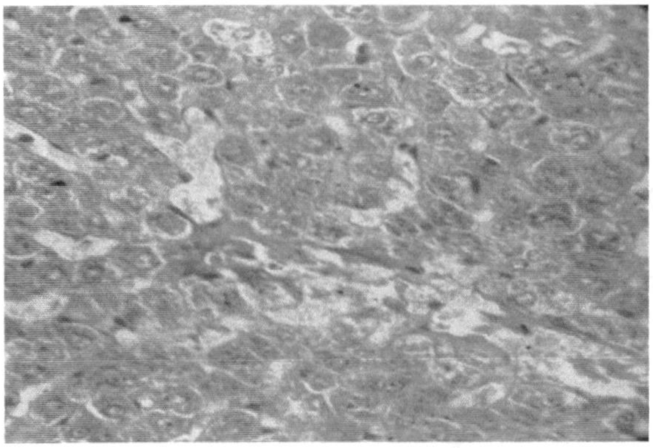

Figure 2. Morphological changes in the liver at doses of 1000 mg / kg: very pronounced plethora of vessels and degeneration of hepatocytes.

In the spleen - a large numbers of black granules were observed in the red pulp with all the doses. Large follicles with germinal centers active were observed at of 250 mg / kg and 500 mg / kg. The predominance of red pulp over white pulp was indicated at a dosage of 1000 mg / kg, white pulp was with indistinct outlines, the follicles were without light centers and large cells were found.

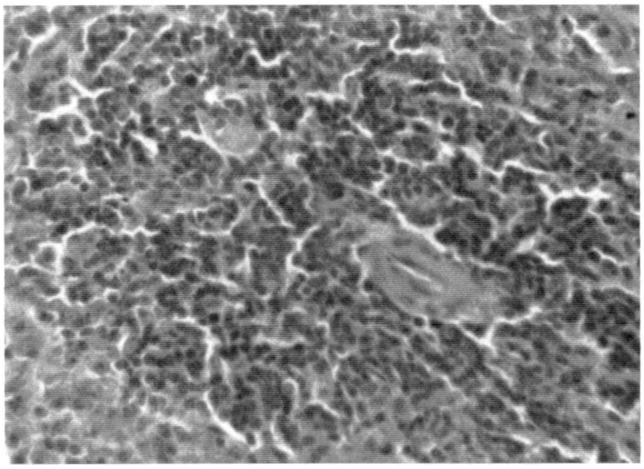

Figure 3. Morphological changes in the spleen at a dosage of 1000 mg / kg were presented a white pulp with indistinct outlines and follicles without light centers, large cells were found.
Nanoparticles were not accumulated in the organs, which was confirmed by the lack of visualization on MRI.

3. Conclusions

Changes in varying degree were revealed in the analysis of micropreparations of the kidneys of laboratory animals of experimental group, depending on the solution concentration of iron nanoparticle. The maximum changes were observed at a dosage of 1000 mkg / kg. The presence of vascular venous plethora was evidence of the reaction on the coming of Fe nanosized particles into the systemic circulation, and degenerative changes in the organs testify to the presence of the damaging effect of nanoparticles. These changes were not observed in the control group.

Acknowledgment

Thus, uncoated iron nanoparticles were not accumulate in the organs, during prolonged oral administration. Morphological changes depend on the concentration of nanoparticles: identified violations of the structure of the liver, kidneys and spleen as venous plethora and degeneration of cells at 250 and

500 mkg / kg dose of nanoparticles are reversible, changes in the organs were pronounced with a dosage of 1000 mkg / kg.

References

[1] Losic D, Rosengarten G, Mitchell JG, et al. Porearchitecture of diatom frustules: potential nanostructuredmembranes for molecular and particle separations. J NanosciNanotechnol 2006; 6: 982–9.

[2] Koger N. Prescribing diatom morphology: towardsgenetic engineering of biological nanomaterials. Curr OpinChem Biol 2007; 11: 662–9.

[3] Vo-Dinh T, eds. Protein Nanotechnology, Protocols,Instrumentation, and Applications. Series: Methods in MolecularBiology; Totowa: Humana Press Inc, 2008; 452 p.

[4] Zschornack G, Grossmann F, Ovsyannikov VP, et al.Highly charged ions for high-tech applications.German-Ukrainian Symposium on Nanoscience and Nanotechnology,abstract book; Essen, 2008: 80.

[5] Kentsch J, Durr M, Schnelle T, et al. Microdevices forseparation, accumulation and analysis of biological micro- andnanoparticles. IEE Proc Nanobiotechnol 2003; 150: 82–9.

[6] Neugebauer S, Muller U, Lochmuller T, et al. Characterizationof nanopore electrode structures as basis for amplifiedelectro chemical assays. Electroanalysis 2006; 18: 1929–36.

[7] Greulich C, Kittler S, Epple M, et al. Studies on thebiocompatibility and the interaction of silver nanoparticles with human mesenchymal stem cells (hMSCs). Langenbecks Arch Surg 2009; 394: 495–502.

[8] Saleem M, Meyer MC, Breitenstein D, et al. Thesurfactant peptide KL4 in lipid monolayers: phase behavior, topography and chemical distribution. J Biol Chem 2008;283: 5195–207.

[9] Chekhun VF, Kulik GI, Todor IN, et al. The influenceof ferromagnetic nanoparticles on antitumor effect of doxorubicinin Ehrlich ascitic carcinoma-bearing mice. German-Ukrainian Symposium on Nanoscience and Nanotechnology,abstract book; Essen, 2008: 72.

[10] Bilyy R, Podhorodecki A, Nyk M, et al. Utilizationof GaN:Eu3+ nanocrystals for the detection of programmedcell death. Physica E: Low-Dimentional Systems and Nanostructures 2008; 40: 2096–9.

[11] Arruebo M, Galan M, Navascues N, et al. Development of magnetic nanostructured silica-based materials as potential vector for drug-delivery application. Chem Materials 2006; 18: 1911–9.

[12] Fernandez-Pacheco R, Marquina C, Valdivia JG, et al. Magnetic nanoparticles for local drug delivery using magnetic implants. J Magn Magn Mater 2007; 311: 318–22.

[13] Galanov AI, Yurmasova TA, Savelyev GG, et al. Development of nanostructured iron-based materials as potential vectors for drug-delivery application. Sib Onkol Zhurn 2008; 3: 50–7 (In Russian).

[14] Darton NJ, Hallmark B, Han X, et al. The in-flow capture of superparamagnetic nanoparticles for targeting therapeutics. Nanomedicine 2008; 4: 19–29.

Design of new drug forms by cryo-nanotechnologies

A G Ogienko[1,2], E V Boldyreva[1,3], V V Boldyrev[1,3], A Y Manakov[1,2], S A Myz[1,3], A A Ogienko[1,4], E G Zevak[1,2], A S Yunoshev[1,5]

[1] Novosibirsk State University, REC-008 "Molecular Design and Ecologically Safe Technologies", ul. Pirogova 2, Novosibirsk 630090, Russia

[2] Nikolaev Institute of Inorganic Chemistry SB RAS, Lavrentieva 3, Novosibirsk 630090, Russia

[3] Institute of Solid State Chemistry and Mechanochemistry SB RAS, ul. Kutateladze 18, Novosibirsk 630128, Russia

[4] Institute of Cytology and Genetics SB RAS, Lavrentieva 10, Novosibirsk 630090, Russia

[5] Lavrentiev Institute of Hydrodynamics SB RAS, Lavrentieva 15, Novosibirsk 630090, Russia

E-mail: andreyogienko@gmail.com

Abstract. A detailed study of processes occurring on annealing of frozen solutions of selected APIs in mixed organic-aqueous solvents allowed us to optimize the ratio of components and the experimental conditions for preparing novel forms of the APIs with improved properties.

1. Introduction

Currently, about 40% of industrially produced pharmaceuticals are poorly water soluble [1], what restricts their usage as orally administered drugs because of poor wetting properties and solubility in liquors of the digestive tract. Development, preparation and introduction of novel drug forms are time- and money-consuming, therefore the search for various approaches to optimizing already known and widely used medicines is highly topical. Improvement of bioavailability of drugs through enhancement of their dissolution kinetics is the task of paramount importance and urgency for pharmaceutical science. Besides, the interest of pharmaceutical companies in developing the technologies of producing micro- and nanoparticles of desirable size is related, among other reasons, to the growing popularity of inhalation therapy [2]. However, scarce solubility of pharmaceutical ingredients in water limits severely the application of spray-drying or freeze-drying due to the necessity of evaporating large volumes of solvents during the process, or of removing a large amount of ice by sublimation. This results in a substantial increase in the duration of an experiment and a large energy consumption for the preparation of even relatively small amounts of a drug. At the same time, it is known that the solubility of APIs can be considerably increased, if water is mixed with an organic liquid having a low boiling point. Many of such systems are characterized by formation of clathrate hydrates at low temperatures, which are stable up to fairly high temperatures, thus making it possible in a certain concentration range to carry out drying at essentially higher temperatures under the conditions which exclude the apperance of liquid phases in the system. Fast cooling giving amorphous phases is followed by solvent removal by sublimation, and since no liquid phases are present, fine particles are formed, which do not grow.

Paracetamol is a wide-spread analgetic drug, exhibiting antiinflammatory and antipyretic effect, and it is also a common model object in many studies on key problems of pharmaceutical science [3-9]. For example, in our recent work [10] we have demonstrated the possibility to use the acetone-water mixture for preparing ultrafine powders of the monoclinic paracetamol suitable for direct compression. At the same time, the tetrahydrofuran (hereafter THF) - water system received much more attention of researchers focused on enhancement of solubility of poorly water soluble APIs (e.g., danazol) [11], what is related, among other things, to the fact that the clathrate hydrate in this system is crystallized on cooling of homogeneous aqueous solutions [12]. However, up to the moment, the search for appropriate solvents and their mixtures lacked systematic approach; little attention was paid to the processes occurring on rapid cooling of solutions and on their freeze-drying. As a result of wrong choices of the ratio of components in the mixed solvents, the consumption of energy needed to maintain low temperature during a prolonged drying was very high [11]. Therefore, our primary task was to demonstrate (with paracetamol example), how a detailed study and understanding of the processes occurring in a chosen mixture of solvents on cooling and subsequent freeze-drying allows one to make an optimum choice of the ratio of components in the mixture, to achieve substantial energy saving and to develop novel APIs with improved properties.

2. Experimental

Tetrahydrofuran (THF) and 1,4-dioxane (purified from peroxides and double distilled), chromatographically pure acetone and distilled water were used as solvents. Ultrafine powders of APIs were obtained by spray freeze-drying of their solutions in the organic volatile liquid - water mixtures.

Bruker D8 Advance equipped with a low-temperature TTK 450 Anton Paar chamber permitting to work under vacuum down to 10^{-3} torr was used for X-ray powder diffraction experiments aimed at optimizing drying conditions. The samples were carefully but gently ground in a mortar at the liquid nitrogen temperature and placed onto a holder, which was preliminary cooled down to liquid nitrogen temperature. If a diffraction pattern was measured at low pressure, the sample was kept in the chamber for 5 minutes prior to the start of data collection for an equilibration. Diffraction patterns were measured in the -140 - +20°C temperature range (2Θ scans in the 5-55 degrees range, at a 0.02 degrees step, about 8 minutes per pattern).

Scanning electron microscopy (SEM) analysis was carried out with TM-1000 (Hitachi) to obtain a visual image and to evaluate particle size, shape, and surface. The specimens were mounted on a metal stub with double sided adhesive tape and coated under vacuum with platinum.

The specific surface area was determined by the BET method from N_2-adsorption/desorption isotherm data using an ASAP-2400 instrument (Micromeritics, USA).

3. Results and discussion

The temperature regimes of drying were tuned on the basis of the X-ray diffraction study of the processes occurring on annealing of frozen solutions (prepared by spraying of the starting solution in a vessel with liquid nitrogen) and of the analysis of the phase diagram of the THF-water system, to guarantee that at every stage of drying only solid phases were present in the system. Paracetamol samples were obtained from the THF-water solutions as very light fluffy powders. Fig. 1 illustrates the difference in the volumes of the same amounts of ultrafine paracetamol obtained in this work and of the commercial sample as purchased. Specific surface area of the paracetamol samples measured by sorption/desorption of N_2 was equal to 2.43 ± 0.06 m^2/g for our samples, and $0.13\pm0.01 m^2/g$ - for the commercial sample. According to SEM data, the samples of ultrafine paracetamol contain porous spherical agglomerates with typical sizes of 30-70 μm. These agglomerates are formed as the solvent is removed from frozen droplets, which are produced by rapid cooling of the starting solution. It was noted that even on very slight mechanical impact (movement of the sample on tumbling of the vial; friction with a spatula) the agglomerates disintegrate into constituting planar particles with linear dimensions 2-4 μm and thickness about 100 nm (Fig. 2).

Fig.1. A comparison of volumes of the same amounts (1.0 g) of ultrafine paracetamol obtained in this work (left) and of the commercial sample as purchased.

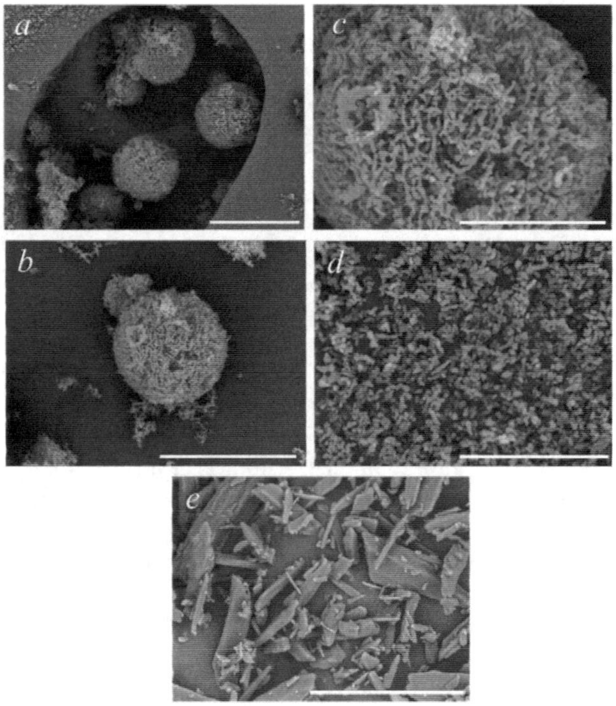

Fig. 2. Samples of the ultrafine paracetamol (*a* - *d*) obtained by spray freeze-drying of a frozen solution of paracetamol in the THF-water mixture. For a comparison: the commercial sample (e). The scale bar: *a, e* –100 μm; *b* – 20 μm; *c, d* –100 μm.

A study of the dissolution dynamics (Fig. 3) has shown that the samples of ultrafine paracetamol dissolve much faster at the initial minutes of dissolution as compared to the commercial sample, yielding the concentration in solution more than five times higher, than that achieved by dissolving a commercial sample during the same time. Besides, the dissolution of paracetamol samples prepared in this work in both water and 0.1M HCl was complete already on the 2nd minute, while the commercial samples could be dissolved in 20 minutes only.

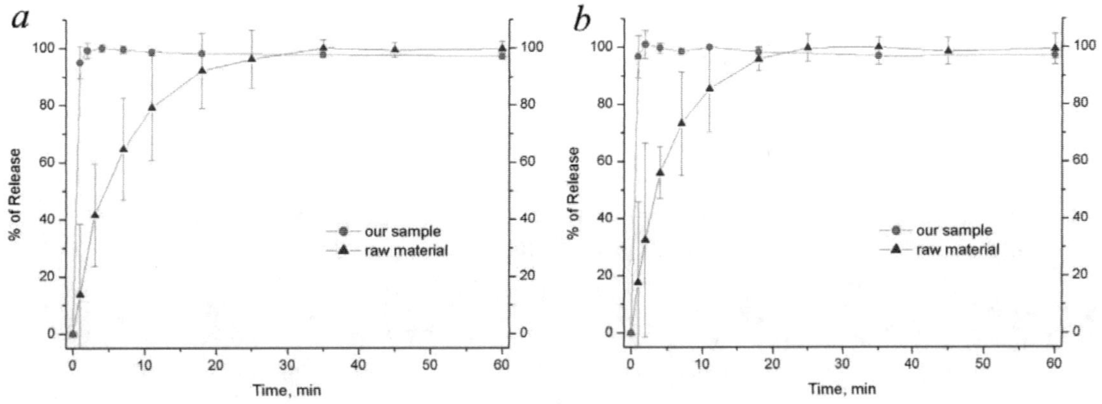

Fig. 3. Dissolution profiles of paracetamol samples obtained in this work, and of the commercial samples ("raw material") in different media (*a*: water; *b*: 0.1 M HCl).

This approach to obtaining fine particles can be applied to practically any solid pharmaceutical. Another example is provided by ibuprofen, which is a very common analgetic and antiinflammatory drug, which shows a bad dissolution behavior because of its hydrophobic structure. A rapid drug release is preferable for analgesic drugs. Decreasing a particle size is a promising way to improve the bioavailability of the drug. According to SEM data (Fig. 4), ibuprofen samples prepared by our technique (from different mixtures of solvents) contain agglomerates with sizes 30-500 μm consisting of flat 1-10 μm particles with the thickness less than 300 nm (depending on the solvent mixture used). Specific surface area values of SFD ibuprofen obtained in this work and of the commercial sample are 2.58±0.11 and 0.05±0.01 m^2/g, respectively. The suggested technique of the micronization of ibuprofen makes it possible to prepare fast-dissolving suspensions, which meet the requirements to analgetic and antipyretic drugs for babies and infants.

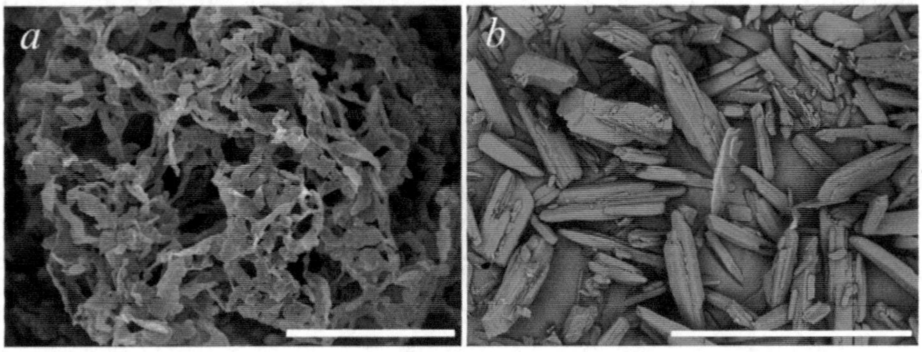

Fig. 4. SEM images of the ibuprofen samples: obtained by spray freeze-drying (*a*), and commercial sample (*b*). The scale bar: *a* –50 μm; *b* – 500 μm.

4. Conclusion

Two-component mixtures "volatile-liquid – water" giving clathrates at low temperatures can be used as solvents for APIs. Flash-cooling and subsequent evaporation of these solutions can give micronized samples with improved dissolution profile, compactability and other properties. If the ratio of the components is chosen correctly, the concentration of the initial solution and the evaporation temperature can be high enough, to reduce energy consumption as the drying time significantly (this is of critical importance for technology), and to avoid the increase in the size of fine particles (since no liquid is present in the system at any stage anymore after cooling of the starting solution). The suggested technique based on freeze-drying of frozen solutions prepared from mixed water-organic solvents allows one to prepare a wide range of other APIs as ultrafine powders.

Acknowlegments

The authors thank "Tatchempharmpreparaty" and "Altaivitaminy" for having kindly provided the drug substances. This work was supported by the Integration project № 109 SB RAS, Projects 21.44 and 5.6.4 of Presidium of RAS, by a grant from RFBR (11-03-12114-ofi-m-2011) and by the Russian Ministry of Education and Science (State Contract of RF № 14.740.11.1023).

References

[1] Lipinski C A, Lombardo F, Dominy B W, Feeney P J. 2001 *Adv. Drug. Deliv. Rev.* **46** 3
[2] Chow A H L, Tong H H Y, Chattopadhyay P, Shekunov B Y 2007 *Pharm. Res.* **24** 411
[3] Joiris E, Di Martino P, Berneron C, Guyot-Hermann A M, Guyot J C 1998 *Pharm. Res.* **15** 1122
[4] Boldyreva E V, Drebushchak V A, Paukov I E, Kovalevskaya Y A, Drebushchak T N 2004 *J. Therm. Anal. Calorim.* **77** 607
[5] Espeau P, Ceolin R, Tamarit J L, Perrin M A, Gauchi J P, Leveiller F 2005 *J. Pharm. Sci.* 9**4** 524
[6] Kachrimanis K, Fucke K, Noisternig M, Siebenhaar B, Griesser U J 2008 *Pharm. Res.* **25** 1440
[7] Karki S, Friscic T, Fabian L, Laity P R, Day G M, Jones W 2009 *Adv. Mater.* **21** 3905
[8] Palmieri G F, Bonacucina G, Di Martino P, Martelli S 2001 *Drug Dev. Ind. Pharm.* **27** 195
[9] Talegaonkar S, Khan A Y, Khar R K, Ahmad F J, Khan Z I 2007 *Iranian J. Pharm. Res.* **6** 95
[10] Ogienko A G, Boldyreva E V, Manakov A Yu, Boldyrev V V, Yunoshev A S, Ogienko A A, Myz S A, Ancharov A I, Achkasov A F, Drebushchak T N 2011 *Pharm. Res.* DOI: 10.1007/s11095-011-0502-x
[11] Rogers T L, Nelsen A C, Sarkari M, Young T J, Johnston K P, Williams R O 2003 *Pharm. Res.* **20** 485
[12] Dyadin Y A, Bondaryuk I V, Zhurko F V. Clathrate hydrates at high pressures. In: Inclusion compounds. V.5, Atwood JL, Davies JED, MacNicol DD. Oxford: Oxford University Press; 1991, p. 214–275.

Journal of Physics: Conference Series **345** (2012) 012045 doi:10.1088/1742-6596/345/1/012045

Technology of Liposomal Tiosens, Cifelin and Lysomustin for Industrial Purposes

E V Sanarova[1], **E A Kotova, AV Lantsova**

N.N.Blokhin Russian Cancer Research Center of the Russian Academy of Medical Sciences

Kaschirskoe highway 24, 115478 Moscow, Russia

E-mail: sanarova8686@mail.ru

Abstract. This work is devoted to the development of national antineoplastic drug (Tiosens, Cifelin, Lysomustin) liposomal dosage form (LDF) circuit technology and their manufacturing technology. In modern oncology liposomes, which are hollow phospholipid vesicles, are used as delivery systems protected drugs from biodegradation, and healthy cells from the toxic effect of chemotherapeutic agents. The technology of their production is stretching and multistage. It is also necessary to give consideration a lot of factors that influence on the finished product quality.

1. Introduction

Tiosens (tetra-3-phenylthiophthalocyanin aluminum hydroxide) refers to the photosensitizer of the second generation and is designed to treat tumors by photodynamic therapy. Tiosens LDF was proposed to increase the selectivity effects on tumor tissue and improve the solubility of the substance. This LDF showed a high antineoplastic activity at the preliminary stages of the study and has been recommended for batch production.

Lysomustin is nitrosoalkylurea derivative based on aminoacid L-homocitrulline. Biological studies have shown the effectiveness of this drug in small cell lung cancer and melanoma. Lysomustin is unstable both in alkaline and in acidic solutions, having the maximum stability of its solutions in a narrow range of pH values, and undergo degradation under influence of visible light at room temperature in aqueous solution. In this connection, to enhance clinical trials a novel lysomustin LDF, provided drug stability during storage, high selectivity for tumor cells and low toxicity were proposed for intravenous administration.

Cifelin is less toxic alkylating agents represented chloraethylamin peptide derivative with such aminoacids as L',D'-phenylalanine and L',D'-valine, synthesized in VONTS USSR AMS. Clinical trials of Cifelin tablets have been revealed significant antineoplastic activity of the drug in several tumors monotherapy, but because of the substance insolubility in the water is needed to apply high daily doses causing a lot of side effects. Cifelin incorporation into liposomal vesicles was one of the possible way to increase its solubilization and bioavailability. During investigation of developed

[1] To whom any correspondence should be addressed.

dosage form specific activity in a number of transplantable tumors in mice by intraperitoneal injection was showed that the dosage form multiply exceeds substance antitumor activity.1.

2. Materials and methods

Preparation of large multilayer liposomes (LML) was performed using the lipid film hydration. Exact samples of phosphatidylcholine (LIPOID E PC S, Germany), cholesterol (SIGMA, Germany), polyethyleneglycol-2000-distearoylphosphatidylethanolamine (LIPOID, Germany), Tiosens substance (SRC "NIOPIK") or Cifelin substance (RCRC n.a. N.N. Blokhin) were dissolved in 200 ml of chloroform (chemically pure, Himmed, Russia) and transferred to a round bottom flask with a capacity of 20 liters. The chloroform solution was evaporated on a rotary evaporator (Heidolph Laborota 20 control, Germany) at water bath of 35±2 ° C to form a lipid thin film, followed by redrying for 2 hours under vacuum before residual solvent have been completely removed. After this, the lipid film was hydrated by water for injection or an aqueous lysomustin solution (for liposomes containing Lysomustin) to complete washing out of the flask walls. The resulting LML emulsion was filtered through membrane filters with pore diameters of 1.2 and 0.45 μm (Pall, Germany), and reduse to a small size in a homogenizer Microfluidizer M-110S (Microfluidics, USA) at a pressure of 40 mm Hg to prepare monolayer vesicles. The resulting liposomes were sterilized by filtration through a polycarbonate membrane filters with pore diameter of 0.22 μm and mixed with sterilized 40% sucrose solution in a 3:1 ratio. Prepared Tiosens, Cifelin and Lysomustin liposomal emulsion were dosed by 6 ml in 25 ml vials and subjected to freeze-dried using a set Edwards Minifast DO.2 "(Ero Electronic SpA, UK). Lyophilization was performed on previously setted mode of cooling before -55ºC and heating shelves to +20ºC.

For studying the chemical and pharmaceutical properties of freeze-dried LDF vial content was rehydrated in 5.8 ml of deionized water to produce homogeneous emulsion. At all stages of manufacturing liposome size was measured by Nicomp 380 Submicron Particle Sizer (U.S.A.), pH value of freeze-dried LDF was potentiometrically determined.

3. Results and discussion

In the course of investigation the formulations were selected and circuit preparation Tiosens, Lysomustin, Cifelin LDF based on making of the lipid film, production of large lipid vesicles, followed by homogenization, sterilizing filtration and freeze-drying of product were developed. All series met the declared quality parameters: visual environment (dry porous mass), identification (the absorption maximum), the average mass of vial content, pH value, weight loss after drying (not more than 3%), particle size (not more than 200 nm) and active substance quantity (mg/vial).

Sufficiently high antitumor activity of Tosens liposomes were shown during a PDT session in transplanted mice tumors by calculating the inhibition of tumor growth values (ITG%). ITG for Ehrlich's tumor was 70 - 76%, for lymphocytic leukemia P-388 was 70%, and for epidermoid Lewis lung carcinoma was 60-64%.

Also a high antitumor effect of Lysomustin nanostructured dosage form was shown on lymphoblastic leukemia L-1210 in all tested doses (125-250 mg / kg) and on the Lewis lung carcinoma maximum Lysomustin antitumor effect occured at doses 225 and 250 mg / kg, curing 71% and 86% of the animals, respectively.

The third studing drug was effective against a tumor model of breast adenocarcinoma (ITG = 84%) and a model of cervical cancer (ITG = 83%). The study of liposomal Cifelin in outbred rats with transplanted Walker carcinoma caused 89% of animal treatment when administered orally, and 100% was for intravenous administration.

It is necessary to ensure liposome manufacturing in order to proceed to the next stages of preclinical research consisted in investigation of Tiosens, Lysomustin and Cifelin nanostructured formulation toxicity, pharmacokinetics, safety assessment and preparing for clinical trials. The manufacturing technology allows to prepare minimum 800 ml of liposomal dispersion to produce a series including at least 100 vials.

4. Conclusions

The manufacturing technology for production of stable and effective Tiosens, Lysomustin, Cifelin LDF was developed and studied its critical points. Obtained drugs can bedirected for indepth pre-clinical and clinical trials.

5. Acknowledgements

The work was supported by the grant of the president of the Russian Federation for state support of young Russian scientific candidates of science.

Work is executed according to the scientific and technical program «Creation and practical development in public healthcare of new methods and means of preventive maintenance, diagnostics and treatment of oncological, infectious and other dangerous diseases» at financial support of the Government of Moscow.

AUTHOR INDEX

AUTHOR INDEX

Institute of Physics
Dirac House, Temple Back
Bristol BS1 6BE UK

CURRAN ASSOCIATES INC
proceedings
.com